Buffalo Production Un
Climatic Regi

Buffalo Production Under Different Climatic Regions

Editors

S S Kundu, Ph.D.
Principal Scientist & Head
Plant Animal Relationship Division
Indian Grassland and Fodder Research Institute
Jhansi - 284 003 (UP) INDIA

A K Misra, Ph.D.
Senior Scientist
Plant Animal Relationship Division
Indian Grassland and Fodder Research Institute
Jhansi - 284 003 (UP) INDIA

P S Pathak, Ph.D.
Director
Indian Grassland and Fodder Research Institute
Jhansi - 284 003 (UP) INDIA

Published in Association with
**Indian Grassland and Fodder Research Institute
(IGFRI), ICAR, Jhansi (UP)**

International Book Distributing Co.
(Publishing Division)

Published by

INTERNATIONAL BOOK DISTRIBUTING CO.
(Publishing Division)
Chaman Studio Building, 2nd Floor,
Charbagh, Lucknow 226 004 U.P. (INDIA)
Tel. : Off. : 2450004, 2450007, 2459058 Fax : 0522-2458629
E-Mail : ibdco@sancharnet.in

First Edition 2004

Price: Rs. 850/-

ISBN 81-8189-065-5

Composed & Designed at :

Panacea Computers
33, Nehru Road,
Sadar Cantt. Lucknow-226 002
Phone : 2481164, 2483354, 3127082, Fax : 0522-2480543
E-mail : prasgupt@rediffmail.com

Printed at:

Army Printing Press
33, Nehru Road,
Sadar Cantt. Lucknow-226 002
Phone : 2481164, 2483354, 3246263, Fax : 0522-2480543
E-mail : armypress@indiatimes.com

Dr. Mangala Rai
Secretary to Government of India
Department of Agriculture Research and Education
& Director General Indian Council of
Agricultural Reserarch
Krishi Bhawan, New Delhi - 110001

FOREWORD

Animal husbandry is an integral part of the Indian Agriculture. Of all the livestock species, buffalo holds the greatest promise for food security and sustainable development in the farming systems of Indo-Gangetic plains and many other agro-climatic regions of the country. Buffaloes also form an integral part of the typical Indian Farming Systems due to its ability to sustain under harsh climatic conditions, scarce feeds and fodder that too coarse and lignified and produce high quality milk, with low cholesterol and higher fat content. Buffaloes have evolved as main dairy animal in independent India however; their domestication has been since the epic period of Ramayana and Mahabharata. Domesticated buffaloes primarily reared for milk production in Kanya kumari to the Bugyals above Sri Kedarnath at the height above 4000 meter in north and J&K, and in arid & semi-arid west to hilly states of North-Eastern Regions.

India possesses the highest resources of buffaloes and the best milch breeds of the world, e.g. Murrah, Jaffarabadi, Nili-Ravi, Bhadawari and Surti, which had their origin in North-West India. These breeds have high potential for milk and fat production besides the meat production. There are several other breeds which have regional importance and economic value, e.g. Surti, Bhadawari, Nagpuri, Pandharpuri, Tarai, Parlakhemundi, Manda, Jerangi, Kalahandi, Samalpur, Toda, South Kanara, Swamp etc.

During past 3-4 decades extensive research programmes have been taken-up on various aspects of buffalo production but there appears to be a need for focused coordination and strengthening the process of exchange of knowledge gained at different regions with respect to different breeds. AT ICAR level, the attempts have been made in this direction and concerted efforts are on for improvement and conservation of buffalo breeds of various part of the country under the Net Work Project on buffaloes.

This compilation entitled "Buffalo Production Under Different Agro-Climatic Regions" is an attempt to bring these issues of buffalo feeding and rearing to the forefront and to identify the problems that have to be solved and opportunities it has to offer to the professionals, scientists entrepreneurs and farmers. I find that the organizers have invited papers from variety of workers engaged in the field of animal production, forage and fodder production and utilization in different agro climatic regions of the country. I am sure this attempt will encourage further research and emphasis in the area of buffalo production focusing the regional opportunities and strengths.

(MANGALA RAI)

Preface

Buffalo is, believed to be, originated in the Indian sub-continent. They were probably domesticated about 5000 years ago and evidences from China are there that they were in use some 4000 years ago. Buffaloes have mainly been classified in two broad categories- Swamp type and River type. Swamp buffaloes are mainly used for work and their meat production potential is also outstanding. The River type buffaloes are predominately dairy animals. Out of 170 million buffaloes of the world, 165 million (97 per cent) are in Asia, of which 57 % are found in India alone. The present population of buffalo in India is about 97 million. Currently the buffaloes in India are providing 47.85 million ton of milk, 1.5 million ton of meat and 0.53 million ton of skin annually besides use of their males for draught purpose. Though, total buffalo population is less than half of cattle yet they contribute 55 percent of the total milk produced in country. India possess some of the best milch breeds of buffaloes in the world e.g. Murrah, Nili-Ravi, Jaffarabadi and Surti. There are other breeds such as Bhadawari, Mehsana Nagpuri, Toda, Pandharpuri and Parlakhemundi, which have evolved in different regions of the country due to natural selection over the centuries to meet out the need of the region. Other breeds have a specific regional importance and economic value, e.g. Tarai, Manda, Jerangi, Kalahandi, Sambalpur, Toda and South Kanara, Banni, Kutanad, Chilika etc. Studies conducted in the country have revealed that a marked improvement can be achieved in the overall productivity of the buffaloes through breed improvement programs, feed and fodder resource development, management and health care. In order to provide a common platform to deliberate, discuss and recommend the opportunities and challenges in different agro-climatic regions of the country in feeding, breeding, management and health of buffaloes for sustainable production. National seminar was planned, book chapters/lectures were invited from planners, scientists, teachers, extension workers. Thirty papers were received covering various aspects of buffalo production have been included. First four chapter covers an overview of buffalo production, feeding and nutrition and forage and grazing resources availability in different agro-climatic regions of the country. Chapter 5 to 11 describe about different breeds prominent in different agro-climatic zones. In chapters 12 to 15 the meat production potential of the buffaloes in the country as whole has been discussed. In chapter 16 to 22 nutritional aspects along with the use of conventional and unconventional feed resources have been taken up. In chapter 23 to 29 reproductive problems and strategies for their management for sustainable buffalo production have been discussed while the role of buffaloes in the economy of Asian countries has been brought out in chapter 30.

We feel that the authors have made a good effort in bringing out issues of buffaloes reared in different regions of the country. The book through these chapters brings out the potentials of buffalo rearing for meeting the challenges of milk and meat production, conservation of regionally important breeds, up-gradation of buffalo production system in different agro-climatic regions and ample opportunities for improved livelihood and social equity.

The encouragement, guidance and foreword by our most reverend Dr Mangala Rai ji, Secretary (DARE) and D.G. (ICAR) are gratefully acknowledged. We appreciate the commitment of our authors and do take the responsibilities for those important issues still missing in the book for an inquisitive reader. The help rendered by scientists of PAR division is thankfully acknowledged.

S S Kundu
A K Misra
P S Pathak

List of Contributors

A. K. Misra
Plant Animal Relationship Division
Indian Grassland and Fodder
Research Institute,
Jhansi-284 003 (UP)

A. K. Pathak
Poshak Feeds India Pvt. Ltd., 71/3
Mile Stone,
G. T. Road, Karnal -132 001
(Haryana)

A. K. Samanta
Indian Grassland and Fodder
Research Institute
Jhansi – 284 003 (UP)

Anil Kumar
Indian Grassland & Fodder Research
Institute, Jhansi

B P Kushwaha
Indian Grassland & Fodder Research
Institute, Jhansi

B.S. Meena
Indian Grassland and Fodder
Research Institute,
Jhansi-284 003 (UP)

C. B. Sachan
Krishi Vigyan Kendra, College of
Agriculture,
Gwalior (MP)

D. Swarup
Indian Veterinary Research Institute,
Izatnagar-243 122 (UP)

D.N. Verma
Narendra Deva University of
Agriculture & Technology,
Kumarganj,
Faizabad - 224 229 (UP)

K. Anilkumar
Kerala Agricultural University,
Mannuthy, Thrissur- 680 651 (Kerala)

K. Kareemulla
National Research Centre for Agroforestry, Jhansi-284 003 (UP)

L K Karnani
Indian Grassland & Fodder Research
Institute, Jhansi

M. M. Roy
Indian Grassland & Fodder Research
Institute,
Jhansi - 284 003 (UP)

M. Muthukumar
National Research Centre on Meat,
CRIDA Campus,
Hyderabad -500 059 (AP)

M.C. Sharma
Indian Veterinary Research Institute,
Izatnagar-234 122 (UP)

M.M. Das
Indian Grassland and Fodder
Research Institute
Jhansi – 284 003 (UP)

N. N. Pathak
Central Institute for Research on
Buffaloes, Sirsa Road,
Hissar - 125 001 (Haryana)

N. P. Singh
Central Sheep and Wool Research
Institute,
Avikanagar- 304 501 (Rajasthan)

Nagendra Sharma
National Dairy Research Institute,
Karnal - 132 001 (Haryana)

O. H. Chaturvedi
Central Sheep and Wool Research
Institute,
Avikanagar- 304 501 (Rajasthan)

O.P. Dhanda
CCS HAU,
Hisar-125 004 (Haryana)

P.S. Pathak
Indian Grassland and Fodder
Research Institute,
Jhansi 284 003 (UP)

R K Sethi
Project Coordinator (Buffalo)
Central Institute for Research on
Buffaloes, Sirsa Road,
Hisar-125 001 (Haryana)

R. Kadirvel
Tamil Nadu Veterinary and Animal
Sciences University, Chennai – 600
051 (Tamilnadu)

R.C. Upadhyay
National Dairy Research Institute,
Karnal -132 001 (Haryana)

R.K. Raikwar
National Dairy Research Institute,
Karnal – 132 001 (Haryana)

R.N. Choubey
Indian Grassland and Fodder
Research Institute,
Jhansi - 284 003 (UP)

S B Maity
Indian Grassland and Fodder
Research Institute,
Jhansi- 284 003 (UP)

S S Kundu
Indian Grassland and Fodder
Research Institute,
Jhansi- 284 003 (UP)

S. A. Karim
Central Sheep and Wool Research
Institute,
Avikanagar- 304 501 (Rajasthan)

S. K. Dash
Department of Animal Breeding and
Genetics, Orissa University of
Agriculture and Technology,
Bhubaneswar (Orissa)

S. S. Zombade
Poshak Feeds India Pvt. Ltd., 71/3
Mile Stone, G. T. Road,
Karnal -132 001 (Haryana)

S. Senani
Central Agricultural Research
Institute,
Port Blair - 744 101 (A&N)

S.K. Tomar
National Dairy Research Institute,
Karnal-132 001 (Haryana)

S.P. Tailor
Livestock Research Station,
Vallabhnagar – Udaipur (Rajasthan)

S.S. Thakur
National Dairy Research Institute,
Karnal – 132 001 (Haryana)

S.V. Singh
National Dairy Research Institute,
Karnal -132 001 (Haryana)

Satish Kumar
Indian Veterinary Research Institute,
Izatnagar-241 322 (U.P.)

Shiv Prasad
Dairy Cattle Breeding Division, NDRI,
Karnal - 132 001 (Haryana)

Sultan Singh
Plant Animal Relationship Division
Indian Grassland and Fodder
Research Institute, Jhansi-284 003
(UP)

Sunil Kumar
Indian Grassland and Fodder
Research Institute,
Jhansi 284 003 (UP)

T.R.K. Murthy
National Research Centre on Meat,
CRIDA Campus,
Hyderabad -500 059 (AP)

Udeybir Singh
Narendra Deva University of
Agriculture & Technology,
Kumarganj, Faizabad - 224 229 (UP)

V.K.Sharma
College of Vet. & Animal Science,
HPKV,
Palampur-176 062 (H.P)

Contents

CHAPTER 1

Forage and Grazing Resources in Different Agro-climatic Regions of India

P.S. Pathak and Sunil Kumar

Indian Grassland and Fodder Research Institute,
Jhansi 284 003 (UP)

Introduction

Livestock constitutes an integral part of the Indian agriculture. They provide livelihood to 70 per cent of the population engaged in agriculture and contribute 6.7 per cent to the gross domestic production of the country. They not only provide products for human sustenance but also serve as a major energy source for draught power in agricultural operations. The country accounts world's 15 percent of the total world livestock population dependent upon 2 per cent of geographical area. Among other factors, the low productivity of livestock is mainly attributed to poor fodder and feed resources. The growing human population pressure demands more allocation for food production. Livestock products also from part of human diet and their requirement is growing at a faster rate. Production efficiency of the Indian livestock in most of the regions of the country is very low as compared to some of the states like Gujarat, Punjab and Haryana. Balanced nutrition through feed and fodder to livestock is available in selected milk shed areas of the country where intensive fodder production systems are practiced. Rest of the farming community maintains uneconomic large herds on grazing supplemented with stall feeding. As per current estimate, the availability of forage (green and dry fodder) is nearly 60 percent (Anonymous, 2001). To meet this challenge, concentrated efforts are to be made for bridging the large gap in demand and supply of the fodder in this country.

Status of fodder demand and supply

The forage resources in India are mainly derived from crop residues, cultivated forages and grazing from pastures and grasslands. However, the crop residues mainly constitute the major feed material for the animals in most of the states. The country has about 4.9% of the total cropped area under cultivated forages. The requirement for fodder is mainly governed by the density of livestock in an area and secondly by management levels and climatic conditions. The fodder resources with present productivity levels are not sufficient to meet the demand of huge livestock population. The growth of livestock population between the period of 1951 to 1992 is about 61 percent which works out to be 1.5 percent annual growth rate. It is difficult to find out fully reliable figures of fodder production in India, however, working estimates are available. According to an estimate, current fodder requirement is 1006 and 560 million tonnes for green fodder and dry matter, respectively (Anonymous, 2001). The present availability of green fodder from cultivated areas and pastures is 388 million tones which includes supplementation from sugarcane tops and seasonal weeds taken out from cultivated fields. Similarly, in case of dry fodder, total availability is 437 million tonnes which constitutes straw of cereal crops like paddy, wheat, barley, maize, sorghum, pearl millet and few legumes like pigeonpea and chickpea and dry grass from grazing lands and forests. It is clear from the projection of demand and supply of forages that deficit of green fodder and dry matter will be more than 65 and 25 per cent by 2025 (Table 1). The data are undoubtedly indicative of high pressure of livestock on available forages especially green fodder. In a span of next twenty years, picture for bridging the gap between demand and supply appears to be gloomy of appropriate measures are not taken. Such a situation calls for serious attempt to improve supply of green fodder in this country for sustaining the ever-increasing millions of herds every year.

Table-1:   Projected demand, supply and deficits of forage in the country

Year	Supply (mt)		Demand (mt)		Deficit as % of demand	
	Green	Dry	Green	Dry	Green	Dry
2003	387.7	437.3	1006	560.1	61.51	21.81
2005	389.8	441.6	1021	568.0	61.83	22.12
2010	395.2	452.7	1057	588.2	62.63	22.91
2015	400.5	464.0	1095	609.2	63.44	23.72
2020	406.0	475.7	1134	630.9	64.26	24.57
2025	411.5	487.6	1174	653.3	65.10	25.44

Source : Draft report on Fodder during the Xth Plan submitted to Planning Commission (2001).

Livestock population –growth trend

Livestock population in India is dependent upon the biomass residues in crop production for animal production and plays a pivotal role in the rural economy. Earlier, number of livestock holding was considered prestigious in society and was linked with status. This trend is still continuing in some parts of the country. Total livestock population of the country as per census (1992) is 470.14 million. However, projected estimates for livestock population in 2002 are 897.9 million (Pathak, 2003). The phenomenal growth trend is visible in next decade with the annual growth rate of 1.7 between 2002 and 2012 (Table 2). Among the livestock, the growth rate of buffalo is increasing linearly. The buffalo population is increasing with 1.7% annual growth rate while in case of cattle, the trend is negative. During the span of ten years, cattle population is projected to be reduced by 0.08 percent. Trend of population growth in sheep and goat is similar to that of buffalo. Apart from this, poultry growth is also quite higher but equine and camels population is liable to be constant in the next decade also.

Table-2: Livestock population – projected estimates (*in million*)

Year	Cattle	Buffalo	Sheep	Goat	Pigs	Poultry	Equine	Camel	Total
2002	201.0	97.0	65.0	122.1	14.8	396.3	0.4	1.3	897.9
2003	200.8	98.6	66.6	123.2	15.1	406.1	0.4	1.3	912.0
2004	200.7	100.1	68.2	124.2	15.5	416.2	0.4	1.3	926.5
2005	200.5	101.7	69.8	125.3	15.8	426.6	0.4	1.3	941.3
2006	200.3	103.3	71.5	126.4	16.1	437.2	0.4	1.3	956.4
2007	200.2	105.0	73.2	127.4	16.4	448.1	0.3	1.3	971.9
2008	200.0	106.7	75.0	128.5	16.8	459.2	0.3	1.3	987.8
2009	199.8	108.4	76.8	129.6	17.1	470.6	0.3	1.3	1004.0
2010	199.6	110.1	78.6	130.7	17.5	482.4	0.3	1.3	1020.5
2011	199.5	111.8	80.5	131.9	17.8	494.4	0.3	1.3	1037.5
2012	199.3	113.6	82.5	133.0	18.2	506.7	0.3	1.3	1054.9

Source : Pathak (2003)

Forage resources

The major fodder and feed for livestock of our country are the variety of herbage from cultivated forages, i.e. cultivated forage crops, crop residues from crop lands, dry fodder from self owned or community land and grazing from wasteland, degraded land and forest grazing lands. In India, cattle of intensive cropped areas obtain only about 25 percent of their feed from grazing in nearby forests and other uncultivated lands, the balance comes from crop residues unsuitable for human consumption.

The total area under cultivated fodder is only 8.4 mha on individual crop basis which is static since last two decades. The scope for further increase in area under forages seems to be very less due to demographic pressure of human population for food crops. The accurate figures of area and production of forage crops is very difficult to find because these crops are grown in small patches which finds less attention. No systematic efforts for collection of data on area of forage crops have been made so far. However, provisional figures of area under important fodder crops have been reported under the ageis of AICRP on Forage Crops (Hazra, 1998). The state-wise area of some of the forage crops has been given in Table 3. Sorghum (3.6 mha) amongst the kharif crops and berseem (1.1 mha) amongst the rabi crops, occupy about 55 percent of the total cultivated area. Among states, Rajasthan (1.6 mha) occupies first place with regard to area under forages followed by Gujarat (1.1mha) and Punjab (1.0 mha). Area under forages is lowest in Himalayan states like Himachal Pradesh (0.3 mha), Jammu and Kashmir (0.7 mha) and other states (mostly North eastern hill states).

In the diverse climate of India, variety of forage species, range grasses and legumes and trees as well shrubs finds their place in the natural vegetation. Various agro-climatic regions possess certain opportunity and threats to different flora of fodder value. The production potential of many of the forage species is quite convincing in these regions. Amongst the annual fodder crops, berseem and lucerne are the highest green forage producer (60-130 t/ha in a season). Amongst perennial cultivated forages, napier and pearl millet hybrid, guinea and para grass produce high tonnage (90-230t/ha/ green fodder annually). The forage resources with their production potential (green fodder-t/ha) in 15 different agro-climatic region is given in Table 4. The suitability of forage species is visible in their production potential at specific zone. Apart from cultivated forages, range grasses supplements the fodder demand in Himalayan region (Zone 1,2) and western region (Zone 13,14). In coastal and Island region (Zone 11,12 & 15), forages are grown under plantation crops in multi-tier system. Fodder trees and shrubs also contribute substantially to the fodder requirement in lean period in almost all 15 agro-climatic zones of the country. The year round fodder production system in different agro-climatic regions indicates the potential of the environment concerned for harvesting maximum tonnage under intensive management system.

Table-3: State-wise   area   (provisional)   of   important
cultivated fodder crops in some Indian states

State	Area ('000 ha)									
	Sorg-hum	Maize	Pearl-millet	Cow-pea	Guar	Ber-seem	Oats	Lucer-ne	Oth-ers	Total
Jammu & Kashmir	-	35	-	-	-	10	10	10	5	70
Himachal Pradesh	-	15	-	5	-	4	5	-	5	34
Haryana	300	-	100	20	20	200	15	-	10	665
Punjab	400	200	20	20	10	350	25	-	15	1040
Uttar Pradesh	600	100	100	50	-	300	20	5	10	1195
Madhya Pradesh	410	200	100	20	-	150	20	-	15	915
Bihar	30	200	-	10	-	50	5	-	10	305
Rajasthan	900	-	600	10	50	50	5	15	10	1640
Gujrat	300	50	150	30	-	-	5	500	15	1050
Maharastra	400	100	50	20	-	-	-	100	20	690
Andhra Pradesh	100	10	50	10	-	-	-	20	10	200
Karnatka	50	50	30	10	-	-	5	50	10	205
Tamil Nadu	80	10	50	10	-	-	-	100	15	265
Other states	50	25	25	2	2	10	5	15	6	140
Total	3620	995	1275	217	82	1124	120	815	166	8414

Source : Hazra (1998)

In different, agro-climatic regions, the productivity of some prominent cultivated forages is highly variable (Table 5). Among the Kharif forage crops. Sorghum, maize, cowpea, NB hybrid and guinea grass have wide amplitude. However, during rabi, the choice is limited up to oat, lucerne and Berseem. Productivity of sorghum (40-55 t/ha) and maize (40-50) is highest in Upper and trans-gangetic plains as well as Western & Southern plateau & hill region, Lowest being in West coast plains & ghats region. The performance of (10-15%) cowpea is not much variable in all agro-climatic zones. High tonnage of NB hybrid is recorded (> 100 t/ha in all cultivated areas of Southern & Western plateau & hill region it produces NB hybrid in the range of 125-300 t/ha which is above the national average. The yield of Oat is poor (20-25 t/ha) in the agro-climatic zones, where winters are short. Productivity of berseem (70-80 t/ha) and Lucerne (80-90 t/ha) is

Table-4: Forage resources in different agro-climatic regions of India

Cultivated forages		Range grasses/legumes		Other minor forage resources	Production potential of intensive fodder crop rotation (t/ha/yr)	
Crops	Productivity (green fodder) (t/ha)	Species	Productivity (green fodder) (t/ha)		Crop rotation	Productivity (green fodder) (t/ha)
1. Western Himalayan Region						
NB Hybrid	110-130	Setaria grass	30-40	Siratro, Pangola Poa,	Maize + Cowpea-Lucerne+Oats-Mustard	85
Maize	20-25	Tall fescue grass	25-35	Desmodium, Cocksfoot,	NB hybrid + Velvet bean – Berseem + Mustard	123
Cowpea	15-20	White clover	15-20	Broad bean,		
Velvet bean	20-25	Red clover	15-20	Birdsfoot trefoil,		
Guinea grass	30-35	Timothy	20-25	Bauhinia,		
Berseem	50-50	Rye grass	15-20	Bhimal,		
Shaftal	50-60			Mulberry,		
Oat	20-25			Himalayan elm,		
Turnip	20-25			Seasonal weeds		
Teosinte	25-30			Other shrubs		
Chinese cabbage	25-35					
Pea	15-20					

2. Eastern Himalayan Region

Maize	20-25	Para grass	60-80	*Desmodium,*	NB hybrid 106
Cowpea	25-30	Centro	15-20	*Desmanthus,* Seasonal weeds,	Maize + Cowpea – Maize + 85 Cowpea – Maize + Cowpea
Ricebean	20-25	Guatemala grass	20-30	Bamboo sp.,	
Coix	20-25	Signal grass	25-30	*Bauhinia*	
Teosinte	35-45				
NB Hybrid	110-140				
Guinea grass	30-40				
Deenanath grass	30-35				
Oat (SC)	30-35				

3. Lower Gangetic Plain Region

Maize	25-30	Para grass	80-90	*Desmanthus, Desmodium,*	Maize + Cowpea – Dinanath 115 grass – Oats
Cowpea	25-30	Stylo	35-50	Rabi weeds, *Gliricidia,*	Maize + Ricebean – Berseem 105 + Mustard
Sorghum	30-40	Signal grass	25-30	*Acacia,*	
Ricebean	20-25			Bamboo sp.,	
Teosinte	35-45			*Bauhinia*	
NB Hybrid	110-140				
Coix	30-40				
Oat	35-45				
Berseem	60-80				

4. Middle Gangetic Plain Region

Deenanath grass	70-75	Signal grass	25-30	Subabool,	Pearlmillet + Cowpea – Maize + Cowpea - Oats	103
NB hybrid	85-100			Desmodium, Desmanthus, Rabi weeds,	Maize + Cowpea – Sorghum + Cowpea – Berseem + Mustard	96
Sorghum	35-40	Paspalum	20-25	Gliricidia, Acacia,		
Ricebean	20-25	Centro	15-20	Sugarcane tops,		
Cowpea	20-25	Para grass	60-80	Doob grass		
Thin Napier	35-45	Rhodes grass	20-30			
Teosinte	15-20	Stylo	40-50			
Maize	25-30					
Oat (SC)	30-35					
Oat (MC)	40-45					
Lucerne	35-40					
Berseem	60-90					

5. Upper Gangetic Plain Region

Crop		Grasses		Other fodder	Cropping system	Value
Maize	45-50	Tall fescue grass	20-25	Pangola, Cocksfoot, Dharaf grass, Vicia, Paspalum, Bhimal, *Hardwickia*, *Bauhinia, Albizia*, Seasonal weeds, Sugarcane tops	Maize + Cowpea – Toria - Oats	177
Sorghum	40-45	Rye grass	20-25		NB Hybrid + Berseem – Cowpea	121
Cowpea	30-35	Setaria grass	30-35			
NB Hybrid	90-100	Timothy	20-25			
Deenanath grass	35-45					
Oat	30-45					
Berseem	70-80					
Guinea	25-30					
Lucerne	40-50					
Barley	25-30					

6. Trans - Gangetic Plain Region

Crop		Grasses		Other fodder	Cropping system	Value
Maize	45-55	Dharaf grass	20-35	Sugarcane tops, Forage chicory, *Acacia* sp., Seasonal weeds	NB hybrid + Berseem	212
Sorghum	45-55	Rhodes grass	20-35		NB hybrid + Lucerne	176
Pearl millet	35-45	Para grass	80-90			
Teosinte	40-45					
Cowpea	25-30					
Cluster bean	25-35					
NB Hybrid	130-150					
Guinea grass	100-120					
Senji	30-35					
Shaftal	20-25					
Berseem	65-75					

Oat (SC)	30-35				
Oat (MC)	40-45				
Chinese cabbage	15-30				
Metha	20-30				
Barley	30-35				

7. Eastern Plateau & Hill Region

Maize (Kharif)	20-25	Para grass	80-90	Subabool, Australian teak, Sabai grass, Rabi weeds	Maize + Cowpea – Dinanath grass - Oats	131
Maize (Rabi)	30-35				Maize + Ricebean – Berseem + Mustard	112
Deenanath grass	25-35	Stylo	35-45			
Pearl millet	35-40	Signal grass	45-60			
Thin Napier	35-40	Setaria grass	50-60			
Guinea grass	30-35	Butterfly pea	20-25			
Cowpea	30-35					
Ricebean	20-25					
Lablab	12.5-15					
Coix	15-20					
NB hybrid	95-105					
Oat	25-30					

8. Central Plateau & Hill Region

Fodder crop		Grass		Trees/Shrubs	Cropping sequence	
Deenanath grass	60-80	Anjan grass	15-25	Wild ber, Subabool, Acacia sp., Madhuca sp., Ficus sp., Khejri, Neem	NB hybrid + Cowpea - Berseem	176
Maize	25-30	Setaria grass	35-50		Sorghum + Cowpea – Berseem + Mustard – Sorghum + Cowpea	169
Sorghum	30-45	Marvel grass	15-25			
Cowpea	25-30	Dharaf grass	25-30			
NB hybrid	90-100					
Guinea grass	80-100					
Oat	30-45	Glycine	20-25			
Berseem	60-80					

9. Western Plateau & Hill Region

Fodder crop		Grass		Trees/Shrubs	Cropping sequence	
Maize	45-50	Stylo	35-45	Subabool, Acacia Desmenthus, Sesbania, Neem	NB hybrid + Cowpea - Lucerne	253
Sorghum	45-50	Dharaf grass	15-20		Maize + Cowpea – Oats – Maize + Cowpea – Maize + Cowpea	168
Pearl millet (SC)	30-35	Marvel grass	20-30			
Pearl millet (MC)	95-105	Anjan grass	15-20			
Ricebean	30-35	Sehima grass	25-30			
Cowpea	30-35	Blue panic grass	25-30			
NB hybrid	120-125	Rhodes grass	20-25			
Guinea grass	110-125					
Oats (SC)	45-50					
Oats (MC)	60-65					
Berseem	80-85					
Lucerne	110-120					

10. Southern Plateau & Hill Region

Sorghum (SC)	30-35	Marvel grass	35-40	Subabool, Shevari, *Acacia*, *Ficus*, Tamarind	NB hybrid + Lucerne	225
Sorghum (MC)	40-45				Sorghum + Cowpea – Maize + Cowpea – Maize + Cowpea	111
Pearl millet	30-35	Buffel grass	45-50			
NB hybrid	300-350	Butterfly pea	25-30			
Guinea grass	200-250	*Desmodium*	65-70			
Deenanath grass	40-45					
Cowpea	25-30					
Desmenthus	80-85					
Leucaena	40-45					
Foxtail millet-R	10-15					
Foxtail millet-I	20-25					
Lucerne	80-90					

10. Southern Plateau & Hill Region

Sorghum (SC)	30-35	Marvel grass	35-40	Subabool,	NB hybrid + Lucerne	225
Sorghum (MC)	40-45			Shevari, Acacia,	Sorghum + Cowpea – Maize +	111
				Ficus, Tamarind	Cowpea – Maize + Cowpea	
Pearl millet	30-35	Buffel grass	45-50			
NB hybrid	300-350	Butterfly pea	25-30			
Guinea grass	200-250	Desmodium	65-70			
Deenanath grass	40-45					
Cowpea	25-30					
Desmenthus	80-85					
Leucaena	40-45					
Foxtail millet-R	10-15					
Foxtail millet-I	20-25					
Lucerne	80-90					

11. East Coast Plains & Hill Region

Crop	Days	Grass	Days	Legume/Tree	Cropping system	Value
Maize	25-35	Stylo	35-45	Hedge lucerne,	NB hybrid	126
Thin Napier	35-40	Para grass	90-100	Subabool,	Maize + Cowpea – Maize + Cowpea – Maize + Cowpea	90
Guinea grass	35-40	Congosignal grass	40-50	Sabai grass		
Deenanath grass	25-35	*Setaria* grass	50-60			
NB Hybrid	95-120	Marval grass	25-30			
Cowpea	30-35	Dasrath grass	25-30			
Ricebean	20-25	Rhodes grass	25-30			
Coix	15-25					
Oats	25-35					

12. West Coast Plains & Ghats Region

Crop	Days	Grass	Days	Legume/Tree	Cropping system	Value
NB hybrid	160-180	Signal grass	95-100	*Flemingia,*	Guinnea grass in coconut plantation	135
Guinea grass	90-100	Gamba grass	30-35	Subabool,	Congosignal grass in coconut plantation	75
Maize	15-20	Congosignal	35-40	*Desmodium,*		
Sorghum	10-15			Gliricidia,		
Pearl millet	15-20	Dasrath grass	55-65	Tamarind,		
Pigeonpea	20-25	*S. hamata*	20-25	Babool white,		
Sesbania	15-20			Sesbania, Neem		
				Babool black,		

15

13. Gujarat Plains & Hill Region

Crop		Grass/Legume		Trees/Others	Cropping sequence	
Sorghum (SC)	35-45	Marvel (R)	6-7	Subabool, Sesbania, Acacia sp., Khejri, Neem, Sunflower, Forage chicory	NB hybrid + Cowpea - Lucerne	215
Sorghum (MC)	65-70	Marvel (I)	10-12		Maize + Cowpea - Oats -	151
Guinea grass	40-60	Dharaf grass	15-20		Maize + Cowpea - Maize +	
Maize	20-25	Buffel grass	15-20		Cowpea	
Pearl millet	30-35	Dasrath grass	80-150			
Cowpea	20-25	Stylo	15-25			
Guar	10-15	*Chrysopogon*	8-15			
Mung	14-16	*Glycine*	20-25			
Horse gram	18-20					
Lablab	8-10					
Pigeonpea	15-20					
Moth	10-15					
Lucerne	80-90					

14. Western Dry Region

Crop		Grass/Legume		Trees/Others	Cropping sequence	
Pearlmillet	30-35	Dhaman grass	15-20	Khejari, Acacia, Neem	NB hybrid + Cowpea - Lucerne	187
Sorghum	35-40	Buffel grass	15-25		Maize + Cowpea - Oats -	151
Maize	25-30	Marvel grass	20-25		Maize + Cowpea - Maize +	
Horse gram	20-25	Dharaf grass	30-35		Cowpea	
Cluster bean	15-25	Sewan grass	7.5-15			
Cowpea	15-25	Stylo	35-45			
Lablab	10-15	*Siratro*	15-25			
Oat	25-30	*Atylosia*	15-25			
Lucerne	80-90					
Fenugreek	15-20					

15. Island Region

Guinea grass	60-80	Congosignal grass	35-40	*Desmodium,*	Guinnea grass in coconut plantation	127
NB hybrid	100-120	Gamba grass	30-35	*Flemingia*	Congosignal grass in coconut plantation	62
Cowpea	25-30	Signal grass	60-70			
Ricebean	20-25	Dasrath grass	35-40			
Coix	15-25	Setaria grass	30-40			
Maize	20-25	Rhodes grass	25-30			
		Para grass	80-90			

SC – Single cut , MC- Multicut, R- Rainfed, I- Irrigated

Source : Based on Annual Reports (Kharif & Rabi), AICRP on Forage Crops

Table-5: Productivity of some widely adopted forage crops in different agro-climatic regionsof India

S. No.	Agroclimatic region	Productivity green fodder (t/ha)							
		Sorghum	Maize	Cowpea	NB hybrid	Guinea grass	Oats	Lucerne	Berseem
1.	Western himalayan region	-	20-25	15-20	110-130	30-35	20-25	-	30-50
2.	Eastern himalayan region	-	20-25	25-30	110-140	30-40	30-35	-	-
3.	Lower gangetic – plain region	30-40	25-30	25-30	110-140	-	35-45	-	60-80
4.	Middle gangetic – plain region	35-40	25-30	20-25	85-100	-	30-45	35-40	60-70
5.	Upper gangetic plain region	40-45	45-50	30-35	90-100	25-30	30-45	40-50	70-80
6.	Trans-gangetic plain region	45-55	45-55	25-30	130-150	100-120	30-45	-	65-75
7.	Eastern plateau & hill region	-	20-35	30-35	95-105	30-35	25-30	-	-
8.	Central plateau & hill region	30-45	25-30	25-30	90-100	80-100	30-45	-	60-80
9.	Western plateau & hill region	45-50	45-50	30-35	120-125	110-125	45-50	110-120	80-85
10.	Southern plateau & hill region	30-35	35-40	25-30	300-350	200-250	-	80-90	-

18

No.	Region								
11.	East coast plains & hill region	-	25-35	30-35	95-120	35-40	25-35	-	-
12.	West coast plans & ghats region	10-15	15-20	-	160-180	90-100	-	-	-
13.	Gujarat plains & hill region	SC-35-45 MC-65-70	20-25	20-25	-	40-60	-	80-90	-
14.	Western dry region	25-30	35-40	15-25	-	-	25-30	80-90	-
15.	Island region	20-25	-	25-30	100-120	-	-	-	-

highest in Upper gangatic plain & Western plateau & hill region and Gujarat plains & hill region and Western dry region, respectively. This calls for developing new varieties of these crops that can tolerate the abiotic stresses and assure higher productivity. Besides, preference of crops and their combinations to each zone with the edaphic limits is to be identified and developed so that the temporal and spatial dimensions of the production are attended. Emphasis has to be made on area specific new crops that can break the yield barriers and meet the challenges of fodder deficits.

Grazing resources and grazing pressure

Animal husbandry evolved through forests with human evolution and livestock domestication. Thus, grazing was major resource in the past. India's huge livestock population still sustains to a larger extent on the grazing resources. Grazing based animal husbandry plays significant role in rural economy of an agrarian country like ours. The grazing activity is mainly dependent on the availability of the grazing resources from pastures and other grazing lands, *viz.* forests, miscellaneous tree crops and groves, cultivable waste lands and fallow lands. According to Wasteland Atlas of India (2000), the total area available for grazing accounts for about 40 per cent of the country' geographical area. The area available for grazing is highly variable (Table 6). Among the states, Rajasthan (13.04 mha) followed by Madhya Pradesh (9.05 mha), Andhra Pradesh (7.47 mha) and Maharashtra (6.83 m ha) occupied major share of total grazing area of the country. States like Jammu & Kashmir (6.82 mha) and Himachal Pradesh (4.05 mha) in Western Himalayan region have more area under grazing. However, the percentage area under grazing to that of total geographical area lies in states like Nagaland, Manipur, Meghalaya and Sikkim in Eastern Himalayan zone.

Most of the states falling in Indo-Gangetic Plains are poor in respect of grazing resources which otherwise are potentially rich in fodder production from arable lands. Population density of livestock per unit geographical area of India is 1.48 which is indicative of high pressure on land. Among the states, West Bengal harbours highest livestock population density (3.95) followed by Bihar (2.76), Uttar

Pradesh (2.20), Haryana (2.07) and Assam (2.05), much above the national average. States with moderate population density are Tamil Nadu, Punjab, Karnatka, Kerala and Tripura. All the North East hill states except Assam bear less pressure of livestock on their respective total geographical area.

If the population is expressed as livestock unit (ACU) comprising only cattle, buffaloes, sheep and goats on the basis of total grazing area available, India has 3.42 ACU/ha which is higher. The situation is not same in all the states as the value ranges from 0.14 (Mizoram) to 25.77 ACU/ha (West Bengal). States like Andhra Pradesh, Arunachal Pradesh, Gujrat, Jammu and Kashmir, NE states and Rajasthan have 0 - 3 ACU/ha, which shows very less pressure of livestock on the available land. Moderate population density (4-7 ACU/ha) exists in Assam, Karnatka, Madhya Pradesh, Maharashtra, Orissa, Tamil Nadu, Tripura and Union territories. The livestock pressure in these states is manageable. Three states (Bihar, Haryana and Uttar Pradesh) have ACU/ha in the range of 7-14, Kerala in the range of 15-21 ACU/ha and the remaining two states (Punjab and West Bengal) have 22-28 ACU/ha. This indicates that in the last group of six states, pressure of livestock on the land is alarming and deserve proper attention (Table 7). Its again indicative of the fact, other than Kerala, five states *viz.* Punjab, West Bengal, Bihar, Haryana and Uttar Pradesh have maximum density of buffaloes of their total livestock population.

Since agriculture provides a major part of feed/ fodder for the livestock, the net area sown was also considered to find ACU/ha. Considering net area sown, India has 2.52 ACU/ha (net area sown) and at state levels the value ranges from 1.14 ACU/ha (net area sown) in Mizoram to 6.27 ACU/ha (net area sown) in Jammu and Kashmir. Higher levels of ACU/ha (net area sown) i.e.> 5.0 are present in Jammu & Kashmir, Manipur, Himachal Pradesh and Arunachal Pradesh. This shows that the area under agriculture is not able to cater to the need of fodder /feed and the intervention of other systems such as agroforestry that can provide tree fodder and grass fodder may resolve the problem up to some extent (Pathak and Dagar,1998).

Table-6: State-wise grazing resources and grazing pressure based on livestock units (ACU)*

State	Geographi-cal Area (m ha)	Total grazing area (m ha)	Livestock (m)	Lives-tock density	Grazing pressure (ACU/ha)	ACU/ha (Net sown area)
Andhra Pradesh	27.51	7.47	32.91	1.20	2.63	2.58
Arunachal Pradesh	8.37	2.19	0.85	0.10	0.14	5.78
Assam	7.84	2.53	16.07	2.05	3.98	4.49
Bihar	17.39	3.42	47.93	2.76	7.71	3.42
Goa	0.37	0.07	0.25	0.67	1.91	1.23
Gujarat	19.60	4.89	18.60	0.95	2.39	1.68
Haryana	4.42	0.52	9.14	2.07	12.11	2.56
Himachal Pradesh	5.57	4.05	5.12	0.92	0.69	6.27
Jammu&Kashmir	10.14	6.82	8.71	0.86	0.56	6.71
Karnataka	19.18	2.92	29.57	1.54	5.73	1.99
Kerala	3.89	0.21	5.84	1.50	17.20	1.92
Madhya Pradesh	44.34	9.05	46.74	1.05	3.72	2.14
Maharashtra	30.77	6.83	36.39	1.18	3.19	1.54
Manipur	2.23	1.36	1.29	0.58	0.55	6.42
Meghalaya	2.24	1.35	1.19	0.53	0.45	3.60
Mizoram	2.11	0.44	0.21	0.10	0.14	1.14
Nagaland	1.66	1.00	1.08	0.65	0.34	4.80
Orissa	15.57	3.14	22.75	1.46	4.53	2.74
Punjab	5.04	0.27	9.45	1.88	28.64	2.60
Rajasthan	34.22	13.04	48.41	1.41	1.57	1.72
Sikkim	0.71	0.46	0.39	0.55	0.41	2.38
Tamil Nadu	13.01	3.28	25.01	1.92	3.66	2.63
Tripura	1.05	0.19	1.59	1.52	4.77	3.80
Uttar Pradesh	29.44	4.26	64.80	2.20	10.18	3.38
West Bengal	8.88	0.69	35.09	3.95	25.77	3.95
Union Territories	1.10	0.09	0.77	0.70	6.05	
Total	316.64	80.51	470.15	1.48	3.42	2.52

* ACU includes cattle, buffaloes, sheep and goats

Source : Anonymous (1997 & 2000 b)

Table-7: Grazing pressure based on livestock units and action for improved forage supply

Grazing pressure (ACU/ha)	States	Action for improved forage supply
0 - 3	Andhra Pradesh, Arunanchal Pradesh, Gujrat, Jammu&Kashmir, NE states, Rajasthan	Effective use of grazing lands.
4-7	Assam, Karnatka, Madhya Pradesh, Maharashtra, Orissa, Tamil Nadu, Tripura, Union territories	Efficient utilization of arable and grazing lands, Strategic management of crop residues
7-14	Bihar, Haryana, Uttar Pradesh	Strategic management of crop residues, Remunerative food - fodder production system
15-21	Kerala	Multi-tier fodder production system under plantation crops
22-28	Punjab, West Bengal	Strategic management of crop residues

Utilization of land under different production systems in a developing country like India have strong manifestation about the sustainability of the whole system. Report of UK Department for International Development (DFID) in collaboration with ILRI, Nairobi have presented appropriate indices under use of land in different production system based on human pressure the and poor livestock keepers (World Bank Index) in India (Thornton *et al*, 2003). The distribution of land area under different production systems in India has been given in (Table 8). The results of this study indicate that maximum poor livestock keepers/area (0.41-0.90) are concentrated in mixed rainfed production system (MRA, MRH & MRT) while mixed irrigated production system (MIA, MIH & MIT) possess moderate pressure of poor livestock keepers/ area (0.25-0.57). Lowest density of poor livestock/area (< 0.10) is available in rangeland-based habitats with livestock only. In all agrarian society, human population compete with livestock population particularly for land resources. High human pressure (>2.5) in mixed irrigated production system with higher poor livestock keepers/area is responsible for the lower production of livestock of this region. Emphasis of suitable type and breed of livestock in a production system keeping in view the human need, may be a long term measure of increased sustainability and profitability of the system.

Table-8: Distribution of land area under different production systems in India

S. No.	Production system	Area (mha)	Human population (m)	Human pressure	Poor livestock keepers (m)	Poor livestock keeper/Area
1	LGA	9.19	2.89	0.31	0.77	0.084
2	LGH	0.79	0.22	0.28	0.06	0.075
3	LGT	2.45	0.17	0.07	0.05	0.018
4	MIA	91.88	416.51	4.53	37.90	0.413
5	MIH	23.57	146.95	6.23	13.37	0.567
6	MIT	0.16	0.43	2.71	0.04	0.247
7	MRA	109.64	262.89	2.40	62.57	0.571
8	MRH	30.52	115.88	3.80	27.58	0.904
9	MRT	4.72	8.08	1.71	1.92	0.407
10	Others	40.59	54.28	1.34	4.94	0.122
	Total	313.51	1008.30	23.39	149.20	3.41

Source : Thornton *et al.* (2003)

Production systems :

LGA   =   Livestock only, rangeland-based arid/semiarid

LGH   =   Livestock only, rangeland-based humid/sub humid

LGT   =   Livestock only, rangeland-based temperate/tropical highland

MIA   =   Mixed irrigated arid/semi-arid

MIH   =   Mixed irrigated humid/subhumid

MIT   =   Mixed irrigated temperate/tropical high land

MRA   =   Mixed rainfed arid /semi-arid

MRH   =   Mixed rainfed humid / sub humid

MRT   =   Mixed rainfed temperate / tropical high land

4.2 The Grass Covers of India

A reconnaissance survey of grasslands of India conducted from 1954 to 1962 revealed 5 major grass covers based on distribution of grass species and communities. The distribution of grasses was primarily governed by climatic and edaphic factors, chiefly by latitudinal influence followed by altitude and topography particularly the soil moisture relationship. The five grass covers identified are, *Sehima – Dichanthium* type, *Dichanthium-Cenchrus-Lasiurus* type, *Phragmites-Saccharum-Imperata* type, *Themeda-Arundinella* type, and Temperate Alpine type.

Sehima-Dichanthium type

The tract favouring the occurrence of this grass cover lies approx between 8° and 28ºN and between 68° and 87° E. This cover type spreads over the whole of Peninsular India, including the central Indian plateau, the Chhota Nagpur Plateau and the Aravali ranges with a potential coverage of approx. 17,40,000 km². The cover is also found in the coastal region. The cover is represented by dominant perennial grasses viz. *Dichanthium annulatum, Sehima nervosum, Bothriochloa pertusa, Chrysopogon fulvus, Heteropogon contortus, Iseilema laxum, Themeda triandra, Cynodon dactylon, Aristida setacea, Cymbopogon*

---

spp.etc. Important associated species are, *Apluda mutica, Bothriochloa intermedia, Arundinella nepalensis, Desmostachya bipinnata, Eragrostis* and *Eragrostiella* spp.

Examination of protected or lightly grazed sites within this cover type shows that the highest expression of grass cover on gravelly soils consist in the establishment of a *Sehima* community with S. *nervosum* as a dominant species

A *Dichanthium* community, with *D. annulatum, D. caricosum* or *D. aristatum* as the principal species, represent the highest development of the grassland. The plant cover of a developed *Dichanthium* community may exceed 80% and hay production may be about 6.3 t/ha. On level soils with increasing moisture availability, the *Dichanthium* community is replaced partially or wholly by an *Iseilema* community, with *I. Laxum* as a chief species when *Sehima-Dichanthium* cover is subjected to grazing these communities are replaced by *Chrysopogon and Bothriochloa* communities and with further grazing at this stage, these communities are replaced by *Heteropogon* and *Eremopogon* communities, with *Heteropogon contortus* and *E. foveolatus* the chief species.

Dichanthium-Cenchrus-Lasiurus type

The tract favouring the occurrence of this grass is situated approximately between 23° and 32° N and 68° and 80°E. This type is associated with sub-tropical arid and semi-arid regions comprising the northern portion of Gujarat, the whole of Rajasthan, excluding the Aravalli ranges in the south, western UP, Delhi state and the Punjab, Haryana with a potential coverage of more than 4,36,000 km². On the southern boundary of this type a mingling of the grass species of the *Sehima-Dichanthium* and the *Dichanthium-Cenchrus-Lasiurus* cover takes place; therefore, this portion could be considered as a transitional zone between two types. Principal perennial grass species are *Cenchrus ciliaris, C. setigerus, D. annulatum, Cymbopogon jawarancusa, Cynodon dactylon, Eleusine compressa, Lasiurus sindicus Sporobolus marginatus, Dactyloctenium sindicum, Desmostachya bipinnata* etc. Important associate species are: *Chloris,*

Desmostachya, Heteropogon contortus, Saccharum bengalense, Vitevaria zyzanioides, etc.

Under grazing condition, *D. annulatum* is first to disappear in a mixed community of this cover, to be replaced in dominance by *Cenchrus* species and *L. sindicus* with continued grazing, the second stage of deterioration is reached, represented by *Cynodon dactylon* in low lying and moist conditions and *Eleusine compressa* on drier soils.

Phragmites-Saccharum-Imperata type

The area covered by this type is situated approximate between 26° and 32°N and 74° to 96°E. This cover type occurs throughout the Gangetic Plain, the Brahmaputra valley and extends westwards into the plains of Punjab. The area comprises approx. 28,00,000 km² in North eastern states, W. Bengal, Bihar, UP, Punjab and Haryana. Principal perennial species in drier regions are: *Imperata cylindrica, Saccharum arundinaceum, S. spontaneum, Phragmites karka, Desmostachya bipinnata.* Other important species of this grass cover throughout the area of its distribution are: *Bothriochloa intermedia, Vitevaria zizanioides, Imperata cylindrica, Chrysopogon aciculatus, Panicum notatum* etc.

This grassland cover consists of tall coarse grasses and favours swampy and wet situation, it is subjected to cutting and burning; the grass being primarily used for thatching purpose. In their undisturbed state, the swampy area invariably shows the dominance of *Phragmitis karka.* As the habitat becomes drier due to cutting and burning, *Phragmitis karka* is the first species to disappear, giving place to communities represented by *Saccharum, Imperata* and *Sclerostachya.* Introduction of grazing in addition to burning at this stage seems to favour the appearance of *Vitevaria zizanioides.* In the humid region grazing and burning induce further regression. The communities representing three changes are *Sporobolus* with *S. indicus* as the principal species. In some sites, with *Paspalum* and *Chrysopogon* in other sites (*P. conjugatum* (ch. *Aciculatus*).

Themeda-Arundinella type

This cover type occurs in the entire northern and north western mountain tract, on an area approximately 2,30,000 km² in the north-eastern states, West Bengal, UP, Punjab, Haryana, HP and J&K. In the west this type is found approximately between 29° and 37°N, and between 73° and 81°E, and in the east approximately between 22° and 28.5°N, and 88° and 97°E. This type is associated with undifferentiated forest and hill soils; and also with undifferentiated forest sub-mountain regional soils. Principal species of grass vegetation is represented by *Arundinella benghalensis, A, nepalensis, Bothriochloa intermedia*, etc. *Chrysopogon fulvus, Cymbopogon jwarancusa, Cynodon dactylon, Heteropogon contortus, Themeda anathera, Eulaliopsis binata, Ischaemum barbatum*. Associated perennial species are: *Apluda mutica, Arundinella khaseana, Pennisetum flaccidum, Chloris, Desmostachya* etc.

Cutting or grazing of the cover shows preponderance of *Themeda anathera*. When *Themeda* community is subjected to an increased degree of grazing, *Themeda anathera* gradually disappears and its place is taken progressively by *Arundinella nepalensis* and *A. bengalensis*. However, with further grazing *Chrysopogon fulvus* appears to become dominant.

Temperate Alpine type

The cover type occurs on the high hills of Uttaranchal, J&K, HP, West Bengal and North-eastern regions. The tract lies approximately between 29° and 37°N, and between 73° and 81°E in the western part of the country. On eastern side, the type is situated approximately. between 27° and 29.5°N, and 88° and 97°E. However, it differs from the *Themeda-Arundinella* type on that it essentially occurs at higher elevation, beyond timberline, approximately above 3,000 m in the west and above 2000 m in the east. The principal perennial species are: *Agropyron conaliculatum, Chrysopogon gryllus, Dactylis glomerata, Danthonia cachemyriana, Phleum alpinum, Carex nubigena, Poa pratensis, Stipa concinna*. Associated species are: *Poa alpina, Festuca lucida, Eragrostis nigra, Bromus ramosus* etc. The production levels and their potential in different grass covers indicates the possibilities of upgrading the production by technological interventions and management (Table 9). It has been found that application of silvipastoral technology in the *Sehima – Dichanthium* zone allowed annual productivity to increase by more than 8 times of the actual production.

Table-9: Production level (t/ha) of varoius grass covers in India

Grass cover	Harvestable biomass	
	Actual	Potential
Themeda - Arundinella (Sub montane region)	2.2	4.0
Phragmitis - Saccharum - Imperata (Sub tropics, high humid)	5.0	5.0
Dichanthium - Cenchrus - Lasiurus (Dry and arid tract)	3.3	5.0
Sehima - Dichanthium (Central and southern semi arid)	3.5	6.0

Grazing and forage feeding systems in different agro-climatic regions of India

Livestock has been an integral component of Indian Agriculture. However, the upkeep of livestock is highly variable with different agro-ecological regions and socio-economic conditions of agrarian community (Table 10). The grazing and forage feeding system in different parts of the country adopted are stall feeding restricted grazing, open grazing or combination of these three practices. The integration of all possible forage resources available is altogether not sufficient to fulfill the requirement of Indian livestock. Almost all regions possess certain constraints with regard to assured forage supply throughout the year (Pathak *et al.* 2000).

Table-10: Grazing and forage feeding systems in different agro-climatic regions of India (Based on Pathak *et al.* 2000)

Agroclimatic region	System followed	Constraints
Western himalayan region	Pasture grazing in Upper & Middle hills, fodder crops grown in limited area, grazing under these crops, and cut and carry systems in middle hills and grazing in forests and community area in foot hills, During summer grasses harvested under cropping of trees at lower altitude and conserved for winter use.	Acute shortage of fodder due to limited area under these crops, Problem is compounded due to nomadic migratory nature of Bakarwals and Gujjars visiting these areas. Declining pasture productivity due to high stocking rate.
Eastern himalayan region	No proper pasture, animals graze in forests having identification marks on them, medium & large farmers take fodder crops, silvipasture in some pockets, stall feeding on crop residues mostly Rice, legumes and leaves of various trees	Due to grazing high pressure resulted in a drain of biodiversity, Shortage of green fodder in some areas Forage quality is poor and needs blending with legumes.
Gangetic - plain region (Lower, Middle & Upper)	Crop residue based and relatively smaller area is allocated to fodder crops, stubble grazing after harvesting of rabi and kharif crops, In Lower - gangetic plain region only buffaloes are stall fed others graze in open while in Upper - gangetic plain Region, controlled grazing from July to March is practiced	Less priority to forage crops due to high pressure on land for food crops Residue quality is poor and needs blending with desirable high nutrition forages.
Trans – gangetic plain region	Stall feeding on crops and residue, Uncontrolled grazing in limited area,	Shortage of fodder due to high pressure of livestock Quality of crop residues is poor.

Eastern plateau & hill	Crop residue, tree leaves, grazing in forest land	Acute shortage of green fodder due to limited area under forage crops, Poor grasslands Silvipasture system is also less productive due to lack of proper varieties.
Central plateau & hills region	Dry grasses and crop residues, Grazing on degraded lands, forage resources come from natural grasslands, forest area and wastelands	Poor supply of forage resource from degraded and wastelands, Less attention to fodder crops Poor quality of crop residues
Western plateau & hill region	Plain zone has more area under forage while Ghat zone has maximum pasture lands and free grazing practiced , cut and carry method for grass from own land (bunds & hills)	Plain zone have much pressure due to less pasture while Ghat zone faces shortage of green fodder Lower area under fodder crops
Southern plateau & hill region	No specific grazing/forage feeding system Controlled grazing from June to March while open grazing in the forest area is common in the scarce rainfall zone of Rayalseema	Shortage of green fodder due to less average area under forage, Degraded grazing lands for livestock, Low quality crop residues
East coast plain & hill	Silvipasture, fodder trees, less area under forages, stall feeding by rice straw	Scarcity of green fodder due to less acerage under fodder crops
West coast plain & ghat region	Pastures, Free or restricted grazing in crop lands (in non-crop season), community lands and forest lands, stall feeding for crossbred animals	Pressure of livestock on limited forage resources available. Quality of crop residues
Gujarat plains & hill region	Stall feeding to cross bred animals controlled grazing during July – March	Poor carrying capacity of grazing lands due to low rainfall, Pressure on cultivated lands for fodder

Western dry region	Small portion under silvipasture/pastoral purposes, mostly rough grazing or to some extent cut for carrying to subsequent seasons, Stall feeding near cities and canal areas with crop residues mixed with green fodder. Improved pastures of *Cenchrus, Lasiurus* with incorporation of perennial legumes like *Stylo, Clitoria*	Poor productivity of rangelands, Quality of crop residues due to degradation Farmers priority for food grain crops Lack of fodder banks to tide over the dry months/drought years.
Island region	Grasses, Tree leaves / Crops and crop residues specially paddy straw	Shortage of forage and feeding system is vogue, Over grazing, Poor productivity of grasslands due to over exploitation.

Strategy to meet the gap in demand and supply of forages

The scenario of forage supply and demand in India needs to be analysed with respect to soil-plant-animal complex vis-à-vis the requirement of human population in particular region. The major focus, hence forth, is to be given for improved carrying capacity of the system in specific regions. Pathak (2002) identified the possibilities of potential increase through improved technologies and policy support from grasslands. Eartier an attempt was also made to identify some of the options that are highlighted in Table 11.

In addition some strategies would also be required to increase the efficiency of the supply of fodder in the various regions of the country. Some of them are outlined in Table 12.

Table-12: Strategies for improving the efficiency of fodder supplies to the areas under acute shortage

Strategy	Anticipated impact
Establishment of fodder banks This should be the strategy for drought and flood prone areas of the country and should aim at transporting *economically baled* dry fodder from *surplus areas-* 8 major river valleys spread over 40 million ha in the country and 40 million population is affected by floods annually. 74 districts in the country in 14 states and 86 million people are affected annually by droughts.	The resource poor livestock owners most often lose their livestock to the fodder scarcity caused by these natural disasters. Establishment of fodder banks will serve the purpose of easing these scarcities. In addition, the surplus forages produced in other parts of the country, currently wasted will also be efficiently utilised and will contribute to reducing *regional imbalances.*
Conversion of fodder into feed blocks	This will support the initiative of setting up fodder banks and will facilitate the *economical transport* of fodder from surplus areas.
Enrichment of straw/stover with urea	This will help in alleviating the nutritional deficiencies to some extent in the absence of availability of adequate quantities of fodder to fulfil the nutritional requirements of the livestock and contribute to increased production and incomes of rural populace.
Hay/Silage demonstrations	This will help in easing the fodder scarcities during lean season and bring about an even distribution of available fodder through out the year.

Epilogue

Thus, it is apparent that forage demands can be adequately met by mobilizing the resources of 'climate, water, soil, species, varieties and the management. Fodder and its importance are realized only in the deficit zones and deficit rainfall years. Technological options are sufficient to meet these challenges. The potential areas requiring attention are as under :

⊙ Develop degraded lands under silvipasture / pasture.

⊙ Develop degraded forests under silvipastures through Joint Forest Management.

⊙ Conserve extra fodder and crop residues, density and store in fodder banks.

⊙ Popularize chaff cutters in those areas where it is not need at present.

⊙ Use improved varieties of fodder crops.

⊙ Emphasize use of perennial crops in deficit zones alongwith crops.

⊙ Emphasize silvipastoral systems, bund forage technology, year round fodder production. and mixed cropping of forages with crops to meet the demand of livestock.

⊙ Develop strategic supplementation for bean periods to promote higher weight grain / production through leaf meal from *Leucaena*, Stylo, *Sesbania, Gliricidia*.

References

Anonymous, 1997. Basic Animal Husbandry Statistics. Ministry of Agriculture, Department of Animal Husbandry and Dairing, New Delhi pp. 43-52.

Anonymous 2000a. Annual Report, Kharif & Rabi, AICRP on Forage Crops, IGFRI, Jhansi.

Anonymous, 2000b. Wastelands Atlas of India, Ministry of Rural Development, Department of Land Resources Environment of

India, New Delhi. Pp. 10

Anonymous 2001. Annual Report, Kharif & Rabi, AICRP on Forage Crops, IGFRI, Jhansi.

Anonymous, 2003a. Forage Research in Gujarat, AICRP on Forage Crops. GAU, Anand Campus.

Anonymous, 2003b. Forage Research in Himachal Pradesh : Achievement of IX Plan, AICRP on Forage Crops, CSK HPKV, Palampur.

Hazra, C.R. 1995. *Advances in Forage Production Technology*, ICAR Publication, AICRP on Forage Crops, IGFRI, Jhansi pp. 33-41.

Hazra, C.R. 1998. Advances in fodder production systems. In : *Strategy for maximization of forage production* Ed (s) A.K. Mukherjee, S. Maiti, & M.K. Nanda, Directorate of Research, BCKV, Kalyani, West Bengal pp. 40-56.

Pathak, P.S. 2002. Potential technological and management interventions for improving the productivity of grasslands. In : Brithal, P. and Parthasarathy Rao (eds.) *Technology options for sustainable livestock production in India* 164-182 : Proc. Of the workshop on documentation adoption and impact of Livestock Technologies in India 18-19 Jan. 2001, ICRISAT, Patencheru, India.

Pathak, P.S. 2003. *Prospects of Feed Crops in India : The Role of CGPRT Crops*. Working Paper Series No. 64, CGPRT Centre, Bogor, Indonesia, XVIII, 60 pp.

Pathak, P.S. and Dagar, J.C. 1998. Agroforestry : A strategy for rehabilitation of degraded lands. In : (Gopal, B. and Pathak, P.S. eds.) *Ecology Today : An Anthology of Contemporary Ecological Research,* International Scientific Pub., New Delhi pp. 333-371.

Pathak, P.S., Pateria, H.M. and Solanki, K.R. 2000. *Agroforestry System in India : A Diagnosis & Design Approach*, AICRP on Agroforestry, ICAR Publication, New Delhi : 134-143.

Thornton, P.K., Kruska, R.L., Henninger, N., Kristjanser, P.M., Reid, R.S., Atieno, F., Odero, A.N. and Ndegwa, T. 2003. Mapping Poverty and Livestock in the Developing World – A report commissioned by the UK Department for International Development (DFID) on behalf of the Inter Agency Group of Donors Supporting Research on Livestock Production & Health in the Developing World, International Livestock Research Institute, PO BOX-30709, Nairobi, Kenya.

LIST OF FORAGES WITH THEIR BOTANICAL NAMES

Name	Botanical name

A. Cultivated forages – I Annual

1. Cereals

Kharif

Maize	*Zea mays* (L.)
Sorghum	*Sorghum bicolor* (L.) Moench.
Pearlmillet	*Pennisetum americanum* (L.) Leeke
Teosinte	*Zea mexicana* Schrad.
Foxtail millet	*Setaria italica* (L.) Beauv.
Coix	*Coix lachryma* Jobi.
Finger millet	*Eleusine corocana*

Rabi

Oat	*Avena sativa* (L)
Barley	*Hordeum vulgare* (L) Emend. Bowden

2. Legumes

Kharif

Cowpea	*Vigna unguiculata* (L) Walp
Ricebean	*Vigna umbellata* (Thumb.) Ohwi & Ohashi
Lablab bean	*Lablab purpureus*
Horse gram	*Dolichos biflorus* Lam.
Clusterbean	*Cyamposis tetragonoloba* (L) Taub.
Velvet bean	*Mucuna deeringiana*
Pigeonpea	*Cajanus cajan* L.
Broadbean	*Vicia faba*
Mung	*Vigna radiata*

Moth	*Phaseolus ovconitifolius*

Rabi

Berseem	*Trifolium alexandrinum* L.
Lucerne	*Medicago sativa* L.
Shaftal	*Trifolium resupinatum* L.
Fenugreek	*Trigonella foenumgraecum* L.
Indian clover	*Melilotus indica* L.

3. Other than legumes

Chinese cabbage	*Brassica pekinensis* (Lour.) Rupr.
Japan rape	*Brassica compestris*
Pea	*Pisum sativum*
Turnip	*Brassica rapa*

4. Grasses

Deenanath grass	*Pennisetum pedicellatum* Trin.

A. Cultivated fodder – II – Perennial

Napier bajra hybrid	*Pennisetum purpureum* Schum.X *Pennisetum americanum* (L)
Leeke	
Guinea grass	*Panicum maximum* Jacq.
Thin napier	*Pennisetum polystachyon*

B. Range grasses/ legumes – I Perennial grasses

1. Range grasses

Buffel grass	*Cenchrus ciliaris* L.
Yellow buffel grass	*Cenchrus setigerus* Vahi.
Setaria grass	*Setaria anceps* Stapfex Massey
Para grass	*Bracharia mutica* (Forsk.) Stapf.

Signal grass	*Bracharia brizantha* (Hochst ex. A. Rich) Stapf.
Congosignal grass	*Bracharia ruziziensis* Germain & Evard
Dharaf grass	*Chrysopogon fulvus* (Spreng) Chiov
Sewan grass	*Lasiurus hirsutus* (Forsk.) Boiss
Rhodes grass	*Chloris gayana* Kunth
Marvel grass	*Dicanthium annulatum* (Forsk.) Stapf.
Little para grass	*Urochloa mosambicensis* (Hack) Dandy
Sehima grass	*Sehima nervosum*
Blue panic grass	*Panicum antidotale*
Cocksfoot grass	*Dactylis glomerata*
Italian Rye grass	*Lolium multiflorum*
Tall Fescue grass	*Festuca arundinacea*
Perennial rye grass	*Lolium perene*
Timothy grass	*Phleum pratense*
Pangola grass	*Digitaria decumbans*
Guatemala grass	*Tripsacum laxum*

2. Range legumes

Stylo	*Stylosanthes hamata* (L) Taub
S. guianensis (Aubl.) SW.	
S. humilis H.B.K.	
Siratro	*Macroptilium atropurpureum* (DC) Urb.
Butterfly pea	*Clitoria ternatea* L.
Poa	*Poa annua* L.
Centro	*Centrosema pubuscens*
Hedge lucerne	*Desmanthus virgatus* (L) Wild.

White clover	*Trifolium repens*
Red clover	*Trifolium pratense*
Glycine	*Glycine wightii* (R.Grah. ex Wight & Arn) Verde
Desmodium (Green leaf)	*Desmodium intortum* (Mill) Urb.
Desmodium (Silver leaf)	*D. uncinatum* (Jacq.) DC

C. Fodder shrubs/trees

Subabool	*Leucaena leucocephala*
Shevari/Agathi	*Sesbania egyptica*
Babul black	*Acacia nilotica*
Babul white	*Acacia arabica*
Khejri native	*Prosopes cineraria*
Khejri exotic	*Prosopis Juliflora*
Neem	*Azadirachta indica*
Madhuca	*Madhuca latifolia*
Gliricidea	*Gliricidia sepium*
Tamarind	*Tamarindus indica*
Flemingia	*Flemingia congesta*
Bhimal	*Grewia optiva*
Bamboo	*Dendrocalamus sikkimensis, Bambusa balcoa*
Anjan	*Hardwickia binata*
Siris	*Albizia amara, A. lebbek*
Pink Bauhinia	*Bauhinia purpurea*
Chinese ber	*Zizyphus mauritiana*
Wild ber	*Zizyphus nummularia*
Mulberry	*Morus alba*

Himalayan elm	*Ulmus wallichiana*
Australian teak	*Acacia mangium L.*
Ficus	*Ficus religiosa, F. bengalensis*
Sesban	*Sesbania sesban*
Calliandra	*Calliandra Calothyrsus*
Mango	*Mangifera indica*

Animal Agro-Climatic Zoning of India - A Perspective

S.S. Kundu, M.M. Das, A. K. Samanta and P.S. Pathak

Indian Grassland and Fodder Research Institute
Jhansi – 284 003 (UP)

India is endowed with the largest bovine population of the world and bestowed with rich domestic animal diversity having 30 breeds of cattle, 10 breeds of buffalo, 42 breeds of sheep, 20 breeds of goat, 9 breeds of camel, 6 breeds of horses, 3 breeds of pigs and 18 breeds of poultry. India ranks first in respect of cattle (219.64 million) and buffalo (94.13 million), second in goat (123.50 million), third in sheep (58.20 million) and seventh in poultry population (413.00 million). The distribution of buffalo through out the world is presented in Table 1.

Total livestock population increased to the tune of 1.6 times from 1951 to 1997. The livestock population data indicate only 27.3 percent increase in cattle population against 104.15 percent of buffalo during same span of time. The population of sheep, goat and pigs has also increased at the tune of 45.26, 155.93 and 209.09 percent respectively (Table 2). The milk production is highest in Uttar Pradesh including Uttaranchal and Punjab was second in milk production followed by Andhra Pradesh, Gujrat, Haryana, Karnataka, Madhya Pradesh including Chattisgarh, Maharastra and Tamil Nadu. However, Milk production, buffalo population and yield per animal were lower in North Eastern states. Karnataka attained the maximum increase in milk production (19.76%) followed by Uttar Pradesh including Uttaranchal (16.62%), Punjab (8.76%), Bihar including Jharkhand (8.77%), Gujrat (6.05%) and Haryana (6.3%) during 1999-2000 to 2001-2002. However, total milk production in the country increase by 7.35 percent during the same period (Table 3).

Table 1 Buffalo population in the world

Unit: 1000 heads

Country	1991		1998		1999		2000		2001		Annual Growth Rate 1991 - 2001	
DEVELOPING COUNTRIES												
Bangladesh	807		820		828	F	830	F	830	F	-0.1	%
Bhutan	4	F	4	F	4	F	4	F	4	F	-0.1	%
Cambodia	755		694		654		694		626		-2.4	%
China	21,712		22,557		22,677		22,599		22,769		0.3	%
India	82,160	*	90,909	*	92,090	*	93,772	*	94,132	*	1.4	%
Indonesia	3,311		2,829		2,504		2,405		2,287		-3.5	%
Iran	440	F	474		474		460	F	460	F	0.8	%
Laos	1,100		1,093		992		1,008		1,008	F	-1.3	%
Malaysia	202		160		155		155	F	155	F	-2.3	%
Myanmar	2,072		2,337		2,391		2,441		2,500		2.0	%
Nepal	3,044		3,389		3,471		3,526		3,624		1.8	%
Pakistan	17,818		21,422		22,000		22,700		23,300		2.7	%
Philippines	2,647		3,013		3,006		3,024		3,066		2.1	%
Sri Lanka	825		721		728		694		690	F	-2.2	%
Thailand	4,919		2,286		1,912		1,900	F	1,900	F	-11.0	%
Viet Nam	2,859		2,951		2,956		2,897		2,950		0.1	%
SUB-TOTAL	144,674		155,659		156,840		159,109		160,302		1.0	%

Unit: 1000 heads

Country	1991	1998	1999	2000	2001	Annual Growth Rate 1991 - 2001	
DEVELOPED COUNTRIES							
Australia	-	-	-	-	-	-	%
Japan	-	-	-	-	-	-	%
New Zealand	-	-	-	-	-	-	%
SUB-TOTAL	-	-	-	-	-	-	
ASIA-PACIFIC TOTAL	144,674	155,659	156,840	159,109	160,302	1.0	%
REST OF WORLD	5,494	5,055	5,227	5,337	5,422	-0.6	%
WORLD	150,168	160,714	162,067	164,446	165,724	0.9	%

Table 2 Livestock Population trends in India

Species	1951	1956	1961	1966	1972	1977	1982	1987	1992	1997*
Cattle	155.3	158.7	175.6	176.2	178.3	180.0	192.5	199.7	204.6	197.7
Adult Female Cattle	54.4	47.3	51.0	51.8	53.4	54.6	59.2	62.1	64.6	NA
Buffalo	43.4	44.9	51.2	53.0	57.4	62.0	69.6	76.0	84.2	88.6
Adult Female buffalo	21.0	21.7	24.3	25.4	28.6	31.3	32.5	39.1	43.8	NA
Total Bovines	198.7	203.6	226.8	229.2	235.7	242.0	262.4	275.8	289.0	286.5
Sheep	39.1	39.3	40.2	42.4	40.0	41.0	48.8	45.7	50.8	56.8
Goats	47.2	55.4	60.9	64.6	67.5	75.6	95.3	110.2	115.3	120.8
Horses & Ponies	1.5	1.5	1.3	1.1	0.9	0.9	0.9	0.8	0.8	0.6
Camels	0.6	0.8	0.9	1.0	1.1	1.1	1.1	1.0	1.0	0.9
Pigs	4.4	4.9	5.2	5.0	6.9	7.6	10.1	10.6	12.8	13.6
Mules	0.1	0.0	0.1	0.1	0.1	0.1	0.1	0.2	0.2	0.1
Donkeys	1.3	1.1	1.1	1.1	1.0	1.0	1.0	1.0	1.0	0.6
Yaks	NC	NC	0.0	0.0	0.0	0.1	0.1	0.0	0.1	0.1
Total Livestock	**292.8**	**306.6**	**335.4**	**344.1**	**353.6**	**369.0**	**419.6**	**445.3**	**470.9**	**479.9**
Poultry	73.5	94.8	114.2	115.4	138.5	159.2	207.7	275.3	307.1	352.0
Dogs	NC	NC	NC	NC	NC	NC	18.5	18.0	21.8	NA

NC: Not Collected
*Based on provisional results from States/UTs
Source: Department Of Animal Husbandry & Dairying, New Delhi

Table-3: State-wise production of milk during 1999-2000 to 2001-02.

State/ UTs	Milk ('000 tones)		
	1999 - 2000	2000 - 2001*	2001 - 2002**
Andhra Pradesh	5122	5521	5145
Arunachal Pradesh	45	46	55
Assam	733	738	894
Bihar including Jharkhand	3740	3878	4068
Goa	43	44	47
Gujarat	5255	5317	5573
Haryana	4679	4849	4976
Himachal Pradesh	741	760	810
Jammu-Kashmir	1286	1037	1088
Karnataka	4473	4598	5357
Kerala	2673	2771	2907
Madhya Pradesh including Chattisgarh	5600	5806	6091
Maharashtra	5706	5850	6024
Manipur	67	69	73
Meghalaya	62	64	71
Mizoram	18	14	11
Nagaland	50	50	54
Orissa	847	875	865
Punjab	7700	7984	8375
Rajasthan	7260	7455	6330
Sikkim	35	35	46
Tamil Nadu	4574	4899	4629
Tripura	49	51	53
Uttar Pradesh including Uttaranchal	14153	14840	16506
West Bengal	3465	3470	4079

A&N Islands	23	24	25
Chandigarh	42	44	46
D & N Haveli	10	10	11
Daman & Diu	1	1	1
Delhi	290	292	321
Lakshdeep	1	1	1
Pondicherry	36	37	38
All India	**78779**	**81430**	**84570**

*Provisional **Anticipated achievement
Source: Department of Animal Husbandry and Drying, Ministry of Agriculture, New Delhi

Livestock figures of 1992 revealed highest population of female buffalo (over 3 yrs) in Uttar Pradesh, followed by Andhra Pradesh., Rajasthan, Punjab, Bihar including Jharkhand, M.P. including Chattisgarh and Maharastra, whereas in North Eastern states the population density is very low. Six states namely Punjab, Delhi, Haryana, A.P., U.P, Gujrat have higher buffalo population than cattle (Table 4). The population dynamics of different breeds of buffaloes indicate that Murrah breed constitutes the major population followed by Mehsana, Surti, Nilli-Ravi (Table 5). The Toda breed is in the verge of extinction and trends also depicts its alarming picture. The annual growth rate for total buffalo as well as female buffalo population is highest in Nagaland followed by Assam, Mizoram and Madhya Pradesh (Table 6). However, the highest population density (number / square kilometer) is in Punjab and Haryana, followed by Uttar Pradesh and Andhra Pradesh (Table 7).

Punjab has the highest per capita milk availability and also achieved increasing trend from 1996 to 2001 whereas availability was lowest in North Eastern states of the country (Table 8). Meat production figures from 1981 to 2000 indicated that the buffalo meat availability increased by 11 times (127 vs. 1421) whereas beef and veal, mutton and lamb, goat meat and pig meat availability increased by 18.5, 1.8, 1.7 and 7.5 times respectively. Total meat production in

Table 4:   State wise milch bovine population   (in thousands)

State/UT's	Female Cattle			Female buffalo over 3 yrs.	Milch bovine total
	Cross bred over 2 ½ yrs.	Indigenous over 3 Yrs.	Total		
Andhra Pradesh	222	2803	3025	4822	7847
Arunachal Pradesh	6	88	94	2	96
Assam	138	3003	3141	328	3469
Bihar including Jharkhand	65	5301	5366	2587	7953
Gujarat	127	2009	2136	3147	5283
Goa	4	26	30	20	50
Haryana	163	555	718	2263	2981
Himachal Pradesh	125	615	740	473	1213
Jammu-Kashmir	252	780	1032	418	1450
Karnataka	298	4288	4586	2361	6947
Kerala	919	840	1759	113	1872
Madhya Pradesh including Chattisgarh	89	8639	8728	3515	12243
Maharashtra	912	4938	5850	3223	9073
Manipur	29	150	179	38	217
Meghalya	10	189	199	10	209
Mizoram	3	20	23	3	26
Nagaland	46	71	117	9	126
Orissa	262	3963	4225	433	4658
Punjab	737	454	1191	3606	4797
Rajasthan	46	4539	4585	4069	8654
Sikkim	16	48	64	1	65
Tamil Nadu	847	2617	3464	1805	5269
Tripura	45	259	304	8	312
Uttar Pradesh including Uttaranchal	658	6353	7011	10143	17154

State/UT's	Female Cattle			Female buffalo over 3 yrs.	Milch bovine total
	Cross bred over 2 ½ yrs.	Indigenous ovⁿ 3 Yrs.	Total		
West Bengal	430	5256	5686	233	5919
A & N Islands	1	18	19	5	24
Chandigarh	5	-	5	15	20
D & N Haveli	-	12	12	2	14
Delhi	8	17	25	151	176
Lakshadweep	-	2	2	-	2
Pondicherry	30	13	43	3	46
Daman & Diu	-	2	2	-	2
All India	6,493	57,868	64,361	43,806	108,167

Source : Livestock Census 1992, GOI

Table-5: Population dynamics of various breeds of buffalo ('000)

Breed	Total population	Breedable females	Females bred pure	Stud bulls	Trend
Murrah	1384	753	600	12.5	Increasing
Mehsana	543	296	207	1.1	Increasing
Surti	472	260	182	1.8	Increasing
Nilli-Ravi	462	230	184	10	NA
Nagpuri	357	154	123	6.5	Increasing
Jaffarabadi	289	161	128	1.8	Increasing
Bhadawari	173	82	58	0.6	Increasing
Toda	6	3	3	0.1	Decreasing

(Tantia *et al.*, 1994)

Table 6: Total Number (in thousands) of buffaloes and annual growth rate

States/UTs	Total Population (1992)	Annual Growth Rate (%) 1987-92	
		Female	Total
Andhra Pradesh	9154	1.72	0.89
Arunachal Pradesh	9	-3.58	-5.59
Assam	958	11.32	8.99
Bihar	5353	1.90	1.90
Gujarat	5268	3.10	3.19
Goa	44	2.90	1.92
Haryana	4373	2.14	2.70
Himachal Pradesh	703	0.27	-2.41
Jammu-Kashmir	732	5.16	5.32
Karnataka	4251	0.55	1.05
Kerala	296	-1.99	-1.97
Madhya Pradesh including Chhatisgarh	7970	1.86	4.60
Maharashtra	5448	2.82	2.77
Manipur	115	-1.86	-3.86
Meghalya	34	2.90	3.96
Mizoram	7	5.92	6.96
Nagaland	34	14.87	17.78
Orissa	1539	0.32	0.43
Punjab	5238	-1.41	-1.24
Rajasthan	7743	3.90	4.07
Sikkim	2	0.00	0.00
Tamil Nadu	2814	0.79	-2.10
Tripura	20	4.10	4.56
Uttar Pradesh	20086	2.28	1.95
West Bengal	1010	-1.97	-2.78

A & N Islands	15	0.00	1.39
Chandigarh	23	2.98	2.83
D & N Haveli	3	0.00	0.00
Delhi	249	-1.64	-2.66
Lakshadweep	0	-	-
Pondicherry	7	-7.79	-6.89
Daman & Diu	1	-	-
All India	**83499**	**1.94**	**1.91**

Source: Directorate of economics and statistics M/o Agriculture (1997)

Table-7: Buffalo population in different states of India (Concentration /sq km) and milk yield

States	Buffaloes / per square Km	Yield per animal in milch per day
Andhra Pradesh	33	2.2
Bihar	31	3.7
Gujarat	27	3.7
Haryana	99	5.3
Himachal Pradesh	13	3.3
Jammu-Kashmir	3	3.5
Karnataka	22	2.3
Kerala	8	3.6
Madhya Pradesh	18	3.0
Maharashtra	18	3.1
Orissa	10	1.4
Punjab	104	5.7
Rajasthan	23	3.9
Sikkim	16	3.3
Tamil Nadu	22	–
Tripura	2	–
Uttar Pradesh	69	3.5

States	Buffaloes / per square Km	Yield per animal in milch per day
West Bengal	11	4.2
Arunachal Pradesh	–	–
Assam	12	1.9
Goa	12	3.3
Manipur	5.0	3.2
Meghalaya	2.0	1.8
Mizoram	–	–
Nagaland	2.0	3.1

Table-8: Per capita availability of milk during 1996-97 to 2000-01 (g/day)

States/UTs	1996-97	1997-98	1998-99	1999-2000*	2000-01
Andhra Pradesh	169	167	179	187	199
Arunachal Pradesh	115	109	110	106	104
Assam	79	78	78	77	77
Bihar including Jharkhand	100	98	97	104	106
Goa	72	71	74	75	75
Gujarat	289	290	294	301	300
Haryana	617	630	641	653	666
Himachal Pradesh	316	314	309	308	309
Jammu-Kashmir	302	345	354	360	285
Karnataka	191	216	227	237	241
Kerala	199	204	209	228	235
Madhya Pradesh including Chattisgarh	192	194	193	195	198
Maharashtra	161	161	172	173	175
Manipur	75	74	75	74	75
Meghalya	74	73	72	71	72
Mizoram	29	53	61	53	40

Nagaland	86	82	82	82	81
Orissa	54	53	57	65	67
Punjab	823	861	877	903	926
Rajasthan	322	348	365	376	379
Sikkim	190	188	179	175	173
Tamil Nadu	183	185	192	204	217
Tripura	36	45	58	36	37
Uttar Pradesh including Uttaranchal	215	221	227	231	237
West Bengal	123	123	122	121	120
A & N Islands	170	169	165	167	169
Chandigarh	148	147	140	132	135
D & N Haveli	89	86	166	200	194
Daman & Diu	16	17	21	15	14
Delhi	61	59	61	58	57
Lakshdweep	107	97	95	91	91
Pondicherry	44	43	61	39	38
Total	**202**	**207**	**213**	**219**	**223**

*Based on provisional estimates of production and projected population

the country also increased by 5.5 times over the last two decades (Table 9). Total meat production is at the tune of 4.0 million tones per annum (Table 10). The slaughter rate per annum in relation to the population of animals is about 6, 10, 99, 31 and 39 percent for cattle, buffalo, pig, sheep and goat respectively.

Agriculture per se has been only second to textile, gems and jewelry in export during last few years. Leather and manufactures with carpets contribute more than 65.5 percent of it. Import of fertilizer has been of the order of 34,074 crores and edible oil (2.6 million tones), bill is around 5932.76 crores. Buffaloes having higher fat can help to slash the burden of edible oil bills substantially (Table 11).

With the ever increasing demand for food, clothing and shelter to support the burgeoning population in the country, there is a strong need to have a systematic appraisal of the soil, climatic and livestock resources to develop an effective alternate land and livestock use

Table-9:   Meat production during 1981 to 2000 (x 1000 tones)

Year	Buffalo	Beef & Veal	Mutton & Lamb	Goat Meat	Pig Meat	Poultry Meat	Total Meat
1981	127	78	125	277	75	120	850
1982	130	80	132	298	80	130	865
1983	132	80	134	302	80	137	1010
1984	148	149	135	346	82	150	1047
1985	152	150	141	358	85	161	1106
1986	163	160	147	370	86	180	1261
1987	207	239	162	380	80	193	1630
1988	290	232	148	378	357	225	2974
1989	936	845	160	385	359	289	3596
1990	1048	1271	173	410	360	334	3710
1991	1176	1185	168	455	364	362	3800
1992	1182	1216	167	456	397	382	3950
1993	1182	1276	169	466	403	454	4052
1994	1204	1292	171	470	408	507	4259
1995	1351	1365	194	450	420	479	4319
1996	1382	1370	218	454	514	479	4417
1997	1403	1378	222	458	533	527	4521
1998	1380	1401	226	462	543	540	4552
1999	1410	1421	228	466	560	559	4644
2000	1421	1442	229	467	560	575	4694

Table 10: Details of production of meat and meat products (MT) from 1994 to 1998

Source of meat	Year				
	1994	1995	1996	1997	1998
Buffalo	1200	1204	1204	1205	1010
Sheep & goat	637	647	669	670	675
Pig	365	420	420	400	420
Cattle	1290	1292	1292	1292	1295
Poultry	442	578	480	580	600
Total	3569	4141	4065	4147	4000

Table 11 Export of principal commodities (Rupees in crores)

Commodities	1999–2000	2000–2001	% Change
1. Agriculture & allied products	15880.01	17666.38	11.25
2. Plantations	3219.52	3164.57	(-) 1.71
3. Leathers & manufacturers	6890.87	8915.23	29.38
4. Carpets	2795.36	2656.03	(-) 4.96
5. Gems and jewellery	32509.43	33760.71	3.85
6. Textile	39733.22	49318.18	24.12

plan. Since the soils and climatic conditions of a particular region largely determine the suitability of different crops and livestock for their yield potential. Intensive efforts have been made to map agro-ecological regions (uniforms in their soil-site characteristics) in identifying suitable crops to develop suitable cropping pattern for increasing crop and livestock production and livestock raising should also be integrated to complete the system.

Climate

Geographical area of India is about 329 million hectare and is criss crossed by a large number of small and big rivers and only 144 million hectare is net sown area. The presence of the great Himalayas in the North and of the ocean in the South are the two major influences operating on the climate of the country. The first poses an impenetrable barrier to the influence of cold winds from central Asia, and gives the sub-continent the elements of tropical type of climate. The second, which is the source of cool moisture-laden winds, gives it the elements of the oceanic type of climate. The country has a great diversity and variety of climate and an even greater variety of weather conditions. The climate ranges from continental to oceanic, from extremes of heat to extremes of cold, from extreme aridity and negligible rainfall to excessive humidity and torrential rainfall. It is, therefore, necessary to avoid any generalization as to the prevalence of any particular kind of climate, not only over the country as a whole but over major areas in it. There are four seasons, (i) Winter (January – February), (ii) Hot weather summer (March – May), (iii) Rainy South – Western monsoon (June – September) and (iv) post monsoon (October – December).

Rainfall

Rainfall in India is dependent in differing degrees on the South - West and North - East monsoons, on shallow cyclonic depressions and disturbances and on violent local storms which form regions where cool humid winds from the sea meet hot dry winds from the land and occasionally reach cyclonic dimension. Most of the rainfall in India takes place due to South West monsoon between June to September except in Tamil Nadu where it is under the influence of North east monsoon during October and November. The rainfall shows wide variability, unequal seasonal distribution, still more unequal geographical distribution and the frequent departures from the normal. It generally exceeds 1000 mm in areas to the East of Longitude 78 degree E. It extends to 2500 mm along almost the entire West Coast and Western ghats and over most of Assam and Sub-Himalayan West Bengal. On the West of the line joining Porbandar to Delhi and then to Ferozpur the rainfall diminishes rapidly from 500 mm to less than 150 mm in the extreme west. The Peninsular has large areas of rainfall less than 600 mm with pockets of even 500 mm.

Temperature:

The variations in temperature are also marked over the Indian sub-continent. During the winter from November to February, due to the effect of continental winds over most of the country, the temperature decreases from South to North. The mean maximum temperature during the coldest months of December and January varies from 29^0 C in some part of the peninsula to about 18^0 C in the North, whereas the mean minimum varies from about 24^0 C in the extreme South to below 5^0 C in the North. March to May is usually a period of continuous and rapid rise in temperature. The highest temperature occurs in North India, particularly in the desert regions of the Northwest where the maximum may exceed 48^0 C. With the advent of South West monsoon in June, there is a rapid fall in the maximum temperature in the central portions of the country. The temperature is almost uniform over the area covering two thirds of the country, which gets good rain. In August/September there is a marked fall in temperature when the monsoon retreat from North in September. In North West India, in the month of November, the mean maximum temperature is below 38^0 C and the mean minimum below 10^0 C. In the extreme North a temperature drop below freezing point.

Evaporation

Evaporation rate closely follow the climate seasons, and reach peak in the summer months of April and May and the central areas of the country display the highest evaporation rates during this period. With the onset of monsoon towards the end of June, there is a marked fall in the rate of evaporation. The annual potential evaporation ranges between 150 to 250 cm over most parts of the country. Monthly potential evaporation over the peninsula increases from 15 cm in December to 40 cm in May. In the North East, it varies from 6 cm in December to 20 cm in May. It rises to 40 cm in June in West Rajasthan. After the onset of monsoon potential evaporation decreases generally all over the country.

Rivers

India is blessed with a number of rivers and as many as 12 of them are classified as major rivers whose total catchments area is 252.8 million hectare. Of the major rivers, the Ganga - Brahamputra Meghana system is the biggest with catchments area of about 110 million hectare that constitute more than 43 percent of the total catchments area of the major rivers in the country. The other major rivers with catchments area (million hectare) more than 10 m.ha. are Indus (32.1), Godavari (313), Krishana, (25.9) and Mahanadi (14.2). Subernarekha with 1.9 m.ha. catchments area is the largest river among the medium rivers in the country. The average run-off in the river system of the country has been assessed as 1869 km^3, of this utilizable portion by conventional storage and diversion is estimated as about 690 km^3. Additionally, India has 632 km^3 substantial replenishnable ground water potential (India, 2002).

Zoning of country

An agro-climatic zone refers to a land unit in terms of its major climates and suitability for a certain range of crops and cultivars. Agro-ecological region is defined as an area of the earth's surface characterized by distinct ecological responses to macroclimates as expressed by soil, flora, fauna and aquatic systems. The agro-ecological region or zone is the land unit carved out of the agro-climatic region when super imposed on land forms and on the kinds of soils and soil conditions that act as modifiers to climate and length of growing period.

The concept of dividing the world into various zones has a rich and varied history. Of particular interest is the FAO's Agro-Ecological Zones Project that set out criteria based on plant growth days, soil types, rainfall, and temperature regimes.

Since the inception of formal planning in India in 1951, a number of attempts have been made to categorize the country into different agro-ecological regions and zones. Many different approaches were made to classify the land area into climatic regions. Carter (1954) divided India into six climatic regions ranging from arid to peri-humid based on the Thornthwaite criteria for climatic classification. Singh (1974) classified the country according to parameters like units of cattle, buffalo, draft animals, milk cows and milk buffalo per 100 ha of cropped land. He also classified the country based on percentage of total crop area planted to fodder crops over the period 1961-1966 as well as cropping pattern zones. Krishan (1988) delineated 40 climatic zones based on major soil types and moisture index. Muthariah (1988) classified the country on the basis of livestock density using the 1977 livestock census as well as per capita milk production for 1985 - 1986.

The Planning Commission divided the country into 15 broad agro-climatic zones based on physiography and climate (Alagh *et al* 1989). This classification is based on physical conditions such as topography, soil type, geographical formation, rainfall, cropping pattern and development of irrigation. Since these zones were too broad to serve the purpose of planning, project teams were set up for each agro-climatic zone to divide each zone into sub-zones under the National Agricultural Research Project (NARP) of Indian Council of Agricultural Research (ICAR) and suggested 120 micro agro-climatic sub-zones after taking into account the rainfall patterns, temperatures, soil types and existing cropping patterns of each state as a unit (Saxena, 1989; Ghosh, 1991).

The major limitation of these approaches has been non-uniform application of criteria used, and the use of states as a unit for sub-divisions. This resulted in the creation of many sub-zones having similar agro-climatic characteristics but occurring in different states. These approaches also did not give adequate consideration to soils. Hence a need was felt to generate an agro-ecological region map of the country giving due importance to soil environments (Table 12 and 13). With this viewpoint, the National Bureau of Soil Survey and Land Use Planning of ICAR brought out an agro-ecological map of the country through several approximations, using physiography, soils, bio-climatic types and growing period. During 1989, the Bureau brought out a 54-zone map, which had too much delineation. The map and the write up were finally discussed by a committee and the agro-ecological region

Table 12: Estimated areas of broad soil groups in India

Soil type	Area in million hectare
1. Alluvium derived soils	74.3
* Inland	63.1
* Deltaic alluvial	6.8
* Coastal alluvial	4.4
2. Black soils	73.2
3.Red, yellow and laterite soils	87.6
4. Soils of the desert region	30.0
5. Soils of Himalayan region	28.7

(Sehgal, 1995)

Table 13: Total and utilizable quantities of plant nutrients (N, P_2O_3, K_2O) in agricultural, rural and urban wastes during the 1991 – 2025 AD*

Waste material	Quantity (million tones)		
	1991	2011	2025
A. Crop residues			
1. Total	3.22	4.82	5.79
2. Utilizable	1.52	2.21	2.66
3. Fertilizer equivalent value	0.76	1.10	1.33
B. Animal Wastes			
1. Total	12.4	18.02	21.64
2. Utilizable	7.4	10.81	12.98
3. Fertilizer equivalent value	3.7	5.40	6.49
C. Municipal wastes			
1. Total	0.09	0.06	0.07
2. Utilizable	0.04	0.06	0.07
3. Fertilizer equivalent value	0.02	0.03	0.04

D. Sewage wastes

1. Total	0.18	0.26	0.32
2. Utilizable	0.18	0.26	0.32
3. Fertilizer equivalent value	0.15	0.21	0.27

E. Total

1. Total	15.95	23.17	27.83
2. Utilizable	9.19	13.35	16.03
3. Fertilizer equivalent value	4.65	6.75	8.13

(Sekhon, 1996)

map with 21 delineations was approved (Sehgal *et al* 1990). It was hoped that the map would serve India's needs in generating agro-technologies and transferring these to other comparable areas. Further, it may help in substantial land use plan.

Sastry (1993) made a good attempt to collect secondary data and information on various aspects of animal husbandry and classified the data agro-climatic region-wise. The agro-climatic regions considered, were 15 as described by Planning Commission. These 15 regions were regrouped into five regions viz., Himalayan regions (1 & 2); Gangetic Plain regions (3 to 6); Plateau and Hill regions (7 to 10); Coastal regions (11 & 12); and Island region to integrate with livestock production. Some of the general conclusions drawn by Sastry (1993) were: -

1. The numbers of various species of livestock are increasing in the country in general with higher growth rates being recorded for buffaloes, goats and pigs and a marginal rate in the case of cattle and sheep.

2. Paradoxically, the numbers of different livestock appear to be more than optimal compared to land, feed and fodder resources, but there is also a deficit in livestock products as (i) there is not a single animal product that is adequate as per requirements and (ii) the number of animals per 1000 humans is at a historically low level for every species.

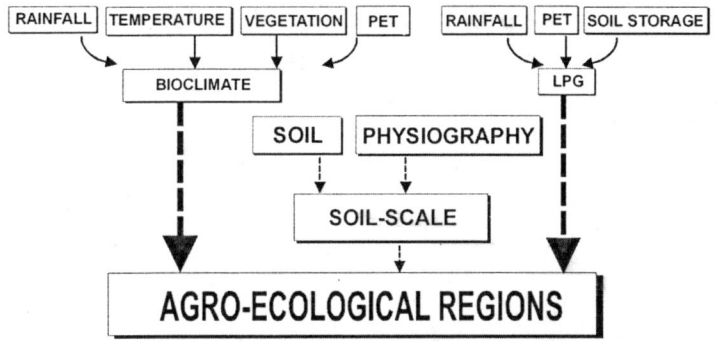

Schematic diagram showing the parameters used for
preparing the map of agro-ecological regions of India

Map-1

Agro-ecological Regions of India

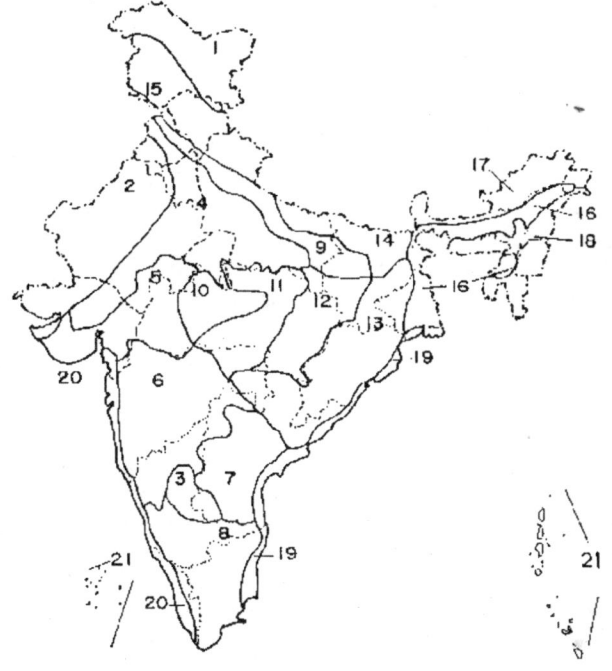

1. Western Himalayas (cold arid)
2. Western plain & Kutch Peninsula
3. Deccan plateau – hot arid
4. Northern plain & central highlands
5. Central highlands & Kathiawar Peninsula
6. Deccan plateau – hot semi arid
7. Deccan plateau & eastern ghat (red & black soil)
8. Eastern ghat & Deccan (red loamy soil)
9. Northern plain
10. Central highlands (Malwa & Bundelkhand)
11. Deccan plateau & Central highlands
12. Eastern plateau (Chhataisgarh)
13. Eastern plateau (Chotanagpur) & eastern ghat
14. Eastern plain
15. Western Himalayas (worm sub humid)
16. Assam Bengal plains
17. Eastern Himalayas
18. North Eastern hills (Purvachal)
19. Eastern coastal plains
20. Westren ghat & coastal plains
21. Islands of Andaman Nicobar & Lakshadweep

Map-2

Agro-Climatic Zones if India

1.	Western Himalayan Region	9.	Western Plateau and Hills Region
2.	Eastern Himalayan Region	10.	Southern Plateau and Hills Region
3.	Lower Gangetic Plain Region	11.	East Coast Plains and Hills Region
4.	Middle Gangetic Plain Region	12.	West Coast Plains and Hills Region
5.	Upper Gangetic Plain Region	13.	Gujrat Plains and Hills Region
6.	Trans-Gangetic Plain Region	14.	Western Dry Region
7.	Eastern Plateau and Hills Region	15.	The Islands Region
8.	Central Plateau and Hills Region		

Map-3

3. The people's needs for animals and their economic importance for the farmers seems to remain as in the past. Only the efficiency of animal production has come down over the years. The most important reason for this is the ever-increasing depletion of traditional livestock feed resources, especially the common land resources in villages.

4. The entire milieu of a region (climate, terrain, soil, ecology, people and through them crop production) determine which species mix of animals are to be reared there.

5. Need for work bullocks still continues to be the prime moving force behind cattle rearing and tractorization has not yet caused any great reduction in bullocks numbers.

6. Shortage of feeds, especially grazing land, is universal and is going to persist for a long time to come.

The conclusions having regional ramifications for dairy production were: -

◉ Cattle are important in all the agro-climatic regions, predominantly as work animals; cattle as dairy animals (crossbreds) are prevalent in the West Coast Plains and Ghats region and in some pockets in the Western Plateau and Hills, Southern Plateau, East coast Plains and Hills in the two Himalayan regions.

◉ Milk production and marketing is well developed in agriculturally better-developed parts of the North Western, Western and Southern parts of the country. The importance of buffalo as a dairy animal decreases as one-move eastwards and southwards, but their concentration in peri-urban areas is generally high.

◉ Excluding the two Himalayan regions, in all the other regions, dairies remain appear to have feed constraints, especially grazing land.

It is evident that India is very diverse with regard to land use patterns, cropping systems and livestock husbandry. Various workers have classified the country into different agro-climatic regions since the start of the planning process in the country. But the suitability of macro-regions from a farming system perspective, particularly livestock farming combined with crop husbandry, needs to be reconsidered. Parameters like livestock density; productivity of milking cattle and buffalo; sheep, goat and pig productivity; feeds and fodder availability, milk availability, resource infrastructure, marketing, etc. need to be considered for any perspective plan in context of livestock productivity. Besides the above parameters, socio-economic parameters like family size, social status, income, literacy rates, customs and traditions also need to be included while classifying the country into different macro and micro regions.

Since, soil and climatic conditions of a region largely determine the suitability of crops growth period and their yield potential; almost same is applicable to livestock husbandry. The degree of correlation/ dependence of animal raising has decreased on soil and climate because of human interventions-

1. Transportation of feeds and fodder towards region of deficit

2. Animal housing systems have been developed to mitigate the vagaries of climate, to a large extent.

Based on above aspects an effort hase been made to integrate and classify livestock production into different agro-climatic regions as identified by planning commission of India. In the proposed zoning (Table 14) i.e. animal agro-climatic regions, minor deviations /changes have been adopted.

1. Trans and upper gangetic plains have been merged because of similar animal husbandry practices of the farmers.

2. Lower gangetic plains region, portion of Eastern plateau and hill region specially the Jharkhand and Orrisa have been regrouped under the name of eastern plateau and plain region keeping in view the status of livestock sector in those particular areas.

Table14: Integration of animal husbandry with the agro climatic zones of India

Planning Commission agro climatic zones	IGFRI perspective
1. Western Himalayan region	1. Western Himalayan region
2. Eastern Himalayan region	2. Eastern Himalayan region
3. Lower gangetic plains region	3. Eastern plateau and plains region
4.Middle gangetic plains	4. Middle gangetic plains
5. Upper gangetic plains region	5. Trans and upper gangetic plains
6. Trans gangetic plains region	
7. Eastern plateau and hills region	
8. Central plateau and hills region	6. Central plateau and hills
9. Western plateau and hills region	7. Western plateau and hills
10. Southern plateau and hills region	8. Southern plateau and hills
11. East coast plains and hills region	
12. West coast plains and ghat region	
13. Gujrat plains and hills region	
14. Western dry region	9. Western dry zone
15. Island region	10. Coastal and Island region

3. East coast plains and hill region, west coast plains and ghat region, Island region are brought under the one region i.e. coastal and Island region because of similar status of livestock enterprise.

4. Dry areas of Gujrat plains and hills region have been merged with the Western dry zone while the rest portion have been placed under the western plateau and hills region.

1. Western Himalayan region

The region encompasses Jammu & Kashmir, hilly areas of Punjab, Himachal Pradesh and Uttaranchal. Indian hill and Tarai zones account for 26.8 million hectare. As a result of high rainfall, the hill soils are exposed to extensive erosion, at higher altitude causing the soil shallow in depth, leached in bases and making them moderate to highly acidic, light in texture with low water holding capacity, and thereby make the soil deficient in certain nutrients. In the cold and arid zones, the soils are neutral to alkaline (Katoch and Singh,2000). The major crops include wheat, rice, maize, minor millets (ragi), pulses, oil seeds etc. Apple is the dominant fruit crop of Western Himalayas comprising of 38 % of cultivated areas of Himachal Pradesh.

Among all the livelihood activities operating in the western Himalayas, dairying is considered as creamy activity, which has lot of potential for household protein security as well as economic up-liftment of the inhabitants. In the low hills areas (Doon vally, Tarai, Bhabar and Shivalik areas), animal husbandry plays a crucial role for milk and meat. The region is therefore, abundance in high yielding milch cattle and buffaloes. In the mid hills region, cattle are reared for milk as well as draught power, buffaloes and goats are maintained for milk production only. In the high altitude of Western Himalayas are predominated by sheep and goat due to abundant availability of alpine pasture. In Uttaranchal, among all the ruminants (43.22 million), cattle constitute 46 % whereas buffalo represented by 19 – 21%, sheep and goat constitute 8.2% and 25% respectively (Vir Singh,2002). The livestock enterprises contribute 20 –40% of the farm levels income; the herd size per family is 1 –2 buffaloes, 0.6 cows, 1.4 bullocks and 3.5 goats.

Buffalo is the main milch animal in the Uttaranchal hills. Tarai breed of buffalo yielding 2 –3 liter of milk are mostly seen in Tarai areas. Presently, improved type of buffaloes (cross of *Murrah* and *Bhadawari*) is well received by the farmers in hills. The milk yield in buffalo is upto 6.0 liter in mid hills and 12.0 liter in *Murrah buffalo* in lower hills has been reported.

Cynodon dactylon and *Ischoemum angustifolium* are major green forage source for ruminants. In addition to that, farmer family also own trees/ shrubs like Grewia optiva, *Celtis australis* or *Melia azaderach* (Mitha neem) on the riser of the crop field to augment the green fodder supply during lean period (Mittal and Rai, 2002). The commonly cultivated fodder in Western Himalaya include Berseem (*Trifolium alexandrinum*), lucerne (*Medicago sativa*), oats (*Avena sativa*), Jowar (*Sorghum bicolor*), Maize (*Zea mays*) etc. and their nutritive value are comparable or marginally higher when cultivated in plains.

2. Eastern Himalayan Region

It includes Assam, Darjeeling district of West Bengal, Sikkim and the states lying beyond Assam in the Far East namely Arunachal Pradesh, Manipur, Mizoram, Tripura, Nagaland, Meghalaya. Soil erosion in the high altitude and flooding are the main problems of the region. Tea gardening and associated activities are important employment source in hilly parts of Assam and West Bengal. The Jhoom cultivation practiced in the nearly one third of the cultivated areas has caused denudation and degradation of soils, with the resultant heavy run off, massive soil erosion and floods in the lower reaches and basins. Rearing of livestock is important livelihood activity in hill farming system. Pig farming is most developed; goat is coming the next for meat production. Co-operatives have been initiated among pig farmers in the states of Nagaland, Meghalaya and other far East states.

Livestock species include cattle, buffalo, goat, pig, mithun, and yak. In Brahamaputra valley of eastern Himalayas, swamp buffaloes (*Kaziranga*) are reared by the local inhabitants for both draught power as well as milk production. The milk production of these buffaloes is very low ranging from 2 -3 liter per head per day. These buffaloes are usually dark grey or black to albinoids in colour and can thrive well in high rainfall and high humid climate. Siri is the native cattle breed of Darjeling district and Sikkim. Milk yield is around 2 – 6 liter per head per day. Ruminant livestock unit (RLU) is maximum in Assam (5.4 million) and least in Mizoram (0.02 million). The cattle population predominates in all the states followed by goat, buffalo and sheep. But in Manipur, buffalo population is higher than goat (Anandan *et al.*, 2001).

Mithun and yak are predominantly reared through grazing on

pasture. The dominant plant species in the grazing field are *Bromus trypogen, Potentilla fulgen, Fimbristylis diphyla, Fragaria vesca* (Basu and Chatterjee, 2001).

Livestock are generally fed locally available raw feed materials like roots, stem, leaves collected from Jungle, kitchen waste etc. Animal are left for grazing in the forest immediately after morning milking and allowed for grazing up to evening and returned back around 4 – 00 PM for the evening milking. At the time of milking the concentrate is offered consisting of boiled maize, rice grain, radish with leaves and mustard cake.

3. Eastern Plateau and Plains Region

It encircles the territorial boundaries of West Bengal, Jharkhand, Orissa. The food habit of the human beings tilted towards fish and poultry farming. Thus, the livestock enterprise received lesser attention for resource allocation by the farmers.

Rice is the staple food for this region. The other crop includes cultivation of wheat, maize, bajra, mustard, til, potato, jute, vegetables etc. The livestock are mostly dominated by non-descript or Zebu cattle. The important buffalo breeds are *Kujang, Manda, Sambalpuri* (Orissa). There is no known breed of cattle in this region however following strains of cattle are found; namely *Ghumsuri, Bingharpuri* (Lesser known strain of Orissa), *Bengali* (a lesser known strain of West Bengal). In some areas of West Bengal like Burdwan, Nadia, 24 parganas, Murshidabad district, crossbred cattle are reared by the farmers. The goats occupy a significant proportion of livestock population because of their characteristics meat quality and its ability to survive under poor management conditions. Black Bengal breed of three categories viz.; Black, grey and white are famous for its meat quality. As a result of demand for milk, crossbred cows as well as *Murrah* buffaloes are reared in the peri urban areas. In certain districts, sheep (*Garol*) are reared both for coarse wool and meat. In rural areas the poorer community rear pigs as a secondary source of income and for meat.

The livestock are maintained mainly on grazing and agricultural crop residues like paddy straw with little access of agricultural byproducts and kitchen waste. The pasture consisted of mainly *Dub* grass (*Cyanodon dactylon*). In addition to grazing mostly on bunds and roadside grasses, the animal receives 4 - 6 kg paddy straw as a bulk roughage (Das *et al.*, 2003) . At the time of milking mustard cake or de-oiled rice bran either singly or along with maize grain or gram chuni are offered to lactating animals. The concentrate supplementation to the animal solely depends on the economic status as well as the productivity of animals. But in the peri-urban areas, lactating animal are solely maintained on paddy straw and concentrate based diet with little green forage like hybrid *Napier*, *Dinanath* grass, *Bracharia* etc.

4. Middle Gangetic Plain

This area encompasses plains of Bihar and eastern parts of Uttar Pradesh. The rainfall is high and 39% of the gross cropped area is irrigated and the cropping intensity is 142%. Rice in Kharif and maize and wheat in Rabi are the main crops. Pulses, oil seed, sugarcane, vegetables, fruit crops are also widely grown. The preference of the livestock is in the order of buffalo followed by cattle and goat. The people of this region rear *Murrah* buffaloes. The native cattle strain of Bihar is *Purnea, Sitamarhi*. Crossbred animals are widely distributed through out the region (Roychaudhary, 1997). The feeding of livestock is mainly from crop residues grown by the farmers. The straw of wheat / paddy constitute the basal dry roughage of the animals. The minor source of dry roughage includes pulse straw.

Keeping in view the relatively higher productivity of the animals, advanced farmer also cultivate green fodder like maize, sorghum, berseem, hybrid napier etc. Fodder production practices are popular by the farmers in Begusarai, Samastipur and Muzaffarpur because of provision of good quality seeds and extension services provided by milk cooperatives (Sohane *et al.*, 2001). The farmers usually chaffed the green fodder and offered to animal for better utilization. The use of concentrate pellets is common by the members of milk producer cooperative societies. The balanced concentrate mixture available in the market is mostly used by the livestock owner of peri-urban / urban areas. In rural areas home made concentrate mixture consisting of oil cakes along with wheat bran / wheat flour are usually offered to animals. The concentrate mixture is mixed with wheat straw and water to make *"Sani"* and offered to animal at the time of milking. The green fodder, if available, is also added in the *"Sani"* to fortify the ration.

In the areas of Uttar Pradesh, the farmer offered more concentrate mixture to milch buffalo during rainy season, followed by autumn, spring, summer and winter. The average milk production of the buffaloes varies from 3.6 to 5.5 liter per day (Verma *et al.*, 2001).

5. Trans and Upper Gangetic Plain

It spreads over the areas of western Utter Pradesh, plains of Punjab, Haryana, Delhi, Sriganganagar district of Rajasthan and Chandigarh. Soils of Uttar Pradesh region are as fertile as those of Punjab and Haryana. Cropping intensity is 144%. *Jamunapuri* goat is famous in this region for its high milk yield. Livestock rearing is an important subsidiary occupation in the rural areas of Uttar Pradesh. Buffalo enterprise plays a crucial role in providing income to the marginal and small farmers. The '*Bhadawari*' breed of buffalo is crossed with *Murrah* for enhancing the milk productivity. The milk and milk products are commonly used by the farmer's family and dung is applied in the crop field to increase the productivity. The buffaloes are maintained on green and dry fodder produced in own farms. The animals are fed with some concentrate like cottonseed cake, mustard cake, barley husk and wheat flour, rice bran, wheat bran, gram churi etc. depending upon their availability. In rest parts of the region, crop production is the important sector of economy. About 60 % of labour force generates 40 % of income from this sector. Cultivation of crops like wheat, rice, pearl millet, cotton, gram, rapeseed, mustard and sugarcane is the most important activity of this region. Buffalo population is higher than cattle by about 40 % (Sastry, 1995). Buffalo breeds are predominantly Murrah. Cattle are mostly crossbred. Besides the list of native cattle breed are *Sahiwal, Haryana, Tharparker* etc. High fat milk is generally preferred. 65 % of total milk production goes for home consumption. The farmers use both dry and green fodder for feeding to their livestock. The concentrate is either home grown or procured from the market. Dry roughage includes wheat straw, sorghum kadbi, pearl millet kadbi etc. A number of livestock owner also allocate their substantial lands for cultivation of green fodder. The commonly grown fodder is berseem, oat, maize, bathua etc. Because of high genetic make up of both cattle and buffalo, the diet of animals shares a huge quantity of concentrate to sustain the milk production. Homegrown concentrate mixture is generally comprised of cottonseed cake, wheat flour, mineral mixture, cluster been seed, mustard cake, common salt etc. (Singh and Kundu, 2003) Most of the farmers fed crop residues and concentrate mixture in the form of 'sani' (Thakur and Tomar, 2001). Livestock are fed on the basis of milk yield but dry stock are fed mainly on wheat / paddy straw with little concentrate mixture. Recently the animal feed manufacturer has also introduced the complete feed block for different milk production levels

6. Central Plateau and Hills

This region lies in the center of Indian subcontinent and spreads over southern part of Uttar Pradesh, south-east Rajasthan, central and northern Madhya Pradesh and Chattisgarh. Hills and ravines are widely spread in this region causing a huge area not available for cultivation. The tract is semi –arid and rainfall varies from 740 – 1200 mm. Wheat, barley, pearl millet, pulses like mung, urd, gram, oilseed like mustard, groundnut, soybean, fruits and vegetables are widely grown in this region. The livestock sector is mainly comprised of buffalo, cattle, sheep and goat. The ravines are predominated by the local buffalo breed *(Bhadawari)*, local goat and *Jaluni* sheep. The local breeds of buffalo are being replaced by crossing with high yielding *Murrah* buffalo. It is also mentioned that *Bhadawari* buffalo is famous for its high fat (14%) secretion in milk. Due to abundance of shrubs *(Jharberi, Pilua* and *Deshi Karaunda)* and trees *(Ruani, Chhonkra, Vilayati Babul)* in the huge tracts of this region, sheep and goat are concentrated and contributed significantly to the income of small and landless farmers.

A system of animal grazing known as *"Chutta Pashu"* grazing system or *"Annapratha"* or *"Stray Animal Grazing"* is very common in Budelkhand region of this zone (Shukla *et al* 2000). The animals are left free for grazing without any monitoring. The owner who have the high yielding cattle and buffalo, raised under stall-feeding system. The high yielding buffaloes as well as cattle are maintained on balanced ration comprising of roughage and branded concentrate mixture. The basal roughage like wheat straw, groundnut haulms, pulses straw are mixed with home made concentrate mixture and water to make the "sani". The home made concentrate mixture usually include oil cake (mustard / cottonseed cake), wheat flour, gram chuni / gram husk.

7. Western Plateau and Hills

This region is comprised of varying topography and covers most of the interior parts of Maharastra, south-western parts of Madhya Pradesh and Gujrat. The annual rainfall varies from 850 – 950 mm. The irrigated area is nearly 12.4 % with canals being the main source. Wheat, bajra, sorghum, maize, guar, mung, sugarcane, cottonseed are widely cultivated.

The zone has made significant strides on co-operative milk production and made the country to be proud for highest milk production in the world. It is the home tract of several important breeds of cattle and buffalo. This include *Surti , Jaffarbadi, Mehsana, Nagpuri, Pandharpuri* breeds for buffalo and *Kankrej, Gir, Khillari, Krishnavally* breed for cattle (Nivsarkar et al., 2000). The white revolution not only strengthens the national milk grid but also presently serving as model for the global co-operative movement for several sectors.

Cattle and sheep are reared by both nomads and stationary farmers. Buffalo keeping is important activity among the local inhabitants. The productivity of cattle and buffalo is generally low. Large-scale slaughterhouse and meat processing center located in this region provokes the farmer for rearing animals for meat.

The dry roughage resources are comprised of wheat straw, groundnut haulms, bajra straw, maize straw, sorghum stover etc (Garg et al., 2003). Green fodder commonly grown in this area is bajra, cowpea, guar. Hybrid napier, local grass, maize, Lucerne, moth, sorghum. Creeper of sweet potato is also fed to animals as green supplement. The major concentrate commonly used for feeding includes oil cakes (cottonseed, coconut, sunflower, mustard), grains (bajra, guar, maize, barley), by-products (wheat bran, gram chuni). Molasses available from sugarcane industries are utilized by compounded feed manufacturer for pelleting and urea supplementation. Different unconventional feed resources like babul pods, corn steep liquor, casitora seed, mahua seed cake, subabool seed are also incorporated in the diet of animals whenever available (Pande and Gupta, 2001)

8.   Southern Plateau and Hills Region

The region includes interior parts of Andhra Pradesh, Karnataka and Tamil Nadu with annual rainfall in range of 670 – 1000 mm. Rainfed farming is mostly adopted in the area and cropping intensity remains around 111%. Major crops are cereals like sorghum, bajra, maize, ragi, paddy, wheat, oilseeds like groundnut, sesame, rape, mustard, linseed, niger, safflower, sunflower etc. (Raju *et al*, 2002, Anandan *et al.*, 2003). The other crops of this region are coconut, cotton and sugarcane. Animal husbandry activity is one of the major components in the overall agricultural production system. Livestock species include buffalo, cattle, sheep and goat. *Zerangi, Peddakimedi, Paralakhemundi* breed of cattle are commonly found in the territory of Andhra Pradesh. In West Godavari district, the dairying is highly developed and the major contribution of milk originates from buffalo. *Godavari* buffalo breed is crossing of native with *Murrah*. In rest of the region, buffalo population outnumbered cows. The reason for higher buffalo is due to higher availability of coarse cereal crop residues and buffalo being better converter of fibrous crop residues than cattle. The native cattle of Karnataka are *Amrit Mahal, Hallikar,* and *Krishna Valley* breed. In Tamil Nadu, native breed of cattle are *Kangayam, Bargur. Toda* breed of buffalo are found in Nilgiri hills of Tamil Nadu. In south Karnataka region, high degree of cross breeding of cattle adopted by the farmers with animal husbandry activities practiced mainly on commercial scale.

The main dry crop residues commonly fed to livestock are ragi straw, sorghum stover, rice straw, sugarcane tops, legume straw from pulses and groundnut haulms also fed to limited extent. The green forages are from the cultivated fodder mainly hybrid napier, lablab and rest from public primary grazing areas consisting of permanent pasture and tree crops like Glyricidia, Neem, Drum stick, Banana, Mulberry, Subabool etc. The oil cakes, brans and chunis constitute the major portion while grain constitutes a small fraction of concentrate mixture. In Nilgiri hills, the buffaloes are solely maintained on grazing. No supplemental feeding of concentrate is done. The grassland is comprised of Kikya grass interspersed with white clover and *Cynodon dactylon*. Fodder tree includes *Acacia melanoxylon, Dendrocalamus strictus* and *Celtis spp.*

9. Western Dry Zone

This region includes mostly desert parts of Rajasthan as well as northern parts of Gujrat. The zone is characterized by hot sandy desert, erratic rainfall and scanty vegetation. The average annual rainfall is only 390 mm with the fluctuation from year to year. Bajra, guar, moth are the main khariff crops while rabi crops include wheat and gram. Livestock assets are of immense important in this region as they provide both inter-year and intra-year stability in income and employment and their impact on desert ecology is substantial. Due to the imbalance between animals and biotic energy, migration of animals is a common phenomenon. The livestock include camel, cattle, sheep and goat. The zone is home tract of cattle breed of Malvi, Mewati, Nagori, Tharperkar, Rathi (Nivsarkar *et al.*, 2000). Among sheep important breed are Chokla, Nali. Sheep are mostly carpet wool producer. Camels are generally used for transport, ploughing of fields, tactive power and can survive under conditions poor / fibrous feed resources. The lists of roughage resources include pala, moth chara, groundnut haulmns, guar phalghati. During summer the leaves of phog, khess, bawali, khejri, neem are commonly fed to animals. In the rainy season, green grasses like bhusal, anjan, sevan, bokoria, gokhru are available

10. Coastal and Island Region

It runs along the peninsular India along its eastern and western fringes and is demarcated as coastal plains, hills and the interior ghats, plateau or plains. Rain fall in western region ranges from 2000 – 4000 mm and in the eastern it varies from 780 – 1300 mm. The island includes the territory of Andaman and Nicober islands, Lakhasadeep, which are typically equatorial. Coconut based cropping system are largely practiced. Fisheries are major livelihood activity among the inhabitants. Paddy, wheat, gram, banana, tubers, tapioca, rubber, sapota (chiku) are widely cultivated in this region. The livestock community is comprised of both large and small ruminants. The density of buffalo is higher in eastern region than western parts. The important breeds of cattle are *Ongole, Vechur* (small breed of cattle in the hills of Kerala). *Vechur* breed of cattle is famous for relatively higher milk production than other local breeds and its adaptation to the hot humid climate of the region. Another important cattle is *Sunandini* (crossbred of Brown Swiss bulls with non descript cows). Buffalo breed in costal Orissa is *Paralakhemundi.* Cross breeding of cattle has got monumental success in western region specially in Kerala while it does not make significant strides in eastern costal region. Sheep are comparatively lesser than goats. In islands, about half of the cropped area is under coconut plantation. Therefore livestock sector is not much developed.

The availability of dry matter in terms of crop residues, concentrate and green are 273, 39.8 and 226 thousands tones in Andaman & Nicobar Islands (Ananthram *et al.*, 2001). Greens accounted for 83 % of total dry matter and crop residues accounted for only 14% of which paddy alone contributes more than 90%. The higher availability of greens in Andaman & Nicobar Islands is due to greater forest cover. The concentrate ingredients in islands include grains, bran and chunnis as well as coconut cake. The ruminant livestock unit (cattle, buffalo, sheep and goat) is 0.54.

Perspective

Agriculture is a livelihood of about 65 percent people and livestock is part and parcel of it. Most of the buffaloes are of nondescript type and only 10 breeds of buffalo have been characterized. The buffalo population is increasing in the irrigated and agriculturally developed areas. Six states namely Haryana, Punjab, Delhi, Andhra Pradesh, Gujrat and Uttar Pradesh having higher buffalo population than cattle. *Murrah* or *Murrah* type animals are available in almost all regions because of their higher milk productivity. The buffalo growth is significant in most part of the country; however, it has registered a decrease in Punjab. Buffalo meat and leather export is increasing and has made a significant stride. The leather and manufactures with carpets contribute 65.5% of export from the agriculture sector. Presently, India is richest in bovine population and attained highest milk production in the world. To achieve the targeted growth of 8%, the animal agriculture sector in general including buffalo production through improved feeding, breeding, management and health care, a regional focus appears to be inevitable. The country is proposed to be classified into 10 Animal Agro Climatic regions because of following reasons:

1. Trans and upper gangetic plains have been merged because of similar animal husbandry practices of the farmers.

2. Lower gangetic plains region, portion of Eastern plateau and hill region specially the Jharkhand and Orrisa have been grouped under the name of Eastern Plateau and Plain Region keeping in view the status of livestock sector in these particular areas.

3. East coast plains and hill region, west coast plains and ghat region, Island region are brought under the one region i.e. coastal and Island region because of comparable status of livestock enterprise.

4. Dry areas of Gujrat plains and hills region have been merged with the Western dry zone while the rest portion has been placed under the western plateau and hills region.

The regions are as follows: 1.Western Himalayan Region, 2. Eastern Himalayan Region, 3. Eastern Plateau and Plains, 4. Middle Gangetic Plains, 5. Trans and Upper Gangetic Plains, 6. Central Plateau and Hills, 7. Western Plateau and Hills, 8. Southern Plateau and Hills, 9. Western Dr zones and 10. Coastal and Islands Region. It is expected for each region, separate species (Table 15) and thrust areas have been suggested for different categories of farmers (Table 16). Specific know-how, dovetailing, the peculiar needs of the region and their resources, are hour of need for equitable and sustainable growth of livestock industry in the country along with its length and breadth.

Table 15: Species density and developmental priorities of various species in different regions of the country

Agro climatic region	Species density				Development priority	Regions product
	Highest	High	Medium	Low		
1. Western Himalayan region	C	BG	S	P	S-G-B-C	Wool-Meat
2. Eastern Himalayan region	C	GP	–	BS	C-G-P	Meat-Draft
3. Eastern plateau & plains	CG	–	BS	P	G-C-P	Draft-Meat
4. Middle gangetic plains	CG	B	SP	–	C-B-G	Milk-Meat
5. Upper & trans gangetic plains	BC	G	S	P	B-C–G	Milk-Meat
6. Central plateau & hills	C	BG	–	SP	C-G-B	Milk-Meat
7. Western plateau & hills	CG	B	S	P	S-G-B-C	Milk – Meat
8. Southern plateau & hills	C	BSG	P	–	C-S-G	Meat-Milk
9. Western dry region	GS	CB	–	P	S-C	Wool –Meat
10. Coastal & island region	C	G	B	SP	All livestock	Milk-Draft-Meat

C= cattle, B= buffalo, S= sheep, G= goat, P= pig

Table 16: Vision for different categories of dairy farmers

Type of livestock rearing	Present status	Future demand / vision
Minimum input	o Grazing land with less grass cover. o Sheep are not prolific *. o High concentration of animal. o Very poor market *.	o More grazing land on forest and problem soils. o Highly prolific breed. o Higher concentration of animals with high growth potential. o Better market (eliminating/ decreasing profits of middleman) o Subsidized institutional support such as veterinary services (vaccination), training etc. o Finance support on easy terms.
Low input	o Problem soils underutilized o Shortage of quality fodder o Good local breeds available o Poor market *	o More grazing land required from forest or problem soils o Availability of good quality feed and fodder. o Improved breeds of livestock for milk and body weight gain. o Chaff cutter o Better market (eliminating / decreasing profits of middleman). o Milk collection facility o Subsidized institutional support such as veterinary services, training etc. o Develop co-operative system.

Medium input

- Feed shortage and less green fodder availability *
- Chaffing practiced *
- Good breeds available*
- Costly veterinary help*
- Poor market *

- Availability of good quality feed and fodder
- Balanced concentrate mixture
- Chaff cutter
- Improved breed of livestock for milk and body weight
- Good veterinary facilities
- Better market (eliminating / decreasing profits of middleman).
- Milk collection facility
- Paid institutional support such as veterinary services, training etc.
- Proper housing of animals

Commercial

- Fodder availability low
- Costly feed
- Breeds with low reproductive efficiency
- High rate of interest on loan
- Costly milking machine
- High rate of power
- Shades available
- Costly veterinary aid
- Rate of milk and milk products low

- Availability of good quality feed and fodder
- Improved breed with high productive (3000 liter per lactation) and reproductive efficiency (elite herd)
- Paid institutional support
- Intensive veterinary care
- Better and assured market
- Balanced concentrate mixture / mineral mixture
- Milking machine
- Facilities for chilling of milk
- Land with irrigation facilities
- Generator for assured electric supply
- Trained manpower availability

* Key response areax

81

References

Alagh,Y.K. 1989. Agro-climatic Regional Planning and Overview. National Planning Commission, Government of India, New Delhi, India.

Anandan, S., Ramchandra, K.S., Raju, S.S. and Angadi, U.B. 2001. Status of livestock and feed resources in the northern states of India. Proceedings of 10th Animal Nutrition Conference held at NDRI, Karnal from November 9 –11, 2001. Abstract no. 096.

Anandan, S., Raju, S.S. and Ramachandra, K.S. 2003. Status of livestock and feed resources in northern Karnataka region. Indian Journal of Dairy Science. **56**: 230 – 234.

Anantahram, K., Raju, S.S., Anandan, S. and Ramchandra, K.S. 2001. Animal feed resources availability in the union territories of Pondicherry and Andaman & Nicobar Islands. Proceedings of 10th Animal Nutrition Conference held at NDRI, Karnal from November 9 –11, 2001. Abstract no. 083.

Basu, A. and Chatterjee, A. 2001. Nutritional evaluation of a pasture land for yaks in Arunachal Pradesh. Proceedings of 10th Animal Nutrition Conference held at NDRI, Karnal from November 9 – 11, 2001. Abstract no. 058.

Carter, D.B. 1954. Climates of Africa and India according to Thornthwaite's (1948) classification. John Hopkins University publication in climatology 7 (4).

Das, A., Ghosh, T.K. and Haldar, S. 2003. Mineral distribution in soil, feeds and grazing cattle of different physiological stages in the red laterite and new alluvial agro climatic zones of West Bengal. Indian Journal of Animal Sciences. **73 (4):** 448 – 454.

Garge, M.R., Bhandari, B.M. and Sherasia, P.L. 2003. Macro-mineral status of feeds and fodder in Kutch district of Gujrat. Animal Nutrition and Feed Technology. **3:** 179 –188.

Ghosh, S.P. 1991. Agro climatic zone specific research – Indian perspective under NARP. ICAR. New Delhi.

India. 2002. A reference Annual Publication Division. Ministry of Information and Broadcasting. Government of India.

Katoch, K.K. and Singh, C.M. 2000. Technologies for managing hill agriculture resources. Indian Farming. **2:** 57 –61

Mittal. R.K. and Rai, S.D. 2002. Farming system approach for sustainable agricultural development in Uttaranchal. Indian Farming. **5:** 9 – 10, 25.

Muthariah, S. 1988. A social and economical Atlas of India. Oxford Publishers, New Delhi, India

Nivsarkar, A.E., Vij, P.K. and Tantia, M.S. 2000. Animal Genetic resources of India–cattle and buffalo. ICAR Publication. New Delhi.

Pande, M.B. and Gupta, R.S. 2001. Availability and utilization of livestock feed resources in different agro climatic zones of Gujrat. Proceedings of 10th animal Nutrition Conference held at NDRI, Karnal from 9 – 11, November 2001. PP 145 – 150.

Raju, S.S., Anandan, S., Angadi, U.B., Anantahram, K., Prasad, C.S. and Ramchandra, K.S. 2002. Assessment of animal feed resource availability in southern Karnataka region. Indian Journal of Animal Sciences. **72 (12):** 1137 – 1140

Roychudhary, P.N. 1997. Milk wave in Bihar. Dairy India. 125 –130.

Sastry, N.S.R. 1995. Livestock sector of India – regional aspects. International Book Distributing Co. Lucknow. Uttar Pradesh.

Sastry, N.S.R. 1993. Animal husbandry in different agro-climatic regions of India-report of a consultancy mission. Swiss development Corporation, New Delhi, India

Saxena, A.P. 1989. Strategies for Agricultural Research and Development- A Zonal Approach. ICAR, New Delhi, India

Sehgal, J. L., Mandal, D.K., Mandal, C., Vadivelu, S. 1990. Agro-Ecological Regions of India. National Bureau of Soil Survey and Land Use Planning (ICAR), NBSS Publ. 24, Nagpur, India

Sehgal, J.L. 1995. Cited in Vision 2020. IISS Perspective plan. ICAR

Sekhon . 1996. Cited in Vision 2020. IISS Perspective plan. ICAR

Shukla, S.K., Solanki, K.R. and Dwivedi, R.P. 2000. Plant multipurpose trees and shrubs for live fence in Bundelkhand region. Indian Farming. **1:** 11 – 15

Singh, Jasbir. 1974. An Agricultural Atlas of India: A Geographical Analysis. Vishal Publications, University of Campus, Kurukshetra.

Singh, Ram and Kundu, S.S. 2003. Quality evaluation of some animal feed stuffs available in Haryana. Animal Nutrition and Feed Technology. **3**: 143 - 150

Sohane, R.K., Jha, P.B., Choudhary, M.K. and Kumari, Asha. 2001. Data base information on feeds, fodders and livestock in north Bihar. Proceedings of 10th Animal Nutrition Conference held at NDRI, Karnal from November 9 -11, 2001. Abstract no. 097.

Tantia, M.S., Kumar, P., Joshi, B.K., Vij, P.K. and Nivsarkar, A.E. 1994. Population status of cattle and buffalo breeds of India. Journal of Dairying, Foods and Home Science. **13**: 192 – 194

Thakur, S.S. and Tomar, S.K. 2001. A study on the feeding practices being followed by dairy farmers of Karnal area. Proceedings of 10th Animal Nutrition Conference held at NDRI, Karnal from November 9 -11, 2001. Abstract no. 086.

Verma, D. N., Singh, S.P., Srivastava, D.K. and Maurya, S.K. 2001. Seasonal variation in feed milk relationship in buffaloes in Gorakhpur region UP. Proceedings of 10th Animal Nutrition Conference held at NDRI, Karnal from November 9 -11, 2001. Abstract no. 088.

Vir Singh. 2002. Small dairy production systems in Uttaranchal hills. Indian Farming. **3**: 23-25.

CHAPTER 3

Buffalo Production: An Overview

Anil Kumar, B P Kushwaha, L K Karnani and S S Kundu

Indian Grassland & Fodder Research Institute, Jhansi

Buffaloes have been a blessing to India. Their huge number, superior genetic potential, docile nature, less demanding in terms of management and nutrition, disease resistance, thriving under stress, their contribution towards milk production, draught and manure and above all their being part of our mythology speaks of their integration with human beings. There are 170 million buffaloes (Table 1) in the world (FAO, 2004) of which 165 million i.e. 97 % are located in Asia. Of the total buffaloes in the world 57 % are found in India alone followed by 15% in Pakistan, 13% in China, 2% each in Nepal, Philippines, Viet Nam and Myanmar, and 1% each in Indonesia, Thailand and Laos. Thus, 95% of all buffaloes in the world are present in 10 countries of Asia. Buffalo population in different countries of the world is presented in table 1.

Table 1.   Buffalo population (,000) in selected countries (2003)

Countries	Population	% of world
World	170,458	100.0
Asia	165,447	97.1
India	96,900	56.8
Pakistan	24,800	14.5
China	22,760	13.4
Nepal	3,750	2.2
Philippines	3,146	1.8

Viet Nam	2,814	1.7
Myanmar	2,600	1.5
Indonesia	2,350	1.4
Thailand	1,800	1.1
Laos	1,080	0.6
Bangladesh	830	0.5
Sri Lanka	635	0.4
Cambodia	626	0.4
Iran	550	0.3
Azerbaijan	312	0.2
Turkey	164	0.1
Malaysia	140	0.1

(FAOSTAT, 2004)

Buffalo is believed to be originated in the Indian sub-continent from the wild arni (*Bubalus arnee*) which are still found in forests of Assam. They were domesticated about 5000 years ago and evidences are there that they were in use about 4000 years ago in China. Buffalo belong to the class Mammalia, sub-class Eutheria, order Ungulata and family Bovidae to which the tribe Bovinae belongs. Wild Asiatic buffalo were found in or near the Brahamaputra valley in Assam, Arunachal Pradesh, Meghalaya, west of Madhya Pradesh - Orissa border in India and Kosi Tappu, Nepal in southern Asia, wild buffaloes are considered to be feral domestic water breed. Translocated or feral buffaloes are also found in Australia, Brazil, the Philippines, Italy and Sri Lanka. Asian wild buffaloes comprise of three distinct species viz. *Bubalus arni, Anoa depressicomis* and *Bubalus mindorensis*. African Buffalo (*Syncerus coffer*) are of two types. S. *caffer* the large black buffalo of Southern Africa, northern to Ethiopia and Somalia and the smaller red type *Syncerus coffer nanus* which occur in western Uganda and ranges west and south through Zaire, Gabon and Congo to north-Angola (FAO, 1995).

Types and breeds

There are two main categories of buffaloes namely, riverine and swamp, depending upon variation in their habitat and genome. River buffaloes are massive in size mostly with curled horns, prefer to enter clear water, have 50 chromosomes, are primarily used for milk production and also used for meat production and draught purpose. Swamp buffaloes are stocky animals with marshy land habitats and have 48 chromosomes. They are primarily used for draught power in paddy fields and haulage and are also used for meat and milk production. Swamp buffaloes are mostly found in South-East Asian countries and few animals are also available in north-eastern states of India. The cape variety of the African buffalo has 52 chromosome. The Congo buffalo has 54 chromosomes (Heck et al 1968, Ulbrich and Fisher 1967).

India has some of the best milch breeds of buffaloes in the world e.g. Murrah, Nili-Ravi, Surti, and Jaffarabadi, which had their origin in North - West India. These breeds have high potential for milk and fat production besides, draughtability and meat production. There are several other breeds which have regional importance and economic value, e.g. Bhadawari, Nagpuri, Pandharpuri, Tarai, Parlakhemundi, Manda, Jerangi, Kalahandi, Sambalpur, Toda and South Kanara, etc. Mehsana and Godavari are the breeds developed from grading up of Surti buffaloes in Mehsana Distt. of Gujarat and local non descript buffaloes in Krishna and Godavari Distt. of Andhra Pradesh, respectively. Sethi (2003) has estimated that the number of pure bred animals of above specified breeds is about 20 - 25 % of the total buffalo population in the country while rest of the buffaloes are non-descript.

Acharya and Bhat (1984) classified Indian buffaloes on the basis of well-defined characters and categorized them into 5 groups (Table 2).

Table 2. Classification of Indian buffaloes.

Group	Breed	Breeding Tract
Murrah type	Murrah	Rohtak, Jind, Hisar, Bhiwani, Sonepat.
	Nili Ravi	Ferozepur distt. in Punjab
Gujarat	Surti	Kaira and Baroda distt.
	Jaffarabadi	Kutch, Junagarh and Jamnagar districts
	Mehsana	Mehsana, Sabarkantha and Banaskantha districts
Uttar Pradesh	Bhadawari	Bhadawar estate, Bah tehsil in Agra distt., Gwlior and Etawah districts.
	Tarai	Tarai region of Uttaranchal and UP
Central India	Nagpuri	Nagpur, Akola and Amravati districts of Maharashtra
	Pandharpuri	South Maharashtra, West Andhra Pradesh and North Kamataka
	Parlakhemundi	Ganjam, Koraput and Plateau region of Orissa.
	Kalahandi	Hilly region of Andhra Pradesh and Orissa
	Sambalpur	Bilaspur district
South India	Toda	Nilgiri hills
	South Kanara	West coast in Kerala

Some important buffalo breeds of different regions of India are described below:

Murrah

Murrah is one of the best breeds of buffaloes. Its home tract lies in Rohtak, Jind, Hisar and Bhiwani district of southern Haryana, the Union Territory of Delhi and Patiala Distt. in Punjab. However, the animals of this breed can be seen throughout the country. Murrah buffaloes are jet black and large in size with long and deep body.

Horns are short and tightly curved in a spiral shape. This breed has derived its name from shape of its horns (Murrah means "twisted horns") Udder is capacious extending from hind legs to just behind naval flap with prominent milk-veins. Murrah buffaloes produce around 2000 kg of milk. Murrah breed is widely used for up-gradation of local buffaloes in many parts of the country. The number of pure bred animals of this breed is declining day by day due to immigration of superior animals from its breeding tract to the milk shed areas and inter breed crossing.

Jafrabadi

Jafrabadi is the heaviest of all the Indian breeds of buffaloes. The animals of this breed are found in Junagarah, Bhavanagar, Nreli and Rajkot districts of Gujrat. The breed is also named as Bhavanagri and Gir. Animals are massive in size. Average weight of a bull is around 1000 kg. The most common colour of Jaffarabadi buffaloes is black but animals of grey and copper colour are also available. Some animals have white spots on forehead, fetlock and tail. Head is drooping and big with bulging forehead. The rest of the face is small and wide. The bone of forehead lens and covers the eyes to give "sleepy eye appearance". Horns emerge side ways downwards compressing the head and are pendulous in nature just falling on both sides of the neck. Body is usually long, massive and fatty. The buffaloes of this breed produce around 2200 kg of milk in a lactation of 305 days. Exceptional buffaloes are also very high producers.

Nili Ravi

Nili- Ravi buffaloes are found in Firozepur, Gurdaspur and Amritsar districts of Punjab. The breeding tract is spread all along the Sutlej River on the Indo-Pak border. This breed is similar to Murrah in almost all characters except for white markings on extremities and walled eyes. Skin and hairs are normally black-brown. Fawn and grey animals are also found. They are usually wall eyed and have white markings on forehead, face, muzzle, legs and tail. The most desired character of the female is the possession of white markings due to which the breed is also known as "Panch Kalyani". Moustache is white. Average milk yield is of 1811-2000 kg in 300 day lactation length.

Bhadawari

Bhadawari buffaloes are known for higher fat content in their milk, which may go upto 13 percent. Animals of this breed are found in the ravines of Yamuna, Chambal and Utangan rivers spread over in *Uttar* Pradesh and Madhya Pradesh. The breeding tract and natural habitat of this breed are Bhadawar estate comprising Bah tehsil of Agra, Chakarnagar and Barhpura blocks of Etawah in Uttar Pradesh, Ambah and Porsa tehsil of Morena and Mahangaon tehsil of Bhind district in Madhya Pradesh. They produce about 1000 kg of milk in a lactation with an average fat of 8 per cent. Fat may vary from 6- 12 percent in their milk. The population of this breed is declining at an alarming rate in the breeding tract. Estimated population in the whole breeding tract is reported between 13500 to 15000 (Kushwaha et.al., 2004).

Pandharpuri

Pandharpuri buffaloes are found in South Maharashtra spread in Kolhapur, Solapur, Pandharpur, Sangli and Satara distt. of Maharashtra. The animals are medium sized having long and narrow face, prominent nasal bone, comparatively narrow frontal bone and long and compact body. Hair colour is grey, tan and blackish. The horns are very long, running backwards, upwards and twisted outward and touching extended beyond shoulder blade. The neck is comparatively longer and thin. The udder is medium in size with well placed teats. They produce around 1500 kg of milk in a lactation and are considered to be consumer friendly, by way of allowing door to door milking for milk quality satisfaction of consumers.

Surti

The animals of this breed are found in Anand, Nadiad, Kaira and Vadodara district. The animals are also commonly found in South Rajasthan extending over to Udaipur, Bhilwara, Rajsamand, Chittorgarh and Dungarpur distts. in Rajasthan. The Surti buffaloes are medium in body size. Coat colour of Surti buffaloes varies from rusty brown to silver grey. There are two white lines on lower side of the neck. Udder is well developed, finely shaped and squarely placed between the hind legs. Surti buffalo produce around 1500 kg of milk in a lactation.

Godavari

Main breeding tract of Godavari buffaloes lies in east and West Godavari districts of Andhra Pradesh. This breed has their origin from interbreed crossing followed by grading of local non-descript buffaloes with Murrah breed. The animals of this breed are also found in areas of Tanuku, Bhimavaram, Narasapur, Ramchandrapuram, Kothapeta, Alamurualuqa and part of Tadepanigudem and Kovvuru, Krishna deltaic areas of Gudlavalluru. The Animals are medium statured with compact body. Colour is predominantly black with spare hair coat of coarse brown hairs. The horns are short, flat, curved, slightly downwards, backwards and then forward with a loose ring at the tip. Udder is medium in size, bowl shaped and well placed medium sized teats. Milk yield is around 2050 kg in a lactation of 305 days.

Swamp

Swamp buffaloes are found in Assam state of India and are spread over entire river bank of Brahmaputra. Animals are reared under semi-domesticated, nomadic (*khutis*) and settled system of management. Grazing of grasses, shrubs and leaves is most common. Body colour of the swamp buffaloes is primarily black with varying shades from dark to slaty black. Albinos are, however, not uncommon in Assam. They produce around 500 kg of milk in a lactation with average fat of 8.48 percent. They are primarily reared for draft purpose. The swamp buffaloes have 48 chromosomes.

Toda

The Toda buffaloes are reared by Toda tribes of Nilgiri hills and got their name after them. The breed had been developed, which could withstand extreme climatic variation of high rainfall and high humid area. In the earlier times breed got its importance not because of its milk production unlike other buffalo breeds, but because of its association with cultural activities of the Todas. Besides Toda tribes, other communities which maintain these buffaloes are Badagos, Kotos and Irulas and the average herd size is 34 buffaloes (Balachandran, 1996). Toda buffaloes are medium in size and are considered to be quite powerful. Coat colour of the calf is generally fawn at birth which varies from grey, light grey and dark grey. In growing calves, at about 2-3 months, the fawn colour changes to ash grey. Body is fairly long with a broad and deep chest. The horns are quite large, set wide apart, outward, upward to form a characteristic semicircle. Udder is small and not so prominent. They produce around 500 kg of milk in a lactation of 200 days. Karthikeyan et al. (2002) reported the total population of the Toda buffaloes in the breeding tract as 3,313 with the break-up details of 1,809 adult females, 70 males young stock, 584 female young stock and 850 calves (as of 1994).

Mehsana

Mehsana buffaloes are found in Mehsana distt. of Gujarat from where it derives its name. It is also common in Banaskantha, Sabarkantha, Gandhinagar and Ahmedabad districts. The breed is considered to have originated from regular inter-breeding between Murrah and Surti. Approximate population of this breed is around 0.4 million and is reputed for high breeding efficiency. The Mehsana buffalo is a medium sized animal. Body colour is mostly black. Horns are generally sickle-shaped with the curve more upward than the Surti breed and less curved than in the Murrah breed. Udder is attached high in the back and carried well. Milk veins are prominent. Mehsana buffaloes produce around 2000 kg of milk in a lactation.

Banni

Home tract of Banni buffaloes is Banni region comparising the Khavada, Bhuj, Bhachau and Anjar area of Kutch district in Gujrat. Banni buffaloes have unique importance, economic value and special adaptation characteristics to the draught hit areas of Kutch region. Population of these animals may be around 2.5 to 3 lakhs. The animals of this breed are maintained under zero input system. Animals go for grazing at 5:00 pm and come back to owner on their own at around

10:00 am. Buffaloes are milked once in 24 hours and average milk yield is 5-6 litres per day (Singh et.al. 2004).

Nagpuri (Berari)

Nagpuri buffalo is a native of the Vidarbha region of Maharashtra. This breed is commonly found in Nagpur and the adjoining districts of Wardha and Berar. This breed has 4 distinct strains, viz. Purnathadi (Akola distt.), Ellichpuri (Ammravati distt.), Gaulani and Nagpuri (Nagpur distt.). All these strains are commonly known as Nagpuri or Berari buffaloes. Nagpuri is a dual purpose breed of buffalo. These buffaloes are used for heavy draught purposes. These buffalo produce around 900 to 1,200 kg of milk. Sirothia et. al. (2004) reported the estimated population of Nagpuri buffaloes as 256851 in Nagpur, 46688 in Amravati, 12762 in Akola and 14736 in Yavatmal district. Around 32.75 percent of the total population of Vidarbha region belong to Nagpuri buffaoes.

Marathwada

Marathwada buffaloes are medium in size and found in the Marathwada region (Parbhani, Nanded, Beed and Latur) of Maharastra and are named after the name of the region. Coat colour varies from greyish-black to jet black, white markings on forehead and on lower parts of the limbs and white switch are common. Horns are medium in length. Udder is medium and milk yield ranges from 850 to 1000 kg in a lactation.

Tarai

The Tarai buffaloes are found in Tarai area of Uttaranchal from Tanakpur to Ramnagar, but originally are a native breed of hilly areas (Kaura, 1961). The animals are moderate in body size with coarse and slightly convex head and prominent nasal bones. Horns are flat and long with broad base and pointed tip. Back is straight in males, slightly shallow in females. Skin and coat colour varies from black to brown. The animals are commonly used for draught purposes in agricultural operations. Buffaloes produce around 2-3 kg daily averaging around 700 kg in a lactation of 250 days.

Kalahandi / Peddakimedi

This breed is known as Kalahandi in Orissa and Peddakimedi in Andhra Pradesh. These buffaloes are found in hilly parts of Andhra Pradesh and adjoining areas of Orissa. This light coloured (grey or ash grey) animal is hardy and is moderate producer. The animals have slightly prominent forehead with broad and half-curved horns running backwards at tip. Animals are docile in temperament, hardy and well suited for draught purposes in hills as well as in plains.

Manda / Parlakkhemundi / Ganjam

These buffaloes are found in hilly areas of border of Andhra Pradesh- Orissa. Skin is brown / gray with yellowish tuft of hair on knees and fetlocks. Tail switch is also yellowish white. Arch like red ring around the chest, 8-9 cm wide is common. Horns are long and semi-circular extending backward and inward. Jaws and nostrils are wide and prominent. Eyes are sharp with broad red margin around the lids. The animals are slow in movement and commonly used for draught purpose with moderate quality of milk yield (Ray et al, 1999).

Jerangi

The buffaloes of this breed are small in size and are found in Jerangi hills of Orissa and adjoining northern Vishakhapatanam and west Ganjam (Andhra Pradesh). The buffaloes have short face and small barrel. Skin is thinner black and coloured. Small conical horns are common. The animal is useful for ploughing water logged paddy fields, with moderate draught capacity.

Sambalpuri / Kimedi / Gowadoo

Sambalpuri buffaloes are found in Orissa and Bilaspur district in Chhatishgarh. The animals are large in size and powerful but docile. The forehead is broad and flat. Horns are short, narrow and curved into a semicircle, extending backwards, upward and then forward at the tip. Tail is long and narrow with white switch. Body and coat colour is black, rarely brown or ash grey. Buffaloes are regular breeder and comparatively more productive breed of the region. Some exceptional buffaloes may yield as high as 2300-2700 kg in about 340 days lactation (Pathak and Singh, 2001).

South Kanara

The buffaloes of this breed are found around Manglore region of Karnataka. Fast and active animal of medium built, famous for agricultural operations and racing at festival occasions. The population of this breed is reported to be declining at a faster rate.

Kuttanad

Kuttanad buffaloes are found only in Kuttanad area of Kerala. It is comprising of two districts of the state namely Kottayam and Alappuzha. The peculiarity of the area is that paddy fields of the Kuttanad are under water for most of the time. The present population of these animals is around 500 numbers. Males are exclusively used for working is paddy fields. Female animals are used as producers of male calves. The milk yield from these animals is very less and fed to the male calves. Milk production of these animals is meagre and amounts to 1 to 2 liters per day. The Kuttanad buffaloes are usually milked once a day.

Chilika

The buffaloes of this breed are found around Chilika lake in Orissa. The animals of this breed are reared with almost zero input. This breed has unique characteristics to adapt in the existing milieu. Chilika buffaloes can eat the sea weeds from under water defecate to the benefit of zooplanktons/fish/prawns and need no housing for their shelter. Buffaloes move into lake by the evening and come back to the farmers (owners) by morning. The population of the pure bred animal is declining. (Sadana and Jain, 2004).

Assamese / Mongoor

Medium sized, conical face, broad forehead, thick coat of hair and usually a black colour are common features of this breed. Low yielder females commonly breed with wild counterpart (Pathak and Singh, 2001).

Sikamese

These buffaloes are the natives of the Sikkim state. This is a small sized hardy buffalo (250-400 kg) with compact body and thick coat, suited for hilly terrain, with black and grey colour (Pathak and Singh, 2001). The buffaloes of this breed are poor milkier.

Information on body measurements, body weight, production and reproduction performance of various breeds is presented Table 3, 4, 5, and 6 respectively.

Table 3. Body measurements (cm) of various breeds of buffaloes

Breed		Male			Female		
		Length	Height	Heart girth	Length	Height	Heart girth
Bhadawari	Mean	116.9	122.8	184.5	115.0	123.1	184.3
	No.	90	90	90	169	169	169
Jaffarabadi	Mean	127.7	126.1	207.7	132.6	129.1	200.9
	No.	58	58	58	1,419	1,419	1,422
Mehsana	Mean	153.7	133.7	200.6	141.7	127.5	189.3
	No.	55	55	55	314	314	314
Murrah	Mean	150.0	142.0	220	148.0	132.7	202.4
	No.				1,372	1,698	1,746
Nagpuri	Mean	180.0	140.0	210.0	128.6	122.8	181.8
	No.				413	413	413
Nili-Ravi	Mean	160.0	140.0	230.0	165.4	134.2	207.7
	No.				150	140	120
Pandharpuri	Mean				132.9	130	192.8
	No.				201	201	201
Surti	Mean	142.0	130.0	190.0	118.8	124.9	184.0
	No.				25	25	25
Toda	Mean				132.7	121.8	180.4
	No.				131	131	129

(Nivsarkar *et al.*, 2000)

Table 4.Body weights (kg) of various breeds of buffaloes

Breed		Birth weight			Adult weight	
		Male	Female	Overall	Male	Female
Bhadawari	Mean	27	25	25.3± 0.23	475	425.71± 7.72
	No.			860		49
	Range			24-27		300-540
Mehsana	Mean	29.51± 0.51	28.51± 0.45	29.0	565.4	484.2
	No.	209	252	461	55	314
	Range	16-44	14-40	14-44	400-602	315-580
Murrah	Mean	31.7	30.0	30.3	567	516
	No.	2,186	6,043	8,574		1,761
	Range	28-34	26-33	26-34	450-800	350-700
Nagpuri	Mean	29.0± 0.32	28.l±: 0.14	28.61± 0.3	520	363.5
	No.			1,028		200
	Range	27-30	25-29	25-30		340-400
Nili-Ravi	Mean	35.1	34.5	34.8	567	454
	No.	182	222	404		
	Range	27-39	27-38	27-39		
Pandharpuri	Mean	28.0± 0.91	25.61 ± 0.74	26.8		416.2±0.10
	No.	17	26	43		11
Surti	Mean	26.3	24.5	25.2	500	382.61± 12.14
	Range	24-30	23-29	23-30		318-414
Toda	Mean	27.91± 0.60	28.01 ± 0. 59	27.91± 0.43	380	380
	No.	57	43	100		

(Nivsarkar *et al.*, 2000)

Table 5. Production performance of various breeds of buffaloes

Breed		Total lactation	First lactation	Lactation length	Dry period
		milk yield (kg)	milk yield (kg)	(days)	(days)
Bhadawari	Mean	903.1	780.0± 25.4	271.9±3.98	190.0
	No.	1,514	931	1,028	421
	Range	658-1,142	699-1,165	140-350	145-295
Jaffarabadi	Mean	2,238.7± 74.87	2,151.3± 130.53	305.1±9.61	144.9±8.4
	No.	70	29	70	57
	Range	2,151-2,336		289-319	
Mehsana	Mean	1,988.0	1 940.4	316.7	166.7
	No.	1,352	713	1,219	1,260
	Range	598-3,597	598-3,221	157-513	14-656
Murrah	Mean	1,751.8	1,678.4	298.7	154.8
	No.	15,765	16,195	16,390	8,665
	Range	1,003-2,057	904-2,041	269-337	127-176
Nagpuri	Mean	1,055.4	933.3	286.4	129.1±4.85
	No.	645	126	996	374
	Range	780-1,520		263-297	80-155
Nili-Ravi	Mean	1,850.2	1,483.4	294.2	150.6
	No.	607	1,124	1,072	2,543
	Range	1,586-1,929	1,268-1,854	263-316	115-202
Pandharpuri	Mean	1,502.3±69.3 5	1,197.3	330.0	144.0±10.54
	No.	39	4	39	34
	Range	1,168-1,680		296-346	108-155
Surti	Mean	1,285.43	1,396.5	344.7	205.4±8.79
	No.	2,274	2,502	900	580
	Range	1,208-2,208	1,208-2,203	280-405	155-289

(Nivsarkar *et al.*, 2000)

Table 6. Reproduction performance of various breeds of buffaloes

Breed		Age at first calving (days)	No. of services/ conception	Service period (days)	Calving interval (days)
Bhadawari	Mean	1,477.44±17.63		178.98±10.60	478.7±11.55
	No.	596		724	715
	Range	1,335-1,550		83-317	390-630
Jaffarabadi	Mean	1,361.7	1.5	93.4±0.69	440.3±14.32
	No.	715	715	715	54
	Range	1,250-1,670	1-2		427-455
Mehsana	Mean	1,265.9	1.93	161.0	475.5
	No.	664	1,074	1,260	1,260
	Range	677-2,500	1.75-2.15	24-646	313-945
Murrah	Mean	1,319.0	2.93	136.3	452.9
	No.	19,991	1,074	6,209	11,083
	Range	1,214-1,647	1.40-3.75	125-187	430-604
Nagpuri	Mean	1,672.0	1.31±0.23	115.7	429.6
	No.	200	374	1,459	557
	Range			34-435	350-721
Nili-Ravi	Mean	1,359	2.38	202.2	487.7
	No.	2,459	1,318	1,739	2,620
	Range	1,216-1,617	2.1-3.4	169-290	313-945
Pandharpuri	Mean	1,255	3.0±0.39	165.0	465.0
	No.		37	34	32
Surti	Mean	1,692.70	2.81	142.6	534.7
	No.	939	13,760	400	641
	Range	1,050-1,770	1.5-3.0	93-164	430-564

(Nivsarkar *et al.*, 2000)

Growth In Buffalo Production

The buffalo population has increased by 33 percent in last two decades growing @1.4% p.a. (Table 7). Among countries having sizeable buffalo population buffalo increased by 96% in Pakistan during the period followed by Egypt, Nepal, India and China. Most of the Asian countries have registered a positive growth in buffalo population. However several other countries witnessed a decline in its number mainly because of their declining role as a draught animal. Buffalo is gaining popularity in Brazil where it is growing @3.7% p.a. and now the population stands at 1.2 million heads. The average herd size of buffaloes all over the world is 2-3 animals except in Italy, Syria, Bulgaria and South America where the herd size is bigger (Ranjhan, 2003). In Asia buffaloes are very well integrated with the cropping system and they are raised mainly under intensive and semi-intensive system of management. State wise buffalo population in India is presented in Table 8 and buffalo population trend in Table 9.

Table 7. **Growth in buffalo population and the rate of growth during two phases**

Country	% increase or decrease		Growth rate % p.a	
	1961-81	1981-2001	1961-81	1981-2001
World	40.4	33.2	1.7	1.4
Asia				
India	31.8	39.5	1.4	1.7
Pakistan	77.9	95.8	2.9	3.4
China	121.9	22.6	4.1	1.0
Nepal	214.5	45.0	5.9	1.9
Egypt	57.9	49.0	2.3	2.0
Philippines	-17.4	7.6	-1.0	0.4
Viet Nam	5.7	18.0	0.3	0.8
Myanmar	87.8	27.1	3.2	1.2
Indonesia	-14.0	-6.2	-0.8	-0.3
Thailand	23.4	-75.1	1.1	-6.7

Country	% increase or decrease		Growth rate % p.a	
	1961-81	1981-2001	1961-81	1981-2001
Brazil	760.3	106.4	11.4	3.7
Laos	109.4	19.5	3.8	0.9
Bangladesh	-4.8	74.4	-0.2	2.8
Sri Lanka	16.4	-26.4	0.8	-1.5
Cambodia	-22.3	55.0	-1.3	2.2
Iran	7.5	90.9	0.4	3.3
Azerbaijan				
Italy	469.4	85.4	9.1	3.1
Turkey	-9.6	-85.8	-0.5	-9.3
Malaysia	-24.2	-43.4	-1.4	-2.8

(FAOSTAT, 2004)

Table 8. State-wise buffalo population in India (,000)

State	1987	1992
Andhra Pradesh	8,758	9,132
Arunachal Pradesh	12	10
Assam	623	958
Bihar	4,872	5,353
Goa	40	45
Gujarat	4,502	5,268
Haryana	3,827	4,373
Himachal Pradesh	794	701
Jammu & Kashmir	565	732
Karnataka	4,035	4,251
Kerala	328	296
Madhya Pradesh	7,351	7,970

Maharshtra	4,753	5,447
Manipur	140	115
Meghalaya	28	34
Mizoram	5	7
Nagaland	15	34
Orissa	1,506	1,509
Punjab	5,575	5,238
Rajasthan	6343	7,775
Sikkim	2	3
Tami Nadu	3,129	2,814
Tripura	16	20
Uttar Pradesh	18,240	20,086
West Bengal	1,163	1,011
UT's	330	299
Lakshadweep		
Pondicherry	10	
ALL INDIA	75,967	83,499

Table 9. Trends in population of cattle and buffaloes, 1951 to 1987 (in millions)

Category	1951	1956	1961	1966	1972	1977	1982	1987
Buffalo								
Young Stock	14.75	16.07	18.45	18.59	20.12	21.79	24.72	29.37
Females over 3 years	21.86	22.34	25.03	26.14	29.24	31.87	32.51	39.14
In milk	10.22	11.81	12.58	12.92	15.07	16.96	17.99	23.15
Dry and not calved	10.79	9.86	11.66	12.59	13.54	14.31	14.40	15.52
Buffaloes used for work		0.67	0.79	0.63	0.63	0.60	0.12	0.47

Category	1951	1956	1961	1966	1972	1977	1982	1987
Males over 3 years	6.80	6.50	7.67	8.19	8.06	8.37	7.95	7.46
Used for breeding	0.31	0.33	0.29	0.33	0.22	0.22	0.63	0.51
Working	6.03	5.95	6.61	6.97	7.01	7.32	-	3.76
Used for breeding /work	NA	NA	0.51	0.62	0.60	0.61	5.97	2.87
Others	0.46	0.22	0.26	0.27	0.23	0.22	1.35	0.32
Total	43.41	44.91	51.15	52.92	57.42	62.03	*69.78	75.97

* Difference of total are due to inclusion of projected figures for Arunachal Pradesh, Punjab and Meghalaya for which data were not available

Buffalo Density / Number of Buffaloes per 1000 Human Beings

Analysis of buffalo distribution indicates that India, Pakistan and Nepal have 25 or more buffaloes per sq. Km. Philippines and Sri Lanka have about 10 and all other countries have less than 10 buffaloes per sq km. But the true association of a livestock species with human beings is reflected in its number per 1000 human beings. Table 10 indicates that Laos with 195 buffaloes every 1000 human being is more closely associated with buffaloes inspite of their small total population (1.08 million). Pakistan and Nepal have over 150, India 91, Egypt, Philippines, Viet Nam, Myanmar, Thailand, Sri Lanka, Cambodia and Azerbaijan between 25 to 50 buffaloes per 1000 human beings.

Table 10. Buffalo density and number per 1000 human beings

Country	Density (No./sq km)	Buff. No. per 1000 human beings
World	1.2	26.9
Asia	5.0	43.1

India	28.6	91.1
Pakistan	29.3	159.5
China	2.4	17.6
Nepal	24.6	150.6
Egypt	3.5	51.1
Philippines	10.2	39.7
Viet Nam	8.5	35.5
Myanmar	3.7	51.9
Indonesia	1.2	10.9
Thailand	3.0	24.8
Brazil	0.1	6.4
Laos	4.4	194.6
Bangladesh	5.8	5.9
Sri Lanka	10.1	35.3
Cambodia	3.5	46.4
Iran	0.3	7.5
Azerbaijan	3.4	36.3
Italy	0.6	3.3
Turkey	0.2	2.1
Malaysia	0.4	6.3

(FAOSTAT, 2004)

Buffalo as a Milch Animal

Buffaloes contribute 12 percent of the world's total milk production. However, their share is much more in the Indian sub-continent where they are mainly reared for milk. Of the 72.6 million tonnes of buffalo milk produced in the world, 97% were produced in Asia. India and Pakistan together contributed 92 % of the total buffalo milk produced in the World. In India buffaloes are less than half the number of cattle, yet they produce 55 percent of the total milk produced in the country. Similarly, 66.7 % of the total milk produced in Pakistan and 65.2 percent in Nepal is contributed by buffaloes. The major buffalo milk producing countries are presented in table 11. In Egypt over 50 percent of the milk produced comes from buffaloes. Similarly, in countries like China, Viet Nam, Myanmar and Malaysia where buffaloes are mainly raised as draught animal, they contribute 15 to 20 percent of the total milk production.

Buffaloes as a milch animal has been the backbone of the operation flood programme in India through which millions of farmers are gainfully employed in the dairy business directly besides many more in the forward and backward linkage. Assuming 2 milch buffaloes being owned by a farmer, 17 million farmers are directly involved with the buffaloes in India. Buffalo milk being rich in fat and people in developing countries having developed a taste for it, commands premium price in the market. Although superior germplasm of milch buffaloes are located in the Indian sub continent, a concerted effort is required for upgrading the millions of non descript buffaloes which require huge number of superior bulls to be provided to the remote villages.

Table 11.Total and buffalo milk production in selected countries (2003)

Country	Total milk production (,000 Mt)	Buffalo milk (,000 Mt)	Buffalo milk as % of total	Growth in total milk production (% p.a) (1961-2003)	Growth in buffalo milk production (% p.a) (1961-2003)
World	600,978	72,616	12.1	1.34	3.40
Asia	185,023	70,394	38.0		
India	86,960	47,850	55.0	3.52	3.54
Pakistan	27,811	18,520	66.6	3.72	3.59
China	17,246	2,650	15.4	5.47	2.56
Egypt	4,085	2,077	50.8	3.05	2.42
Nepal	1,237	807	65.2	1.96	2.08

Country	Total milk production (,000 Mt)	Buffalo milk (,000 Mt)	Buffalo milk as % of total	Growth in total milk production (% p.a) (1961-2003)	Growth in buffalo milk production (% p.a) (1961-2003)
Iran	5,954	230	3.9	3.21	2.94
Italy	12,042	140	1.2	0.30	5.64
Myanmar	650	116	17.8	4.17	4.51
Sri Lanka	298	68	22.8	2.54	2.44
Turkey	9,496	63	0.7	0.90	-3.20
Viet Nam	158	31	19.7	5.60	3.28
Iraq	539	28	5.1	0.16	-0.64
Bangladesh	2,139	22	1.0	2.04	0.80
Malaysia	46	7	15.9	0.76	-1.81
Bulgaria	1,389	5	0.3	0.37	-5.94
Syria	1,767	2	0.1	3.99	1.00

(FAOSTAT, 2004)

In major buffalo producing countries, the proportion of milch buffaloes to the total ranges from 30-40 (Table 12). However, Italy had the highest proportion (69.7%) of milch buffaloes.

Table 12. Milch buffaloes in different countries (2003)

Country	No. of milch buffaloes (,000)	Milch buffaloes as % of total
World	52378	30.7
Asia	50699	30.6
India	34000	35.1
Pakistan	9700	39.1
China	5250	23.1
Egypt	1550	43.5
Nepal	959	25.6
Iran	200	36.4
Italy	124	69.7
Myanmar	296	11.4
Sri Lanka	105	16.5
Turkey	65	39.9
Viet Nam	31	1.1

(FAOSTAT, 2004)

Buffalo as a Meat Animal

Although buffalo remain a non significant player in the global meat scenario, their importance in Asian and some other countries is tremendous. The share of buffalo meat to the total meat in countries like India, Pakstan, Egypt and Laos is between 20 to 27 percent. But in Nepal it contributes over 50 percent of the total meat produced in the country. Buffalo meat has been growing over 2 percent p.a. in most of the countries. The outstanding performance is being shown (Table 13) by countries like Pakistan (3.7 % p.a.), China (7.36), Nepal (5.18) Philippines (4.5) and Cambodia (8.3% p.a.). In countries where buffaloes have traditionally been looked upon as dairy animal, its potential as a meat animal largely remain unexploited, especially in India, in spite, of the fact that unlike cattle there is no taboo attached to their felling. The male progeny are not taken care of and they die young. The peri-urban dairies bring their fresh stock of buffaloes from villages and upon completion of lactation these animals are slaughtered thereby losing the superior genetic stock. These city dairies do not raise buffalo calves (male and female both) because they remain focussed on getting more milk and hence very few heifers could be seen at such dairies and almost no male. But the demand at the slaughter house points toward an insatiate demand for buffalo meat. Ranjhan (2003) has reported that about 5.5 million male buffalo calves are removed due to intentional killing incurring a loss of about US$ 11 million per annum. If these male buffalo calves are reared as meat animal, they could be a good source of employment and income to many. Buffalo as meat animal is also gaining popularity even in Western countries because of its superiority over beef. Compared to beef, the buffalo meat contains 92% less saturated fat, 45% less calories, 67% less cholesterol, 11-30% more protein and 10% more minerals. The buffalo meat has better water holding property and ideally suited muscle fibre. Buffalo meat has muscle pH of 5.4 and shrinkage on chilling is 2%. Buffalo fat is white and meat darker in colour than cattle beef because of more pigmentation and less intracellular fat (2-3% marbling than 3-4% in beef). Moreover, the deadly Mad Cow disease (BSE) has not yet been reported in buffalo. Therefore, the demand for buffalo meat is expected to grow in near future.

Table 13. Total and buffalo meat production in selected countries (2003)

Country	Total meat production (,000 Mt)	Buffalo meat (,000 Mt)	Buffalo meat as % of total	Growth in total meat production (% p.a) (1961-2003)	Growth in buffalo meat production (% p.a) (1961-2003)
World	249851	3180	1.3	3.03	2.61
Asia	102156	2872	2.8		
India	6038	1471	24.4	3.07	2.36
Pakistan	1892	509	26.9	3.99	3.70
China	69715	396	0.6	8.20	7.36
Egypt	1445	307	21.2	3.85	3.16
Nepal	251	130	51.8	3.37	5.18
Viet Nam	2487	99	4.0	4.61	1.97
Philippines	2130	81	3.8	4.51	4.45
Thailand	2197	53	2.4	4.02	-2.04
Indonesia	1917	45	2.4	4.22	0.27
Myanmar	553	22	4.1	3.98	2.22
Laos	94	18	19.5	3.82	3.64
Cambodia	183	13	7.3	4.64	8.26
Iran	1600	12	0.7	3.90	1.13
Sri Lanka	124	5	4.3	2.78	0.45
Turkey	1348	5	0.4	2.50	-2.42
Malaysia	1090	4	0.3	5.80	-2.22
Bangladesh	433	4	0.8	2.27	1.47

(FAOSTAT, 2004)

Number of buffaloes slaughtered (Table 14) for meat production is 13% in the World. In Pakistan, China and Nepal the slaughter % ranges between 15-17 while in India it is 11%. It has been estimated that extraction rate could be safely increased to 30% in India without affecting the milk production or draught production, thereby increasing meat production three folds to 3.6 million by the year 2010 AD.

Table 14.Number of buffaloes slaughtered (2003)

Country	No. of animals slaughtered (,000)	% animal slaughtered
World	22701	13
Asia	20942	13
India	10660	11
Pakistan	3800	15
China	3956	17
Nepal	595	16
Egypt	1750	49
Philippines	425	14
Viet Nam	460	16
Myanmar	132	5
Indonesia	209	9
Thailand	210	12
Laos	166	15
Bangladesh	46	6
Sri Lanka	47	7
Cambodia	84	13
Iran, Islamic Rep of	79	14
Italy	8	4
Bhutan	28	17
Malaysia	20	14

(FAOSTAT, 2004)

Buffalo Hide

India is one of the largest bovine leather producer in the world. Of the 0.857 million (Table 15) tonnes of buffalo hide produced in the world in 2003, 61% were produced in India. China, Pakistan, Egypt, Nepal and Viet Nam are other major buffalo hide producing countries. The highest growth rate is seen in China where it is growing @7.6%

p.a. Modernising the leather industry and encouraging export of finished leather products will help earn the valuable foreign exchange.

Table 15. Buffalo hide production (2003)

Country	Buffalo hide (,000 Mt)	% increase /decrease (1981-2003)	Growth rate % p.a. (1981-2003)
World	856.8	70.4	2.5
Asia	821.4		
India	525.0	47.1	1.8
Pakistan	85.8	147.8	4.2
China	118.8	401.7	7.6
Nepal	33.3	95.3	3.1
Egypt	35.0	142.4	4.1
Philippines	9.4	169.0	4.6
Viet Nam	18.4	50.8	1.9
Myanmar	5.0	26.9	1.1
Indonesia	6.9	4.3	0.2
Thailand	6.3	-67.7	-5.0
Laos	2.7	141.1	4.1
Bangladesh	2.3	85.2	2.8
Sri Lanka	1.6	4.4	0.2
Cambodia	2.8	130.8	3.9
Iran, Islamic Rep of	1.7	48.1	1.8
Italy	0.2	-12.7	-0.6
Bhutan	0.0	-56.3	-3.7
Malaysia	0.5	-48.0	-2.9

(FAOSTAT, 2004)

Conclusion

Buffaloes, which have been the backbone of operation flood programme in India and have a major contribution in the economy of Pakistan, Nepal, Egypt, China and several other South-East Asian countries will continue to dominate the livestock scenario in view of the diminishing role of traditional zebu cattle. With the increasing human population, the requirement for milk and meat products will continue to grow in near future and hence buffalo, as milk, meat and draught animal will continue to serve the mankind. Different breeds of buffaloes which have adapted to their local environment need to be conserved and promoted. Simultaneously, the vast non-descript buffalo population need to be upgraded for increasing their productivity.

References

Acharya R M and Bhat P N (1984). Livestock and poultry genetic resources of India. Research Bulletin No. 1. Indian Veterinary Research Institute, Bareilly.

Balachandran, S. 1996. Final Report (1993 -1996) of ICAR Adhoc Scheme on "Evaluation and Conservation of Toda Buffaloes. Sheep Breeding Research Station, Sandynallah, The Nilgiris, Tamil Nadu.

Cockrill, W. Ross, 1985. Present and future status of buffaloes in world. Proceedings of first buffalo congress, Dec. 27-31, Cairo, Egypt.

FAOSTAT (2004). www.fao.org

Heck, H; Wurster, D and Benirschke, K (1968). Chromosome studies of members of sub families Caprinae and Bovinae family. Bovidae, the musk ox, ibex, aoudad, Congo buffaloes and Gaur. Zutschrift fur saugetierkunde **33**: 172- 9.

Karthikeyan, M.K., Iyue, M., Kandasamy, N. and Paneerselvam, P. (2002). Characteristics and performance of Toda buffaloes of the Nilgiris, India. I. Habitat, morphology and morphometry. Buffalo J., **18**: 303-313.

Kaura, RL. 1965. Indian Breeds of Livestock (including Pakistan breeds). Prem Publishers. Gopalganj, Lucknow.

Kushwaha B P; Anil Kumar; Sultan Singh, Kundu S S, Maity S B and Shrama P (2004). Current status of Bhadawari buffaloes in the country. Abstract in proceedings of National Symposium of Livestock biodiversity *Vis-a-Vis* resurce exploitation: An introspection. February 11-12, 2004 Karnal (Haryana) INDIA.

Nivsarkar, A.E., Vij, P.K. and Tantia, M.S. (2000). Animal Genetic Resources of India :Cattle and Buffalo. ICAR, New Delhi.

Pathak NN and Singh, Inderjeet. 2001. Buffalo Genetic Resources in India. Summer School on Recent Advances in characterizationi and conservation of Animal Genetic Resources. July 16 - August 6, NBAGR, Kamal.

Ranjhan S K (2003). Changing role of buffalo production in the third millennium. Paper presented in 4th Asian buffalo congress held at New Delhi February 25-28, 2003.

Ray SK, Rao, SV, Mohapatra, PK and Morrenhof, J. 1999. The Paralakhemundi buffalo, its description, evolution and conservation for milk, draught and horn industry. Indian J Animal Production. **31:** 53 - 58.

Report (1997). Basic Animal Husbandry Stastistics. Department of Animal Husbandruy and Dairying, Ministry of Animal Husbandry and Dairying, Govt. of India.

Sadana D K and Jain A (2004). Livestock population of Orissa as animal genetic resources. Abstract in proceedings of National Symposium of Livestock biodiversity *Vis-a-Vis* resurce exploitation: An introspection. February 11-12, 2004 Karnal (Haryana) INDIA.

Sethi R K (2003). Buffalo breeds of India. Paper presented in 4th Asian buffalo congress held at New Delhi February 25-28, 2003.

Singh K P; K R Tajane; D P Pandey and A P Chaudhary (2004). Banni buffaloes: A unique and valuable genetic Resources of Gujrat. Abstract in proceedings of National Symposium of Livestock biodiversity Vis- a- Vis resurce exploitation: An introspection. February 11-12, 2004 Karnal (Haryana) INDIA.

Sirothia A R; N H Fuke; Gurmej Singh; P K Singh; K A Sirothia and T L Katre (2004). Present status of Nagpuri buffaloes: The pride of

Viderbha. Abstract in proceedings of National Symposium of Livestock biodiversity *Vis-a-Vis* resurce exploitation: An introspection. February 11-12, 2004 Karnal (Haryana) INDIA.

Ulbrich F and Fisher A (1967). The chromosome of the Asiatic buffalo (bubalis bubalis) and the African buffalo Syncerus caffer, Zeitschrift fur tierzuchtung und Zuchtungsbiologies, **83**: 219-23.

CHAPTER 4

Fundamentals of Buffalo Nutrition and Feeding

N. N. Pathak

Former Director
Central Institute for Research on Buffaloes, Sirsa Road,
Hissar - 125 001 (Haryana)

Buffaloes were recognised long back as the main dairy animal of the Indian subcontinent as evident from an age old folk lore of eastern region of India saying:

"Gai hai lakshmi, bail Mahadeva,

Bhains hai suari, bahut doodh deya".

Buffalo was also recognised as a meat animal during the Ramayan period (about 6000 to 5000 B.C. earlier) as described in the lines of Ramcharitra Manas of Goswami Tulsidas Ji. Due to successive defeat and loss of large number of main warriors, desperate Rawan went to awake his brother Kumbhakaran before due date. However at initial stage Kumbhakaran did not like to fight Shri Ram but clever Rawan ordered the supply of thousands pitchers of wine and a large number of buffaloes for the meals of Kumbhakaran as shown in the following lines: -

"Rawan Mangeu Koti ghut mad aru mahish anek".

Then Kunbhakaran thundered like very strong lightening after eating buffaloes and drinking wine and then arrogant Kumbhakaran walked alone to battle field from fort as evident from the following lines: -

"Mahish Khai Kari Madira Pana, garja bajraghat samana:
Kumbhakaran durmad run ranga, chala durg taji sen na sanga".

Sacrifice of male buffaloes as offering to Hindu Goddess 'Kali', practiced in the Indian subcontinent since the pre-historic period. The flesh of sacrificed offerings are eaten by devotees as "Prasad" (food left by the deity).

Use of male buffaloes as draft animal for riding is used by Lord Yamraj, the Hindu God of death, whereas furious form of male buffalo is evident from the fierce fighting between Goddess Durga and Buffalo Demon-Mahishasur. The Mysore town of Karnataka state is believed to be constructed on the chest of Mahishasur (Mahish means buffalo and Ur means chest).

All these mythological and historical statements provided ample evidence that buffaloes were domesticated much later than the many other farm animals.

Although, buffalo has been domesticated and evolved as a multi-purpose animal since long, indeed birth of male calves in buffaloes that too in the month of Bhadrapad (July-August-30 days) and female human baby was considered inauspicious even during the puranic period as stated by Surapal. The situation is more or less same even today except that the male buffalo calves have gained significant importance as the beast of burden and recently as a good source of high quality edible flesh.

Evolution of Buffalo Feeding Systems in India

Domestication of buffaloes was probably initiated during the epoc period of Ramayan and Mahabharat about 6000 to 5000 B.C. in the Indian subcontinent. The wild buffaloes were widely distributed in the hills, forests, river valleys and almost every region with plenty of water sources as the animal has great affinity with water and swamp. Among the two major genera of buffaloes– *Bubalis* and *Syncerus* only former could be domesticated. Probably India was the home land of Bubalis from where it spread to some parts of Northern Africa (Egypt) and Southern part of Europe. Even today protected herds of wild buffalo *(Bubalus bubalis. Var. arni)* are present in good number in the forests of north-east states with concentration along the Brahmaputra and its tributaries and Manas river. Another group of well-protected wild buffaloes is maintained in the wild life reserve of Kolhapur district in

Maharashtra yet unidentified herds of wild buffaloes are present in the forests of Madhya Pradesh, Orissa, Jharkhand and West Bengal. There are also small herds of tamed buffaloes in Manipur state.

In beginning domesticated buffaloes were reared exclusively on grazing by the nomadic and tribal communities. Such systems of buffalo rearing are still prevalent in some parts of India, viz. rearing of Zaffarabadi buffaloes in Gir forest, Toda buffalo in Nilgiri region, migratory herd of Himalayan region and shifting herd of river valleys of Assam and other states.

In second step the rearing systems became probably more organized and use of supplement feeding was introduced. Perhaps seasonal scarcity of feeds on natural grassland and pastures created the conditions for the supplemental feeding as even today hungry animals attempt to feed on stored grasses, fodders and other edible ingredients. With the passage of time ownership of animals established and owners started to counting and herding their animals at a protected place for protecting the buffaloes from predators like big cats (Lion, Tiger, Panther etc.) in the vicinity of jungles and from crocodiles along the water sources and swamps.

In the third successive step the systems of rearing and feeding was further improved and buffaloes were made a complete domesticated animal. The feeding systems were developed around the main crops and their processing and milling byproducts under close confinements. Perhaps by this time importance of buffaloes for milk production increased due to increased requirement for human consumption. The buffalo became the main dairy animal and its economic contribution showed probably much better rising trend than the cows. Higher yield of butterfat also encouraged buffalo rearing for subsidiary income and sustainability as the "ghee" consumption and consumption of variety milk products became the status parameter of Indian families.

Modern Systems of Buffalo Feeding

The systems of buffalo feeding in vogue in India from the medieval period are:

1. Various regional systems of buffalo feeding developed from the roughages and concentrates produced in the region.

2. Official rations of farm animals developed on the basis of experience and certain observations were followed on the livestock farms of states during the Mughal period and at state farms and military dairy farms during the British period. During British period feeding system of buffaloes was scheduled on the basis of feeding standards for dairy cattle.

However, the rations used for the feeding of dairy cows, buffaloes and equines became deficient in minerals supply and nutritional deficiency disorders of different kinds particularly rickets, stillbirth, osteomalacia and night blindness appeared in the animals of military dairy farms and equine breeding studs during the terminal years of the nineteenth century and early years of the twentieth century. To solve these nutritional problems Laboratory of Physiological Chemist was established by Dr. F.J. Warth at Science College, Pusa in February 1921. Later on this laboratory travelling from Imperial Agriculture Research Institute, Pusa, Muzzafarapur, Dairy Research Institute Bangalore, and Imperial Bacteriological Institute, Mukteswar to settle at IVRI Izatnagar in 1936. At this stage Laboratory of Physiological Chemist was named as Animal Nutrition section and dimension of research was expanded. Later on it was made Animal Nutrition Division of IVRI and functioned with its four regional research stations which were dissociated from IVRI during the Third Five Years Plan period. Animal Nutrition Division of IVRI was the national reference laboratory but gradually this status vanished due to wanton disregard of its importance. Since Eighth Plan period efforts are being made to scrap the role of Animal Nutrition Division of IVRI, Izatnagar and this reduction in the role of this Division will be an unrepairable damage to this discipline and infrastructure developed will vanish gradually without use.

Sir F. Ware, Animal Husbandry Commissioner with the Government of India called a meeting of the crops and soils wing of the Board of Agriculture and Animal Husbandry at Delhi in February, 1942, where it was decided to prepare five manuals on important aspects of Animal Husbandry, viz. (i). The breeding of Farm Animals in India, (ii). The feeding of Farm Animals in India, (iii). The Management of Farm Animals in India, (iv). The preparation of Animals and Animal Products for the Indian Market and (v). Dairying in India. Dr. P.E. Lander was first amongst the five groups to prepare this manual which was published in 1949 by Macmillan and Co. Limited, Calcutta. Dr. Lander made survey of buffalo feeding in some villages of Punjab which he has described as "It is well known that village buffaloes in India are fed better than cows, because the zamindar pays more attention to them as they are his main source of milk and ghee". He has also given an example of common rations of buffaloes.

He compared this ration with the scientifically formulated

Ingredients	Quantity (lb)	DM (lb)	TDN (lb)	DCP (lb)
Jowar & guar	70	12.00	10.00	0.70
Wheat Bhusa	15	14.90	6.75	-
Cotton Seed	2	1.92	1.40	0.24
Toria Cake	1	0.92	0.74	0.30
Gram	1	0.92	0.80	0.12
		30.44	19.69	1.36

balanced ration by him for a similar buffalo of about 1400 lb body weight yielding 24 lb milk of 7% butter fat content, viz.

Comparison of the two rations clearly shows that the rations fed

Ingredients	Quantity (lb)	DM (lb)	TDN (lb)	DCP (lb)
Maize Green	105	20.05	15.75	1.050
Wheat Bhusa	5	4.60	2.25	-
Wheat bran	5	4.65	3.55	0.450
Gram	1	0.92	0.80	0.120
Toria cake	3	2.76	2.22	0.900
		34.98	24.57	2.520
		15.90 Kg	11.17 Kg	1.145 Kg

in the village of buffaloes were deficient in both energy and protein. This was the status of buffalo feeding about half century earlier in the state considered most advanced in dairy husbandry.

A feeding system was also suggested by Sen in 1938. He used mid Morrison values for evolving a feeding standard, which was modified to meet the higher nutrient requirements of lactating buffaloes yielding milk of much higher butter fat content. Subsequently this standard was revised several times by Sen and Ray (1946, 1955, 1958), Sen, Ray and Ranjhan (1978) and Ranjhan (1999).

During this period a feeding standard for buffaloes was also suggested by Kearl (1982). The standard suggested by Kearl (1982) is a compilation of information collected from different sources including that of Ranjhan and Pathak (1979). Probably some extrapolation of data has also been used for the estimation of nutrients requirements, but many values appear to be quite irrelevant specially for energy requirement and its concentration in the rations. The use of this standard for buffalo feeding is limited, and at many places impracticable.

Another group of nutritionists of IVRI developed feeding standard for buffaloes utilizing most of the published data emerged from the experiments conducted in different laboratories of India (Pathak, 1988). This was further improved by utilizing more data from time to time (Pathak and Verma, 1993). The latest publication (Mandal, Paul and Pathak, 2003) provides requirements of feed (dry matter), crude protein, digestible crude protein and energy for growth and lactation worked out with the use of prediction equations evolved from the experimental data. The requirements thus estimated were also compared with the various experimental findings before adoption for the book.

Despite so much informations generated still we are lacking on many aspects of buffalo feeding behaviour and physiological aspects of nutrients utilization. The main reason for this is the consideration of buffalo close to cattle since the beginning of nutritional experiments on this species.

Some Main Aspects Omitted by the Researchers During Experiments

Numerically thousands of experiments have been conducted on various aspects of nutrition and feeding of buffaloes in different countries, but it appears that in none of the experiments some important aspects of buffalo behaviour was taken into consideration. A few of such behaviour having significant effect on feed intake and nutrients utilization are:-

1. Buffalo calves are lethargy at birth and need special attention for feeding in early life.

2. Weaning has adverse effects on lactating buffaloes and in most of the breeds let down becomes very difficult. In many studies of weaning although success has been shown but large scale use of pituitary hormone (Oxytocin) for let down of milk has not been mentioned at all and in many cases despite use it has been denied.

3. Buffaloes use different means for reducing energy expenditure. A few to be mentioned are :-

    (i)   Wallowing for several hours significantly reduced energy loss from the body.

    (ii) The buffaloes are relatively less selective than the cows and spent more time for biting the grasses at one place.

    (iii) The buffaloes spent more time for lying and relaxing.

The most significant method of energy conservation by long duration wallowing almost in still position is not accounted during digestion and metabolism studies, but this may have significant effect on the metabolism and estimates, thus, made in various experiments might have over estimated the energy requirements.

Studies for the Estimation of Requirements

A comparison of information so far available on the requirement of dry matter, protein and energy for the maintenance, growth and milk production (Table-1) shows great variations. There is still need of further refinement in the feeding standard of buffalo. Although recent requirements worked out by Mandal *et al.* (2003) has much greater application than most of the values as far requirements of feed, protein and energy for maintenance, growth and lactation is concerned. For the estimation of requirements for gestation, fattening and work available data are little. So far little efforts have been made for the estimation of minerals and vitamins requirements of the buffaloes and these aspects require more attention for experimentation.

Table-1: Comparison of DM intake between growing and lactation buffaloes

Attributes	DMI (Kg % bwt)		DMI (g.kg 0.75)	
Bwt. (kgO	Buffalo	Cattle	Buffalo	Cattle
90	2.98 (22)	2.79 (28).	92.4± 4.1	86.1±2.8
115	2.78 (28)	2.96 (28)	89.0±1.8	97.1±2.9
140	2.82 (33)	2.85 (53)	94.3±2.3	93.6±3.0
165	2.52 (55)	2.65 (46)	89.9±1.8	96.0±2.6
190	2.35 (25)	2.41 (36)	87.6±3.0	90.4±1.8
210	2.34 (29)	2.30 (15)	88.9±1.3	88.8±2.6
230	2.40 (10)	2.44 (26)	93.8±2.7	97.4±2.8
260	2.26 (33)	2.23 (29)	92.5±3.1	86.9±4.7
320	2.10 (21)	2.08 (11)	88.0±2.3	94.1±3.3
Milk yield (FCM, Kg)				
<9	2.32 (16)	2.79 (42)	107.0±5.1	119.0±2.3
9-11	2.67 (19)	3.67 (13)	122.7±3.3	151.8±8.8
>11	2.67 (20)	3.29 (24)	125.4±2.5	143.9±3.9
Overall	2.54 (55)	3.09 (79)	119.1±2.3	132.0±2.7

The estimates presented in Table-1 are although based on meagre data, but show a little less intake of feed in growing buffaloes than the growing cattle. The differences in case of lactating buffaloes and cows at each level of fat corrected milk (FCM) are highly significant (P<0.001). However, for the generation of more correct data, there is a need of conducting more experiments without any difference in the age, sex, as far possible body weight, season, quality of feeds, types and ratio of feed ingredients, duration of feeding and management practices. Only data from such experiment may precisely show differences, if any, between the buffaloes and cattle for the intake capacity of feeds and utilization of nutrients for various physiological functions like maintenance, growth, reproduction, lactation and quality of milk and meat.

2. Comparison of digestibility of nutrients between the buffaloes and cattle:-

As usual there is lack of adequate data to make a conclusive differentiation between the digestion ability of the two species, some informations available on the comparison of digestibility coefficients (Jang and Majumdar, 1962, Chaturvedi *et al.*, 1973, Pathak *et al.*, 1973, Upadhyay *et al.*, 1973, Jaikishan, 1974, Moran *et al.*, 1983 and Sangwan *et al.*, 1987) are summarized in Table-2.

Table-2: **Mean digestibility coefficients of feeds and nutrients in buffaloes and cattle**

Diets	Sp	Digestibility coefficient				
		DM	CP	EE	CF	NFE
Spear grass	B	54.1	47.5	74.1	62.0	53.2
(Jang & Majumdar, 1962)	H	53.5	49.5	62.9	61.6	52.9
Wheat Bhusa	B	52.9	53.8	69.0	64.8	53.9
(Chaturvedi *et al.*, 1973)	H	52.9	60.7	75.4	64.4	52.8
Wheat Straw +UMLF	B	50.0	46.4	49.0	44.3	72.0
(Pathak *et al.*, 1973)	H	48.7	49.3	44.6	41.9	70.2
Green Cowpea	B	60.1	76.2	71.0	70.1	55.4
(Upadhayay *et al.*, 1973)	H	57.4	74.4	68.1	72.7	49.7
Green Cowpea	B	59.9	71.3	65.5	65.2	65.3
(Jaikishan, 1974)	H	58.2	68.1	64.5	64.8	64.0
Green Maize	B	60.9	63.6	68.3	70.2	62.5
	H	60.7	65.6	53.6	65.5	65.6
Green Oats	B	66.5	66.2	58.3	67.6	72.5
	H	66.4	66.2	55.0	65.0	74.5

Diets	Sp	Digestibility coefficient				
		DM	CP	EE	CF	NFE
Oat Hay	B	54.2	30.3	55.0	61.8	63.0
	H	54.8	31.9	57.6	57.2	64.7
Sole rice Straw	S	36.6	20.5	-37.6	53.1	-
(Moran *et al.*, 1983)	O	37.6	24.4	-42.8	55.4	-
Alkali Treated RS	S	49.7	25.2	27.4	71.1	-
	O	46.9	20.4	12.3	69.8	-
Elephant Grass	S	52.7	61.8	54.5	55.7	-
	O	52.7	64.3	58.5	53.9	-
Wheat Bhusa (65)+GNC (35)	B	49.4	71.3	74.5	59.3	49.7
(Sangwan *et al.*, 1987)	CC	50.1	61.5	73.6	51.9	50.6
Oat Hay (68)+ GNC (32)	B	60.1	74.8	72.7	64.6	60.3
	CC	59.3	62.7	76.5	57.6	58.9
Wheat Bhusa(67)+Guar meal (33)	B	54.5	70.9	71.3	56.8	54.3
	CC	54.7	72.4	73.4	52.8	54.7
Oat Hay (67)+ Guar meal (33)	B	66.4	76.4	76.1	71.0	66.5
	CC	64.3	70.3	78.0	65.3	63.1

NB:- B= Murrah Buffalo; S= Swamp Buffalo
C= Haryana Cattle; O= Ongole Cattle;
CC= Crossbred Cattle

The perusal of values in Table-2 does not show any definite pattern. Since in most of the experiments conditions and management are not described, it is difficult to make precise assessment of the species effects on feed intake and nutrients utilization. The informations so far available on the digestibility of feeds and nutrients do not show any significant superiority of buffaloes over the cattle whether Taurus, crossbred or Zebu.

The situations do not permit to commit on the digestibility values presented in the Table-2. Indeed it can be seen that there is no sign of superiority for feed and fibre digestibility of buffaloes compared to cattle. In some other experiments there is some little improvement that has been recorded in buffaloes, whereas in some other experiments results are almost comparable in both the species.

3. Comparison of nutrients utilization efficiency between buffaloes and cattle for growth production

Mean values of utilization efficiency of DM, CP and TDN per kg body weight gain in grazing buffaloes and cattle worked out by Udeybir *et al.,* (2000) are presented in Table-3.

Table-3: Comparison of nutrients utilization efficiency in growing buffaloes and cattle.

Live weight range (Kg)	Buffalo	Cattle	
	DMI/Kg bwt. gain (Kg)		
100 (70-150)	8.22±0.32 (63)	9.13±0.27 (103)	P<0.05
200 (151-250)	9.43±0.31 (65)	10.54±0.35 (89)	P<0.05
	PI (Kg)/ Kg bwt. Gain		
100 (70-150)	1.10±0.05 (57)	1.26±0.03 (98)	P<0.01
200(151-250)	1.15±0.03 (53)	1.41±0.03 (69)	P<0.01
	TDNI (Kg)/ Kg bwt. Gain		
100 (70-150)	5.05±0.019 (66)	5.66±0.15 (105)	P<0.01
200 (151-250)	5.70±0.12 (65)	6.14±0.19 (83)	P<0.01

Comparison Of Nutrients Utilization Efficiency For Milk Production In Buffaloes And Cows

There appears to be no studies on lactating buffaloes in terms of calorie output in milk for calorie input in diets. Gross and net efficiency of energy and protein utilization in buffaloes and cows for milk production worked out by Paul *et al.* (2003) has been summarized in Table-4.

This has been further utilized for evolving feeding standards for lactating buffaloes (Paul *et al.*, 2002).

Table-4: Comparison of nutrients utilization efficiency for milk production in buffaloes and cows.

Feed/ Nutrient	Buffalo	Cow
DMI (Kg)/ Kg FCM	1.175±0.02 (53)	1.243±0.03 (71) NS
DCPI (g)/ Kg FCM	80.36±2.1 (52)	93.88±2.33 (40) P<0.01
TDNI (g)/Kg FCM	707.6±19 (54)	774.8±17.0 (74) P<0.01
GE efficiency, %	25.19±0.55 (54)	23.17±0.49 (77) P<0.01
NE efficiency, %	60.70±1.57 (44)	59.86±1.18 (73) P<0.001

The values presented in Table-4 clearly indicate the superiority of buffaloes over cattle for nutrient utilization efficiency for milk production. From this physiological event economy of nutrients utilization for maintenance in buffaloes appears to be one of the main factor. However, it needs verification through paired experiments conducted simultaneously on almost similar animals of the two species kept under same management.

Prevalent Buffalo Feeding Systems in India

Migration of lactating buffaloes to stalls of urban and periurban areas is a common practice. Indeed large numbers of buffaloes are reared in the rural areas. There is great variation in the systems of feeding and management of buffaloes in different parts of the country. An attempt has been made to enlist some of the practices of buffaloes feeding:-

1. Stall feeding in crowded housing with little scope of outing in urban areas.

2. Stall feeding with very little outing (1-3 hours) in periurban areas. Buffaloes lying in dirty waters and mud slurry and grazing on scrubs in mango orchards, scrub land and along the railway tract and road sides are commonly seen.

3. Grazing in 'Diara' Land.

4. Grazing in Kacch region- Salt rich herbage.

5. Grazing along water, sources in hilly region.

6. Migratory grazing on high hills of Himalayas during summer months.

7. Grazing on submerged herbage in the coastal areas in the rivers and lakes (Like Chilka lake in Orissa) joining the sea. These are swamp buffaloes.

8. Feeding of buffaloes on rations containing plantain residences mixed with other ingredients in the plantain plantation area of Bihar.

9. Grazing in the plantation lands and along the water streams in the states of Kerala and Tamil Nadu.

Points to be Considered for Conducting Nutritional Studies on Buffaloes Vs. Cattle

Buffaloes appear to possess better efficiency of energy economizing and this trait is probably due to conspicuous differences between the behaviour of the two species recorded by Pathak (1999) and presented in Table-5.

Table-5: Comparison of Some District Behavioural Trails of Buffaloes and Cattle.

Behaviour trails	Buffaloes	Cattle
Response to solar radiation	Preferably move to water source for wallowing in hot sun. in absence of water lie in mud. In absence of both, rest beneath the tree. These are measures of energy conservation.	Start running aimlessly in hot sun. Rarely enter in water & dislike lying in mud. Results are severe panting and more loss of energy.
Grazing behaviour	Moves with a slow pace. Spent more time for biting grasses from an area spending much less energy for movement.	Impatient while grazing. Bites less herbage from a place & spend more energy.
Rumination	More than cattle.	Less than buffaloes

Behaviour trails	Buffaloes	Cattle
Locomotion	Slow pace, easy going.	Aimless, irregular and fast with more body movements.
Aggressiveness	Mostly docile and timid.	Mostly less docile and aggressive.
Social behaviour	May line in crowded herd with little fight.	Require more space and take more time for living together with new mates.
Milk production sustainability	Much better than the cows and less variable (6 months to more than a year).	Highly variable (3 to 12 months).

Conclusion

More experiments are required for understanding the nutrition of buffaloes, viz.

(i)   Exploration of problems associated with rearing of weaned buffaloes calves.

(ii) To determine the causes of disorders of 'let down' of milk after weaning of birth and their amicable remedies.

(iii) Identification of all-important factors associated with energy conservation efficiency of buffaloes for understanding the sequence of physiological reactions.

(iv) Comparative studies on buffaloes and cattle under identical conditions to remove many myths associated with the feeding and nutrients utilization efficiency of buffaloes.

(v)   More precise experiments for the refinement of requirements of energy and protein for various stages of life cycle.

(vi) More experiments for the estimation of minerals and vitamins requirements of the buffaloes for different physiological functions.

(vii) Separate feeding standards are required for the feeding of different type of buffaloes viz.

(a) Dairy (riverine buffaloes yielding milk of less than 8% average milk fat content.

(b) Dairy buffaloes yielding milk of average more than 8.5 % milk fat content.

(c) Swamp buffaloes feeding on fodders of non-saline lands.

(d) Swamp buffaloes feeding on coastal grass and sea weeds, containing high level of salts.

(e) Buffaloes at high altitude.

(f) Working buffaloes.

References

Chaturvedi, M.L., Singh, U.B. and Ranjhan, S.K. (1973). Indian Journal of Animal Sciences, **43:** 387, 676, 1034.

Jaikishan (1974). Ph.D. thesis. Agra University, Agra.

Kearl, L.C. (1982). Nutrient Requirements of Ruminants in Developing countries. International Feed Stuffs Institute. Utah Agriculture Experiment Station. Utah State University, Logon, Utah, U.S.A.

Lander, P.E. (1949). The Feeding of Farm Animals in India. MacMillan and Co. Ltd. Calcutta.

Mandal, A.B., Paul, S.S. and Pathak, N.N. (2003). Nutrient Requirement and Feeding of Buffaloes and Cattle. International Book Distribution Co. Charbagh, Lucknow.

Moran, J.B., Satato, K.B. and Dawson, J.E. (1983). Australian Journal of Agriculture Research. **34:** 73.

Pathak, N.N. (1988). Feeding standard of buffaloes, Indian Veterinary Research Institute, Izatnagar, Publication.

Pathak, N.N. (1999). Proceeding of workshop for evolving strategy for solving feed scarcity. International Livestock Research Institute, Adis Ababa, Ethiiopia, June 13-18.

Pathak, N.N. and Verma, D.N. (1993). Nutrient Requirements of Buffaloes. International Book Distribution Co. Charbagh, Lucknow.

Pathak, N.N., Singh, U.B., Kumar, P., Verma, D.N., Ranjhan, S.K. and Srivastava, R.V.N. (1973). Indian Journal of Animal Science, **43:** 819.

Paul, Shyam S., Asit, B. Mandal and Nitya N. Pathak. (2002). Journal of Dairy Research **69:** 173.

Paul, Shyam, S., Asit, B. Mandal, Alagarsamy Kannan, Guru P. Mandal and Nitya N. Pathak (2003). Journal of Science of Food and Agriculture. **83:** 258.

Ranjhan, S.K. and Pathak, N.N. (1979). Management and Feeding of Buffaloes. Ist Edi, Vikas Publishing House Private Limited, New Delhi.

Sangwan, D.C., Pradhan, K. and Vidya Sagar (1987). Indian Journal of Animal Sciences, **57:** 562.

Sen, K.C. (1938). Nutritive value of Feeds and feeding of Animals. Bulletin No. 25, ICAR, New Delhi.

Sen, K.C. and Ray, S.N. (1946, 1955, 1958, 1964). Nutritive value of Indian Cattle Feeds and Feeding of Farm Animals. Revisions of Bulletin No. 25, ICAR New Delhi.

Sen, K.C., Ray, S.N. and Ranjhan, S.K. (1978). Nutritive value of Indian Feeds and Feeding of Farm Animals. Indian Council of Agricultural Research, New Delhi.

Upadhyay, R.S., Singh, U.B. and Ranjhan, S.K., (1973). Indian Journal of Animal Sciences, **43:** 583.

CHAPTER 5

Toda Buffalo Rearing in Southern Plateau and Hill Regions of Tamil Nadu

R. KADIRVEL

Vice-Chancellor
Tamil Nadu Veterinary and Animal Sciences
University, Chennai – 600 051 (Tamilnadu)

The Toda buffalo is a unique breed, confined to the Nilgiris and is reared by the Toda tribe who are among the most famous aboriginal inhabitants of this country. The life style of this tribal people is largely centred around these buffaloes; hence the breed forms an integral part of their sustenance and cultural heritage. All the anthropological studies on Toda tribals included a detailed account of the buffaloes and their cultural associations. This paper describes the unique characteristics and husbandry of this genetically isolated group of buffaloes inhabiting the "Queen of Hill Stations", the Nilgiris in Tamil Nadu.

Origin and Habitat

The existence of these buffaloes in the hilly terrain could be possibly traced back to 1603 from the report of Finicio, who wrote of the Toda:

> "They have no crops of any kind, and no occupation but the breeding of buffaloes, on whose milk and butter they live"

The possible relationship of this fine species of buffaloes with the Chokatti buffalo of the Krishna River area in Karnataka was stated by Grigg (1880). In 1906, Rivers reported that the Toda buffalo was a variety of the Indian water buffalo and their life on the Nilgiri hills produced a much finer animal than those on the plains. Gunn (1909) and Littlewood (1936) suggested that the Toda buffaloes were indigenous to the hills alone. Several other reports also supported the uniqueness of the breed and their distinguishable characteristics from other breeds of riverine buffaloes in India. Hence, the Toda buffaloes are exclusive and confined to the Nilgiri hills alone. Perhaps, this is the only breed of buffalo, adapted itself to the high rainfall and high humid zone of the plateau of the Nilgiris.

The Nilgiri district extends over an area of 2,545.5 sq. km. and is situated between latitudes 11°15' and 11°30' N and longitudes 76°15' and 77° E. The entire district is hilly with the elevation ranging from 668 to 2,634 metre above mean sea level (MSL). But the Toda buffaloes are predominantly distributed at altitude of more than 2000 metre above MSL. The climatic environment of the hills is temperate and the average rainfall in the area is around 1,200 mm per annum. The meteorological data of the hilly terrain of the Nilgiris is furnished in Table – 1.

Table-1: **Meteorological data of the Nilgiris**

Climatic Variable	Annual Mean	Range
Total rainfall (mm)	1166.1	835.8 – 1477.2
Number of rainy days	88.8	77.0 – 97.0
Maximum temperature (°C)	19.2	18.8 – 19.4
Minimum temperature (°C)	10.6	10.2 – 10.8
Relative humidity (%)		
Morning (7.23 h)	81.5	79.7 – 83.2
Evening (14.23 h)	67.9	64.7 – 69.6
Wind velocity (km/h)	4.4	1.3 – 6.9

Source: Karthikeyan (1995).

Population Status

The census figures of Toda buffaloes do not show any appreciable increase in the population since 1848 to 1986.

Years	Population (in no.)	Reference
1848	2171	Reviewed by Walker (1986)
1930	1619	
1960	2186	
1975	2650	
1986	2002	Nair *et al.* (1986)

However, Mason (1974) reported the decline in the Toda buffalo population due to negative effect of urbanization. On the contrary, Karthikeyan *et al.* (2002a) reported the total population of the Toda buffaloes in the breeding tract as 3,313 with the break-up details of 1,809 adult females, 70 males youngstock, 584 female youngstock and 850 calves (as of 1994). Further they also concluded that the present status of the population is highly vulnerable with respect to conservation.

Distribution

The Toda buffaloes are distributed throughout the plateaus of Nilgiri hills with varying densities. More number of herds are found in Udhagamandalam taluk, followed by Gudalur, Coonor and Kotagiri taluks (Karthikeyan, 1995). The buffaloes are predominantly owned by the Toda tribe; but non-Todas such as Badagas, Kotas and other communities also maintain them.

Morphological Features

Toda buffaloes are medium-sized animals with a fairly long body and a broad and deep chest. A peculiar pattern of change of coat colour from calves to adult is noticed in Toda buffaloes. The coat colour of majority of the new born calves is found to be fawn and the remaining with light grey-coloured coat. As the age advances, the fawn colour charges to ash grey and thick long course hairs also become sparsely distributed. The adult animals are mostly ash grey in colour; however, few light cream-coloured animals are also noticed in the herds.

Other important distinguishable features of the Toda buffaloes are the type of horns and chevron markings. The head is surmounted by slate-coloured, flat horns, marked with distinct rings. They are characteristically crescent shaped or semi-circle in shape, pointing with the sharp tip. In general, two chevron markings are noticed; one just around the jowl and the other below the neck, just above the brisket region.

The average weight of the calves at birth was 27.94 ± 0.43 kg and the body weight at 6 months of age averaged 66 kg (Final Report, 1996).

Usually, Todas do not worship any idol, rather they worship light, fire, mountains, trees, sun and moon. In addition, they are always much fascinated by the horns and their sacred buffaloes.

Behaviour

The Toda buffaloes are aggressive in nature to strangers and could not be easily tractable. But, they could be handled only by the herdsmen. They are not fully domesticated. The buffaloes are gregarious in nature. They do have the habit of wallowing in puddles in marshy lands.

Performance Characteristics

Milk Yield

Dairying is the chief and traditional occupation of the Toda tribe. Milking is usually done after allowing the calves to initiate let down of milk. Sometimes, butter is applied to the teat for ease of milking. Karthikeyan *et al.* (2002b) reported the least-squares means of the daily milk yield in Toda buffaloes (Table-2), as 2.53 ± 0.06 kg. But a maximum daily yield of 6.65 kg is also observed. Though the average daily milk yield appears to be low; it is satisfactory, considering the primitive level of management and their maintenance on grazing alone. No concentrate/any additional feeding is followed by the herdsmen. Buffaloes with more than 10 calvings are also observed in herds.

This reflects their longevity and long productive life of this breed of buffaloes

Table-2: Least-squares means of daily milk yield (kg) in Toda buffaloes

Parity	No. of observations	Mean ± S.E.
1	80	2.439 ± 0.126
2	76	3.027 ± 0.131
3	165	2.639 ± 0.093
4	111	2.767 ± 0.109
5	84	2.180 ± 0.129
6 and above	62	2.123 ± 0.150
Overall	578	2.529 ± 0.062

Constituents of Milk

This breed is well known for its high fat content in the milk. The mean fat content in the milk is estimated as high as 8.27 per cent with the range of 4.8 to a maximum of 14 per cent. The other breed in India, which produces high fat content in milk, is Bhadawari buffaloes (8.0 per cent; Singh and Desai, 1962). The total protein and total solids in the milk of Toda buffalo are 4.29±0.05 and 16.06 ± 0.14 per cent respectively.

Reproduction

In Toda buffaloes, the average age at first calving is 46.92 ± 0.36 month. There seems to be a seasonality existing in the calving. The main months of calving are from July to October and a peak calving occurred in August. Another important reproductive characteristics of the breed is regularity in calving, indicated by the low average calving interval of 14.26 ± 0.36 months. This regularity in calving reflects the reproductive efficiency of this breed of buffalo, which is not the case in other well-recognized buffalo breeds in India such as Murrah and Surti.

Husbandry Practices

Housing

Buffalo calves are housed in pens made of wooden stakes only during night. Calf pens are simply constructed under the spreading stems of a tree, keeping the lateral spreading stems as side walls. The side walls are smeared with a mixture of clay and mud to strengthen the structure and to fill up the gaps to prevent the entry of chilly wind blowing through the slits into the pen. But adults are kept in a circular enclosure of uneven and unhygienic floor guarded by loose stones or wooden stakes without roof, thereby exposing them to the extreme climates of the Nilgiri hills.

Breeding

No males are maintained in the herd. But stud bulls, which are roaming in the forest, mingle with the herds for a few days in different spells to sire all the buffalo cows. So, only natural mating occurred.

Feeding

No separate/supplementary feeding practices are followed. Neither concentrates nor any forages are fed to these buffaloes. These buffaloes are exclusively thriving on grazing alone. Calves are usually allowed for grazing after 3 weeks of birth. Usually young calves are allowed for grazing in the vicinity of the hamlet and not allowed to mingle with adults for the first six months to prevent them from suckling. The young calves are taken to the grasslands a little later in the day while adult buffaloes are let out soon after milking. As the suckling calves are retained, the adult buffaloes return to the hamlet for feeding their calves and evening milking. Otherwise, they remain in the forest and sometimes return next morning. Hence, after 6 to 8 months of calving, only one-time milking is carried out. Sometimes, the calf of dried up dam would be allowed to suckle milk from foster mothers. Similarly, when a calf has died, a surrogate is quickly introduced to keep its mother in milking. A ritual practice of giving common salt twice a year to all the buffalo cows is practised.

The land of entire hilly tract is full of grasses and fresh vegetation. With the onset of frost towards the end of December, the grasses would be burnt and would become coarse and yellow in colour. This change in the greenery continues until the first shower occurs during the month of April. During these months (end of December to March), the animals are deprived of its normal feeding and the buffaloes spent most of its time on grazing in the dense forest and trek longer distances in search of forages. In this process, they remain in the forest continuously for longer period (1 to 3 months); thereby skip milking and ultimately go dry. In recently calved animals, this would shorten their lactation length. Sometimes migration of buffaloes from one hamlet to the other occurs in search of fresh pasturage.

Commonly available vegetation in the pasture is Kikiyu grass (*Pennisetum Clandestinum*), which is abundantly distributed throughout the plateau. Other vegetation includes White clover (*Trifolium repens*) and Hariyali grass (*Cynodon Dactylon*). Some wild grasses and weeds *viz.* Naval kodi (*Oxalis corniculata*), Kuppai Keerai (*Amaranthus Paniculatus*), Mukuthi kachada (*Centella asiatica*), Pal thazhai (*Sonchus branchyotus*), Palkodi (*Embelia gardenaria*), Bothaipil (*Briza minor*) and Panaithangai (*Cotula australis*). In addition, Bracken fern (*Pteridium aquilinum*) is also interspersed in the pasture of the hilly terrain. Fodder trees such as *Acacia melanoxylon, Acacia dealbata, Acacia decurrens,* Dendrocalamus strictus and Celtis species are also present in the region (Report, 1989)

Cultural Practices

The religion of the Toda is a highly ritualized buffalo-cult. Naming of the buffaloes individually, when their first calf is born, is followed as a routine ritual practice. The individual buffaloes do respond to the call by their names. Exchange of buffalo cows along with calves is also practised during marriage ceremony. Occasionally, buffalo cows are slaughtered for ceremonial purposes to propitiate the dead relative of the clan.

Disease Prevalence

Parasitism is the most common cause of calf mortality in this breed during dry season. Dermatomycosis is also found to be a common menace in calves. Sometimes, outbreak of foot and mouth disease (FMD) is noticed in the Toda herds. But in spite of the unhygienic maintenance, there was no report of mastitis in the breed.

Improvement Measures Taken

- ⦿ All India Co-ordinated Research Project on "Transfer of Technology" was carried out through "Tribal Area Research" from 1982 to 1992

- ⦿ An ICAR *ad-hoc* scheme on evaluation and conservation of Toda buffaloes was functioning in TANUVAS for evaluating the performance characteristics and to study the status of the population in the Nilgiri hills.

- ⦿ Biochemical and cytogenetic profiles of the breed were also completed through ICAR-NBAGR Network Project on Core Laboratory.

- ⦿ Molecular characterization of the breed through microsatellite markers has been taken up.

- ⦿ Further, a project on *in-situ* conservation of Toda buffaloes to conserve this unique germplasm is also carried out in sheep Breeding Research Station, Sandynallah, the Nilgiris, Tamil Nadu.

Conclusion

The Toda buffaloes are traditionally sacred dairy animals of the Toda pastoralism. The geographical separation (hills separated from surrounding low lands) had helped in the evolution of this fine genetically isolated breed. High fat content in milk, regularity in calving, longevity and adaptability to the extremes of climate in the hilly tract are specific characters of the breed. Considering these facts, Toda buffaloes are self-sustainable in the low / nil input system of rearing.

References

Finicia, Y. (1603). Letter to the Jesuit Vice-provincial in Calicut (in Portuguese) British Museum. Add. Ms. 9853. (Translated by A.de alberti) Reprinted in: The Todas. (Ed.) Rivers, W.H.R. (1906). Macmillan and Co. Ltd., England. p.727.

Grigg, H.B. (1880). A Manual of the Nilgiri District in the Madras Presidency, Government Press, Madras. (Cited by Walker, 1986)

Gunn, W.D. (1909). Cattle of Southern India. Vol.III. Bulletin No. 60. Department of Agriculture, Madras. pp.50-58.

Karthikeyan, M.K. (1995). Evaluation of Production and Reproduction Performance of Toda buffaloes. M.V.Sc. Thesis Submitted to the Tamil Nadu Veterinary and Animal Sciences University, Chennai–600 007.

Karthikeyan, M.K., Iyue, M., Kandasamy, N. and Paneerselvam, P. (2002a). Characteristics and performance of Toda buffaloes of the Nilgiris, India. I. Habitat, morphology and morphometry. Buffalo J., **18**: 303-313.

Karthikeyan, M.K., Iyue, M., Kandasamy, N. and Paneerselvam, P. (2002b). Characteristics and performance of Toda buffaloes of the Nilgiris, India. II. Production and reproduction performance. Buffalo J., **18**: 315-320

Littlewood, R.W. (1936). Livestock of Southern India. Government of Madras, Madras. pp. 150-159.

Mason, I.L. (1974). Species, types and breeds. *In:* The Husbandry and Health of the Domestic Buffalo, (Ed.). Cockrill, W.R. Food and Agriculture Organization of the United Nations, Rome, pp. 1-47.

Nair, P.G., Balakrishnan, M. and Yadav, B.R. (1986). The Toda buffaloes of Nilgiris. Buffalo J., **2**: 167-178.

Report, (1989). Mineral Status of Plant and Animals and its Relation to the Incidence of Haematuria in Hill cattle. Final Report of ICAR *ad-hoc* scheme, Sheep Breeding Research Station, Sandynallah, Tamil Nadu.

Report, (1996). Final Report of the ICAR *ad-hoc* scheme on Evaluation and Conservation of Toda Buffaloes, Sheep Breeding Research Station, Sandynallah, Tamil Nadu.

Singh, S.B. and Desai, R.N. (1962). Production characters of Bhadawari buffalo cows. Indian Vet. J., **39**: 332-343.

Walker, A.R. (1986). The Toda of South India: A New Look. Hindustan Publishing Corporation, Delhi, pp. 98-118.

Murrah Breed of Buffalo : Its Strength, Weakness and Rearing in Trans-Gangetic Region

R K Sethi

Project Coordinator (Buffalo)
Central Institute for Research on Buffaloes, Sirsa Road,
Hisar-125 001 (Haryana)

Murrah is the most fancied and preferred breed, among various breeds of buffaloes found in India as well as in the world. Breeding tract of Murrah buffaloes stretches around the southern part of Haryana comprising the districts of Rohtak, Jind, Hisar and Bhiwani, the Union Territory of Delhi, Patiala Distt. in Panjab and some part of the western Uttar Pradesh. However, this breed has spread to almost all parts of the country and is being bred either in pure form or is being used for grading up of local nondescript buffaloes. The breed has remarkable ability to adjust quickly to varying environmental conditions and perform equally good in almost all agro-ecological conditions prevailing in the country as well as in various countries.

Murrah buffaloes are jet black and large in size with long and deep body. Head of females is short, fine and clear cut. Bulls are heavy and broad with prominent cushion of short and dense hair. Horns are short and tightly curved in a spiral form. Eyes are bright, active and prominent in females but slightly shrunken in males. Ears are short, thin and alert. Neck is long and thin in females and thick and massive in males. Hips are broad. Fore and hind quarters are drooping. Tail is long reaching below the hock up to fetlock and in some animals ending in a white switch. The skin is thin, soft and pliable with very little hair on the body in adult animals. The colour of the skin is jet black, but animals with fawn / grey hair are commonly found. Udder is capacious extending from hind legs to just behind naval flap with prominent milk-veins. Teats are long and placed uniformly wide apart. Rear teats are generally longer than the fore teats.

Population

India has over 94.13 million buffaloes and they number to approximately 56.6 per cent of the total world buffalo population. Their number in India is increasing at faster rate than the increase in any other country. During the last decade world buffalo population increased by approximately 13.3 millions showing annual increase of about 0.87%. The per cent increase in India was at 1.18 % as compared to the 0.88 % in Asia and 0.43 % in rest of the world.

In India the number of purebred animals of well specified breeds is expected to be about 20 to 25 % of the total buffalo population in the country. Rest of the buffaloes are nondescript in type and have extremely variable composition being either non descript or crosses among various breeds and cannot be categorized in any other well-established breed. Murrah breed being the most prominent and versatile, animals of this breed are found in almost all parts of the country besides their thick presence in the breeding tract as defined above. They number to approximately 6 to 7 millions in the country of which approximately 4.5 to 5 millions are found in Haryana. In addition to the above, large population of graded buffaloes from Murrah parentage is also available in different parts of the country.

Performance   Characteristics

Growth rate: Maturity of an individual is influenced by the growth rate especially in the early age. Growth potential of the animal is primarily dictated by the genetic make up of the animal and can be exploited only when *ad lib* and intensive feeding regimes are adopted. Calves reared under free choice feeding registered growth rate from birth to one month of 472 g per day. In the subsequent months from 1 month to 24 months it ranged from 401 g to 500 g per day (Table-1). Highest growth (500 g per day) was observed between 18 to 24 months whereas after the age of 24 months it was minimum. Body weight at maturity (30 months of age) was recorded as 392±9.9 kg. (Sethi and Chopra, 1995). However, body weight gain at the rate of approx. 1000 g per day has also been reported (Ranjhan, 2003) under feed lot system for fattening of males prior to slaughter of animals.

Table-1: Average body weights and growth rate at different stages in Murrah buffaloes

Age	Body weight (kg)	Growth rate (g/day)
Birth weight	28.7±1.1	-
1 month	43.1±0.5	472±42
6 months	104±5.3	401±26
12 months	179±6.8	415±13
18 months	259±7.1	440±16
24 months	350±7.6	500±22
30 months	392±9.9	233±22

Source: Sethi and Chopra, 1994

Average body weight at maturity in Murrah buffaloes varies from 450 to 550 kg however some of the adult buffaloes weigh as high as 600 kg. Average body weight at first calving reported from various participating centers of Network Project on Buffalo Improvement over the years range from 435 to 512 kg at PAU Ludhiana and 439 to 494 kg at HAU Hisar.

Body Measurements: Calves at birth measure by about 58 cm by the age of maturity gain by about 75 cm. By 2 years of age body length is almost close to the maximum adult size. The height of the animal is the outcome of structural stature of the bone development than deposition of any component. Average height at birth is 59.8±0.8 cm. Calves gain height much faster from birth to 1 year of age (47 cm) than the increase from 1 to 2 years of age (19 cm). At maturity average height in females range from 150 to 165 cm while in males

Table-2: Average body measurements at different stages of growth in buffaloes

Age	Body length (cm)	Height (cm)	Heart girth (cm)	Tail length (cm)	Horn length		Horn circumference		
					Outside	Inside	Base	Tip	
Birth	58.0±0.5	59.8±0.8	64.0±0.8	37.5±0.5	-	-	-	-	
6 months	81.0±1.0	84.3±1.2	109.3±1.2	53.5±1.0	-	-	-	-	
12 months	106.8±1.5	106.8±1.2	136.3±1.8	67.5±1.8	13.1±0.5	8.3±0.3	17.8±0.5	6.9±0.2	
18 months	117.8±1.2	115.0±1.0	158.5±1.8	78.0±1.5	19.3±0.7	11.0±0.3	17.9±0.4	5.6±0.2	
24 months	131.5±1.0	122.5±0.8	174.0±1.5	84.3±1.8	25.8±0.7	12.5±0.6	20.0±0.4	5.6±0.2	
30 months	133.0±1.3	125.8±0.8	184.8±1.8	88.5±1.5	30.4±0.8	14.5±0.7	20.1±0.4	5.3±0.2	

(Sethi and Chopra, 1994)

from 165 to 175 cm. Average heart girth at birth is estimated as 64.0±0.8 cm and at maturity as 184.8±1.8 cm. Total increase in hearth girth from birth to maturity is 120.8 cm and this increased consistently in all the age groups thereby indicating that increase in body weight at later stage is primarily due to the increase in heart girth as is also evident from the very high correlation coefficient (0.92) among the two traits. Average tail length at birth is 37.5±0.5 cm and at maturity (2.5 years of age) 88.5±1.5 cm. Outside and inside lengths of the horn at 12 months of age are 13.1±0.5 and 8.3±0.3 cm, and at maturity the average length is 30.4±0.8 and 14.5±0.7 cm, respectively. The circumference at the base of the horn is 17.8±0.5 cm the tip is 6.9±0.2 cm. The corresponding estimates at maturity are 20.1±0.4 and 5.3±0.2 cm respectively. Tip of the horn sharpen with the advancement of age.

Age at first calving: Weighted average of age at first calving (AFC) estimated from various participating centres of Network Project on Buffalo for Murrah breed from 1997-98 onwards are presented in Table-3. Overall weighted average (average of all the participating centres) during the year 2002–03 was estimated as 45.47 months from 166 calving. Data over the years suggest age at first calving in this breed to stabilize around 45 months. However efforts are needed to bring down to the target AFC of about 40 months in this breed. Wide diversity is apparent in the population from the fact that quite a large number of heifers calve for the first time as early as 27 to 30 months of age.

305 days or less milk yield: Consistent increase in 305 days or less lactation milk yield has been observed in the weighted average estimated from the participating centers of Network Project on Buffalo from 1997-98 to 2002-03 as indicated in Table-3. Overall weighted average during the year 2002-03 has been estimated as 2056 kg from 511 lactations spread over all the herds of Murrah breed. Consistent increase in weighted average in milk yield is apparent as a result of selective breeding. In the process of associated progeny testing several daughters of number of bulls have produced as high as 3000 kg in their first lactation. List of daughters of second set of bulls which produced more than 2500 kg in their first lactation of 305 days or less is presented in Table-4. Best Lactation milk yield to the extent of 3500 to 4000 kg has been recorded in several buffaloes at different locations with peak yield as high as 25 kg.

Haryana Livestock Development Board has identified more than 1400 buffaloes yielding more than 15 kg on the basis of their peak yield under Murrah breed conservation programme.

Table-3: Weighted averages of various performance traits in Murrah breed

Year	Age at first calving (months)	305 days or less lact. Milk yield (kg)	Service period (days)	Calving interval (days)
1997-98	45.6 (162)	1973 (455)	167 (325)	431 (338)
1998-99	47.6 (153)	1928 (412)	159 (323)	469 (331)
1999-00	44.0 (143)	1981 (375)	162 (310)	463 (295)
2000-01	48.6 (176)	1973 (422)	153 (370)	468 (427)
2001-02	46.4 (154)	2017 (505)	150 (397)	460 (374)
2002-03	45.47 (166)	2056 (511)	135.4 (408)	460.5 (383)

Figures in parenthesis are number of observations
Source : Network Project on Buffalo Improvement

Service Period : There has been substantial improvement in service period at all the centres of Network project and the average service period is estimated as 135 days.. Shortest average service period has been reported from NDRI (107±14 days), however in individual cases service period as low as 70 to 75 days has been observed in number of animals and such animals calve regularly with a calf a year. The target value of average service period of 90 to 100 days has been fixed for all the participating centers for this breed which can be achieved by proper reproductive management and monitoring.

Calving Interval: Weighted average of calving interval for the year 2002–03 was estimated as 460 days (Table-4) and it is more or less consistent over the years. Data over the years indicate that this trait is more or less stabilized around 450 days at various participating centres. However still there is scope for improvement of this trait by providing better management so as to improve this trait further to about 400 days.

Table-4: List of outstanding high milk producing daughters of second set of bulls under Network Project on Buffalo for Murrah breed.

Bull number	Location	Daughter number	Location	305 days or less first lact. Yield (kg)
759	CIRB	1921	PAU	2522
761	CIRB	4402	NDRI	2524
761	CIRB	1835	PAU	2702
761	CIRB	1180	HAU	2776
829	CIRB	4409	NDRI	3027
3551	NDRI	1817	CIRB	2752
3551	NDRI	1901	PAU	2709
3736	NDRI	4461	NDRI	2730
3736	NDRI	1104	HAU	2711
1290	PAU	1093	HAU	2723

Murrah as Triple Purpose Animal

Meat Production: Buffalo meat production accounts for about 30% of the total meat production in the country. Buffalo meat is generally produced from spent buffaloes and emaciated young male calves. Dressing percentage in such animals is low and varies between 40 to 45%. However, dressing percentage in these animals can be substantially increased by proper feeding prior to slaughter and in such animals growth rate as high as 1000 gm per day has been reported under feed lot system (Ranjhan, 2003).

In experimental trials daily body weight gain up to 541 gms per day has been achieved under conventional concentrate feeding system. Weight gain during summer months is higher than in hot humid months. (Bharadwaj and Sethi, 1994). Dry matter intake of such animals is generally more than 3% of body weight and dressing percentage as high as 55 to 60% can be achieved. Buffalo veal from young calves of about 4 weeks is considered to be a delicacy and has great export potential. Average cost of rearing was estimated as Rs 40 per kg by Bhardwaj and Sethi, 1994.

Draught Capacity: Buffaloes in most part of the country are primarily used for milk, spent buffaloes and young and emaciated males are sent to slaughter houses for meat production while adult males are generally used for work. Buffalo males of Murrah breed have efficient work capacity especially for load pulling and ploughing in rice cultivation. Though, buffalo bulls are slower in movement than cattle but they can pull heavier load and cover about 3.2 Km per hour as compared to 4.8 to 6.4 km per hour by draft bullocks (Taneja, 1999). Though large buffalo bulls are able to pull heavy load but are not maneuverable like small and compact animals and hence move slowly (Upadhyay, 1999).

Murrah as Improver Breed

Riverine breeds: It has also been advocated since the beginning of First Five Year Plan to use male progeny of superior breeds on low producing and non-descript buffaloes with the objective of upgrading and to put in milk producing capability in them as for as possible, in different parts of the country. In an experiment undertaken at NDRI Karnal on crossbreeding of Murrah and Surti breeds it was found that in crossbreds there was some improvement in milk production and reproductive performance over the Surti breed. Milk yield per day in crossbreds was 3.4 kg while in Murrah buffaloes reared in the same environment it was 3.2 kg. (Basu and Sharma, 1982).

In Mehsana and Banaskantha districts of Gujarat extensive breeding of Surti buffaloes was undertaken with Murrah bulls by farmers in the cooperative sector. Mehsana breed is said to have evolved from above crossing. Mehsana District Cooperative Milk Producers Union in Mehsana district initiated improvement programme during 1987 in association with NDDB. Under this programme so far 95 bulls in 8 sets have completed test mating. The overall average of 305 day first lactation yield of daughters born under the program was 1933 lt. (Namjoshi and Trivedi, 2002).

In Andhra Pradesh extensive breeding of local nondescript buffaloes with Murrah by farmers in the Krishna deltaic region has resulted into the origin of new strain of buffaloes popularly known as Godavari. Performance recording and improvement of this breed has been initiated at ANGRAU Venkataramanagudem, Hyderabad under the Network Project on Buffalo of ICAR. (Sire Directory, 2002).

In Bulgaria crossing of native Bulgarian buffaloes with Murrah imported from India, performance of crossbreds (Bulgarian X Murrah) was close to Pure Murrah in body weights and mid parent value in terms of milk yield (1649 kg) with fat content as high as 7.48 %. However, after few generations of selective breeding in the crossbred population popularly known as Bulgarian Murrah, milk production increased significantly. 6.34 % of the buffaloes produced more than 2500 kg in 305 days or less lactation with average of 2791.8 kg and fat of 7.3 %, 1.13 % buffaloes produced more than 3000 kg with average lactation milk yield of 3226.7 kg and fat 7.16 %. 7 buffaloes produced more than 3900 kg with lactation average of 4011 kg. (Peeva T, 1990).

In Brazil Murrah and Jaffarabadi buffaloes from India were imported for meat and milk production. Average milk production of these buffaloes was about 6 kg with7.2% fat. The work of interbreeding of these buffaloes with the local buffaloes is supervised by the Brazilian Association of Buffalo Breeders (ABCB) and their registration for further breeding. Efforts have been made to establish purebred herds of above breeds as well as of the crossbreds in Brazil and some of the other Latin American countries e.g. Venezuela, Trinidad, Peru and Guyana.

Swamp Buffaloes: About 30% of the world buffalo population is classified as swamp type, which is primarily known as draught animal suited for smallholder paddy fields. It has been reported that swamp animals can plough with higher speed and work on an average 86 to 122 days a year in the Philippines and Thailand (Ranjhan, 1987). The rest of the time animal is rested and used for milk production of about 500 kg per lactation. Spent and surplus animals are generally used for meat production. Crossbreeding with Murrah and Nili Ravi undertaken in the Philippines, Thailand, China, Vietnam and other Asian countries indicate this system of transforming the population as successful. Crossbreeding results between the riverine and swamp buffaloes show more than double the increase in milk yield in first generation. From 483 to 1032 kg in the Philippines (Ranjhan, 1987), 441 to 1154 kg in China (Youngzuo, 1988) and triple crossbred (Swamp / Murrah x Nili Ravi) producing as high as1913.7 kg with fat % of 6.69 in China (Jianxin W, 1990). This has also increased the live weight gain on similar type of rations showing that crossbred animals perform equally well as native buffaloes in these countries.

In Philippines 854 Murrah and Nili-Ravi buffaloes were imported from India between 1917 to 1956. These animals were distributed to a number of institutions which at present maintain only 150 pure bred animals (Ranjhan *et al.,* 1987). This stock was used for the production of Phil-Murrah and Phil-Ravi for combining draught, milk and meat production characteristics. Grades with 50% and above exotic Murrah blood are jet black and show no chevron. Grades with 25% and below exotic blood have the appearance of native animals.

In Vietnam a Murrah herd of 650 buffaloes was established in 1978 at the Buffalo and Forage Research Centre. In due course of time the performance of pure Murrah herd was low probably due to inadequate nutrition and they produced around 1400 kg per lactation as compared to the 1195 kg produced by crossbreds with local buffaloes (Mudgal, 1987).

There is continuing upsurge interest for Murrah breed in various parts of the country as well as abroad in order to upgrade the vast non-descript population indifferent states of India and also to upgrade the locally available low producing buffaloes in various countries e.g. Brazil, Italy, Philippines, Bulgaria and so on.

Germplasm Requirement and Availability

Requirement: In order to cover large breed able population of approximately 45 million buffaloes through AI and or natural service, major share of superior breeding bulls and germplasm lies on Murrah breed as improver breed besides the requirement of superior germplasm of this breed for selective breeding. As a result, large number of genetically superior bulls are needed every year to meet the demand of Murrah bulls. Rough estimate of breeding bulls requirement to cover the breed able population of approximately 45 millions in India is presented below:

Table-5: Requirement of superior germplasm for breeding of buffaloes in India

Proportion of population bred through		Number of frozen semen doses required (millions)	Number of breeding bulls required per annum for NS as replacement
AI	NS		
0	100	0	12600
10	90	7.875 (985)	11340
20	80	15.750 (1970)	10080
30	70	23.625 (2953)	8820
40	60	31.500 (3940)	7560
50	50	39.375 (4922)	6300

Breedable population = 45 million.   Replacement rate of bulls per annum = 20%

Breeding efficiency   = 70%   No. of AI per conception = 2.5
No. of natural services per bull per annum = 500
No. of doses frozen from each bull per annum = 8000

Considering the all India status of AI activities (2001–02) it is estimated that about 25 million artificial inseminations are performed annually covering the entire cattle and buffalo population under AI With greater proportion of AI being done on crossbreds and cattle, it is expected that buffalo share for AI is likely to be about 25% (approximately 6 millions) of the total inseminations. This shows that by the modest expectation only 10 % buffaloes of the total breedable population in the country are covered by AI while 90 % through natural service. To meet this population we require about 12,000 superior breeding bulls (Table-5) to be selected and reared for replacement either for semen production or to be used for breeding through natural service. Major share of these bulls (about 60-70 %) has to be from Murrah being the improver breed for selective breeding / grading up of nondescript local stock.

Availability: There several breeding farms in different parts of the country which maintain buffaloes of Murrah breed. The Murrah bulls generated from these farms are extensively used for improvement of local nondescript buffaloes. National project on cattle and buffalo improvement is being implemented by the livestock development boards in the respective states. Under this programme intensive efforts are being made to establish semen freezing and storage facilities in the state, procure superior quality germplasm and their multiplication either through natural service or adopting artificial insemination. Breed improvement programme is also being undertaken by developmental agencies like NDDB, BAIF and several other non-governmental institutions.

At the Network Project on buffalo improvement, associated progeny testing in Murrah buffaloes is being undertaken associating the herds at PAU Ludhiana, HAU Hisar, CIRB Hisar, NDRI Karnal, IVRI Izatnagar, CCBF Alamadi and NDUAT Faizabad. Through this approach since 1993, 8 sets of breeding bulls have been selected for test mating in these herds as well as in the field units attached to these centers.

Brief information on the number of bulls selected from each centre, their average age at the time of selection, average of the Dam's best yield and Highest Dam's yield in each set and the duration of each set is presented in Table-6.

Table-6: Eight sets of Murrah breeding bulls used under Network Project on Buffalo since July 1993.

Set No.	Duration	Centrewise No. of bulls				Av. age of bulls at selection (months)	Av. of 305 day or less dams best yield (kg)	Highest dam 305 day yield (kg)	305 day or less herd average/No. of records
		CIRB	NDRI	PAU	HAL				
1.	July, 93 to Dec., 94	2	9	0	-	60.5	3050	4114	1820/—
2.	Jan., 95 to June, 96	4	5	6	-	47.5	3002	3898	1920/487
3.	July, 96 to Dec., 97	8	5	2	-	44.0	2876	3275	2053/476
4.	Jan., 98 to June, 99	5	4	5	-	43.5	2999	3401	1973/457
5.	July, 99 to Dec., 2000	6	5	4	-	46.5	3120	3898	1943/551
6.	Jan.,2001 to June 2002	5	5	4	2	40.2	3055	3898	1972/562
7	July 2002 to Dec., 2003	5	2	4	1	34.3	2928	3544	2017/505
8.	Jan. 2004 to June 2005	6	5	5	2	38.4	2931	3690	2056/511

Dissemination of superior bulls / germplasm for breeding purpose: From various centres surplus bulls and frozen semen doses are disseminated every year to various agencies associated with buffalo development and improvement. The number of breeding bulls and frozen semen doses sold from the centres of Network Project on Buffalo for Murrah breed are presented in Table-7 which shows that the availability of superior germplasm is much less than the requirement..

Table-7: Superior germplasm disseminated from various participating centres.

Year	CIRB		PAU		NDRI	
	Bulls	Semen	Bulls	Semen	Bulls	Semen
1998-99	32	50	10	6000	15	1740
1999-00	26	100	22	5847	11	1320
2000-01	16	70	33	3449	9	2230
2001-02	18	21648	18	8579	8	5030
2002-03	18	2270	7	3205	9	2655

Feed Intake and Cost of Rearing

Highest per day intake was registered between 18 and 24 months old for green fodder, i.e. 22.0±0.4 kg. Higher quantity of dry fodder intake (0.88±0.1 kg) and concentrate intake (2.12±0.1 kg) per day was recorded during 2 to 2.5 years old. The overall cost of rearing buffalo calves upto the age of maturity was Rs 8631.5 and the average cost per kg body weight gain was estimated at Rs 26.50. Cost of feeding from 6 to 12 months of age for each kg of body weight gain was lower (Rs 16.80) than the cost estimated at later.

Table-6: Average feed and fodder intake and cost of feeding

Age (month)	Green	Dry	Concentrate	Average cost of feeding per animal (Rs)	Average cost per kg body weight gain (Rs)
Birth-6	460.0±1.2		120.0±1.3	2329.2*	30.90
6-12	1726.9±1.9	57.1±0.5	336.7±1.2	1254.3	16.80
12-18	2218.9±2.3	30.3±0.6	335.7±1.1	1370.3	17.20
18-24	4004.8±4.7	71.7±1.3	342.7±1.4	1965.7	21.60
24-30	2548.0±3.8	159.5±1.7	386.0±1.4	1711.9	40.40

Source : Sethi and Chopra,1994.

Farmers in the breeding tract as well as in other parts of the country follow traditional system of feeding by feeding the locally available feed ingredients. Survey in the breeding tract revealed that medium and large farmers feed more concentrate and follow stall feeding by providing green fodder and dry roughage. Landless farmers who maintain majority of the stock depend upon grazing and limited quantity of concentrate and dry roughage as and when available. Details of feeding system followed by the farmers is presented in Table-7.

Sensitivity to Feeding and Environment

Analysis of data at CIRB revealed that Murrah buffaloes on an average loose 24.6 kg body weight during their first month of lactation and they continue loosing more weight up to fifth month of lactation (27.4 kg). However, by the end of lactation buffaloes regained most of the weight lost during the early months of lactation. Younger buffaloes loose lesser proportion of body weight during the early part of lactation and then gain at a faster rate as compared to the older buffaloes. High milk producing buffaloes lost significantly more body weight than low producing buffaloes during lactation. It is therefore essential to keep the feeding regime at the maximum specially in the early part of lactation so that animals do not loose body weight and in

Table-11: Feeding system of milch buffaloes by different categories of farmers

Category	No. of buffaloes	Mean Body wt. (kg)	Av. daily milk yield (kg)	Feed offered (kg) on fresh basis / day	
I. Landless labourers	108	450	2-4	a) grazing	6-8
				b) wheat straw	2-3
				c) wheat floor	1
II. Small & marginal	324	500	5-9	a) green fodder*	15-20
				b) cotton seed cake	2-4
				c) wheat floor	1.0
				d) wheat straw	4-6
III. Medium and rich farmers	203	500	10-14	a) green fodder	15-20
				b) cotton seed cake	3-4
				c) bajra / wheat	1.5-2.0
				d) cotton seeds	1-2
				e) black gram / 'dal churri'	0-1
				f) wheat straw	4-6

Source: Technology Generation Transfer and Impact. CIRB Hisar

turn the milk yield.

Buffaloes need to be provided appropriate shelter to protect them from excessive heat and cold stress. It is generally observed that younger buffaloes have more heat tolerance capacity than older buffaloes (Sethi *et al.*, 1994).

Seasonality of calving is noticed in many herds as more than 70% calving take place during September to February while only 30% calving occur during summer months i.e. March to July. However, appropriate feeding management and protection during summer reduces the seasonality in calving and uniform spread of breeding is expected.

References

Basu S B and Sharma P A. 1982. Crossbreeding in buffaloes. A Review of Agro Animal Sciences and Health, **6**: 465-466.

Jianxin Wu. 1990 . Water Buffalo in China. Proceedings of the FAO workshop on Open Nucleus Breeding Systems held in Shumen, Bulgaria, November 18-23, 1990.

Namjoshi M and Trivedi KR . 2002. Experiences of Implementing Progeny Testing Programme in Gujarat. National Seminar on Sire Selection for Milk Production Enhancement in Tropics. November 20–22, 2002, Mattupati Kerala.

Peeva Tzonka 1990. The buffalo breeding in Bulgaria. Proceedings of the FAO workshop on Open Nucleus Breeding System held in Shumen, Bulgaria. November 18–23, 1990

Ranjhan SK Faylon PS Momongan VG and Cruz LC. 1987. Husbandry of Swamp buffalo in the Philippines. Philippines Council for Agriculture, Forestry and Natural Resources. Research and Development, Los Banos, Laguna, Philippines.

Ranjhan S K. 2003. Changing role of buffalo production in the third millennium. 4th Asian Buffalo Congress, New Delhi.

Sethi RK and Chopra SC. 1994. Evaluation of growth potential in Murrah buffaloes under free choice feeding system. Indian J Anim. Science. **65**: 438- 441.

Sethi RK and Khatkar MS. 1996. Body weight changes during lactation and their relationship with parity and milk production in buffaloes. Indian J Anim. Science. **66**: 159-162.

Sethi RK, Bharadwaj A and Chopra SC. 1994. Effect of heat stress on buffaloes under different shelter strategies. Indian J Anim. Science. **64**: 1282-1285.

Sire Directory 2002. Network Project on Buffalo. Central Institute for Research on Buffaloes, Hisar, India.

Trivedi, K.R. 1992. Proc. National Seminar on Progeny Testing of Bulls in Tropics. Thurvantapura, Kerala. Feb. 20-222, 1992. pp 59-65.

Upadhyay R C. 1999. Draft potential in buffaloes and its optimum utilization. National Seminar on 'Sustainable development of buffaloes for milk, meat and draft', NDRI, Karnal.

CHAPTER 7

Breeding and Feeding Management of Surti Breed of Buffalo in Southern Rajasthan

S.P.Tailor

Livestock Research Station, Vallabhnagar – Udaipur (Rajasthan)

Livestock sector plays an important role in Indian economy. Besides being an integral part of day to day lives, their contribution to national income is valuable. Acharya (1988) considered buffalo as "milk machine" and "back bone" of dairy industry. Further buffalo occupies third place as meat producing animal and contributes 30% towards total meat production of the country. Buffalo in the Asia is also being used as draft animal. Thus the inherent potential of buffaloes for milk, meat and draft have made them a versatile triple purpose animal in the country. Moreover, buffalo has unique ability to utilize coarse roughages and convert them into better quality animal protein i.e. milk and meat so buffalo fits rather well in the agrarian situations of limited crop livestock production system of the third world countries.

Cockrill *(1974)* described as many as fifteen breeds of buffalo in India, of which seven recognized breeds are Murrah, Nili-Ravi, Jaffarabadi (large-sized), Bhadawari, Mehsana, Nagpuri and Surti (medium-sized). Surti buffalo being lighter in body weight as compared to heavy breeds, consumes less feed, thrives well on limited or no green and produce milk with higher fat and SNF contents. Therefore, Surti can be easily maintained by landless, small and marginal farmers.

Erratic and uneven distribution of rainfall results in frequent famines, small holding with less percentage of irrigated land, low production potential of land due to problematic type of soil, high human and animal pressure on land and uneven topography in southern Rajasthan compels farming community to keep small-sized animals, making Surti as best suited buffalo in this region.

Origin and Distribution

The main home tract of Surti buffalo is southern-western part of Gujarat (Anand, Nadiad, Khara and Baroda districts). Due to forced breeding with Murrah, the Surti breed is disappearing day by day in Gujarat. However it spread up to Southern part of Rajasthan. The breeding tract of Surti in Rajasthan consists of Udaipur, Dungurpur, Banswara, Rajsamand, Chittorgarh, Bhilwara and Sirohi districts.

Typical Characteristics of Surti Buffalo

Medium sized buffalo with wedge shaped barrel, head bread, round between the horns, convex fore head, eye prominent with bright intelligent look and frequently a white streak of hairs over the eyebrows. Ears of medium size with reddish colour inside lower borders of ear are commonly fringed with white hair. Horns are sickle shaped, moderately long and flat, band downward and backward direction, then turn up ward at the tip in the form of a hook. The neck is long in female while thick and heavy in male with two white collars. Medium sized legs, wide and deep hindquarters, pin bones and hipbones are wide apart and tail fairly long ending in a switch.

Udder is well defined, developed with well set teats of medium size and squarely placed.

Livestock Status of Surti Breeding Tract of Rajasthan

The Table-1 depicts land area, human and livestock density and growth rate of buffalo and livestock in different districts of southern Rajasthan as compared to total Rajasthan and other parts of Rajasthan. These seven districts have land area of 55129 sq. km., comprising of 16.11% of total geographical area of the state (342239 sq. km.) and contains 19.66% (8.7 million) and 20.94% (11.4 million) human and livestock population of the state respectively. The density per sq. km. with respect to livestock, human and buffaloes in southern districts of Rajasthan are 206, 157 and 36, which is considerably higher than other part of Rajasthan and all over Rajasthan. Similarly, in these districts the human and livestock ratio is higher (1:31) as compared to other parts (1:21) and all over Rajasthan (1:24) indicating the reason for livestock rearing as main occupation of farmers of the state.

Field Survey

The survey was conducted in 44 villages of Vallabhnagar, Mavli and Girwa tehsils of Udaipur district covering 5493 households. In all 14479 buffaloes were considered for breed age and sex-wise classification. During complete enumeration, feeding, and breeding managemental practices followed by the buffalo breeders were collected through appropriate questionnaires.

Livestock Holding

Out of total livestock (cattle, buffalo, sheep and goat), maximum at 45.83% (14479) were buffaloes followed by 34.19% deshi cattle, 11.65% goat, 6.34% sheep, and 2% cross-bred cattle. The respective values for the state were 18.35, 22.47, 31.86, 26.92 and 0.40%. The livestock distribution according to land holding showed that among the total livestock, only 2.07% were kept by landless, 37.70% by marginal, 30.50% by small, 18.19% by medium and 11.54% by large farmers.

In the area surveyed the proportion of large ruminant (cattle and buffalo) and small ruminant (sheep and goat) was almost equal in landless farmer. The proportion of large ruminant increased by 76.81% in marginal, 81.90% in small, 87.91% in medium and 95.64% in large farmers. It indicated that as the land holding of farmers increased the proportion of cattle and buffaloes i.e. large ruminant, maintained by farmers increased because of the requirement of feed and fodder of large animals are easily met out of agricultural byproducts.

Among the cattle (Deshi + cross bred) 1.87, 35.47, 30.65, 19.42 and 12.60% were reared by landless, marginal, small, medium and large farmers. The respective value for buffaloes and goats were 0.76,

Table-1: Population Status of Southern and whole Rajasthan

Districts	Total area (sq.km)	Human population in (1000)	Livestock population	Density Human	Livestock Density	Buffalo Density	Growth rate over 1992		Ratio H:L
							Livestock	Buffalo	
Banswara	5037	1156	1179315	229	234	45	+13.37	+23.80	1:02
Bhilwara	10455	1593	2710293	152	259	36	-2.57	+13.73	1:70
Chittorgarh	10856	1484	1733665	137	160	36	-6.36	+5.05	1:17
Dungurpur	3770	875	1025940	232	272	47	+4.20	+17.30	1:17
Rajsamand	4768	823	1239630	172	260	47	+0.50	+9.92	1:51
Sirohi	5136	654	953266	127	186	25	-1.10	+35.01	1:46
Udaipur	12511	2067	2559265	165	203	37	+3.19	+11.75	1:23
Total Southern Rajasthan	55129 (16.11)	8652 (19.66)	11381374 (20.94)	157	206	36	+0.57	+19.98	1:31
Other parts of Rajasthan	287110 (83.89)	35354 (80.34)	42967327 (79.06)	123	150	31	+17.86	+29.58	1:21
Total Rajasthan	342239	44006	54348701	129	159	32	+13.76	+25.94	1:24

Figures parenthesis indicate of total Rajasthan

Table-2: Average livestock holding according to different categories of farmers

Categories	Cattle			Buffaloes			Large ruminant	Sheep	Goat	Small Ruminant	Total livestock
	Deshi	Crossbred	Total	In milk	Dry	Total					
Land holding											
Landless	3.04[a]	0.29[a]	3.33	0.89[a]	1.11[a]	3.54	6.87	0.71[a]	0.61[a]	1.32	8.19
Marginal	2.69[a]	0.14[a]	2.83	0.96[a]	0.85[a]	3.57	6.40	0.47[a]	0.89[a]	1.36	7.76
Small	2.59[a]	0.18[a]	2.77	1.00[a]	0.78[a]	3.70	6.47	0.56[a]	0.79[a]	1.35	7.82
Medium	2.61[a]	0.18[a]	2.79	1.02[a]	0.88[a]	3.80	6.59	0.34[a]	0.61[a]	0.95	7.54
Large	2.52[a]	0.17[a]	2.69	1.09[a]	0.76[a]	3.88	6.57	0.93[a]	0.58[a]	1.51	8.08
Castes											
SC	1.68[a]	0.05[a]	1.73	0.64[b]	0.60[a]	2.70[b]	4.43	0.10[ab]	1.70[b]	1.80[b]	6.23
ST	3.38[b]	0.07[a]	3.45	0.32[a]	0.53[a]	1.92[a]	5.37	0.43[ab]	3.10[c]	3.53[a]	8.90
OBC	2.93[b]	0.14[a]	3.07	1.03[c]	0.89[a]	3.92[a]	6.99	0.76[ab]	0.68[a]	1.44[c]	8.43
General	2.02[b]	0.26[b]	2.28	1.04[c]	0.75[a]	3.47[c]	5.75	0.02[a]	0.58[a]	0.60[c]	6.35
Overall	2.63	0.16	2.79	0.99	0.83	3.67	6.46	0.52	0.78	1.30	7.76

Means superscripted by different letters differed significantly

35.53, 30.53, 19.46 and 13.72 and 8.75, 50.79, 26.01, 10.14 and 4.32. The sheep is maintained only by marginal (44.61%), small (39.31%) and medium farmers (16.61%). The sex ratio of male to female of present stock in deshi cattle, cross-bred, buffalo, sheep and goat were 55:45, 19:81, 10:90, 3:97 12:88 with an overall average ratio of 25:75 . The results indicated that farmers are least interested in maintaining males of all the livestock species except deshi cattle because these are reared for draft purpose.

In all 5.03, 4.64, 69.54 and 20.79% of total livestock maintained by the farmers of Shedule Caste (SC), Schedule-Tribe (ST), other backward caste (OBC) and forward castes. The average numbers of livestock per household in SC, ST, OBC and general castes were 5.11, 7.56, 8.37 and 5.84 with an overall average of 7.43.

In the average number of total livestock maintained by farmers were 7.76, out of which 6.46 units of large and 1.30 units of small ruminants. Among the large ruminants there were 2.79 cattle (2.63 deshi +0.16 cross-bred cattle) and 3.67 buffaloes (0.99 milking +0.83 dry +1.85 followers). Out of small ruminants there were 0.52 sheep and remaining 0.78 goats (Table 2).

Breed Composition

In all 14,479 buffaloes were observed for breed composition with following traits. The criteria of classification were as follows:

Breeds	Criteria
Surti	Buffaloes having white collar on neck, sickle shape horn and medium sized body.
Surti type	Buffaloes having any one of the above characters like white collar or sickle shape horn with medium sized body.
Mehsana/Murrah type	Buffaloes having heavy body size with slightly curled horns.
Non descript	Buffaloes do not have any characters but of medium sized.

In all 9.25% (1339) of stock were conformed to pure Surti breed, 32.69% (4733) conformed to Surti type, 4.11% (595) conformed to Mehsana/ Murrah type and remaining 53.95% (7,812) could not conformed to any known breed. The proportion was also similar for growing as well as breedable stock. It indicated that majority of animals (95.89 %) were of medium size which resembled to Surti buffaloes and only 4.11% animals having heavy body size which was not pure Mehsana and Murrah but according to body size, they were heavier than that of Surti and lighter than that of Murrah (Table-3). Further, it also suggested that Surti (medium sized) was the choice of buffalo breed of this region.

Table-3: Breed composition of buffaloes in the area

Breed	Stock		
	Growing	Breedable	Total
Medium sized			
Surti	552 (9.37)	787 (9.26)	1339 (9.25)
Surti type	2015 (34.20)	2718 (31.72)	4733 (32.69)
Nondescript	3092 (52.48)	4720 (55.19)	7812 (53.95)
Sub total	5659 (96.05)	8225 (96.17)	13884 (95.89)
Heavy sized			
Mehsana/Murrah type	233 (3.95)	362 (4.23)	595 (4.11)
Grand total	5892	8553	14479

Figures in parenthesis indicate percentage

Age and Sex-wise Composition

Age and sex-wise composition of buffaloes surveyed have been given in Table-4. In all 17.29% (2504) buffaloes were between age group of 0-6 months, 14.58% (2111) between 6-24 months, 8.82% (1277) between 2-3 years, 9.78% (1416) above 3 years and remaining 49.53% (7171) were adult buffaloes. Out of total stock 40.69% were growing stock and remaining 59.31% were breedable stock.

Table-4: Age and sex-wise composition of buffaloes

Age groups	Sexes		Total
	Male	Female	
0-6 Months	1045 (66.90)	1459 (11.30)	2504 (17.29)
6-24 Months	436 (27.91)	1675 (12.96)	2111 (14.58)
2-3 Years	63 (4.03)	1214 (9.40)	1277 (8.82)
Growing Stock (up to 3 Yrs.)	1544 (98.85)	4348 (33.66)	5892 (40.69)
Above 3 years	18 (1.15)	1398 (10.82)	1416 (9.78)
Buffaloes	--	7171 (55.52)	7171 (49.53)
Breedable group	18 (1.15)	8569 (66.34)	8587 (59.31)
Grand total	1562	12917	14479

Figures in parenthesis indicate percentage

Among females 33.66% were growing stock (11.30% of 0-6 months; 12.96% of 6-24 months; 9.40% of 2-3 years) and remaining 66.34% were breedable group (10.82% of above 3 years and 55.52% adult buffaloes). Most of males (94.75%) were disposed off up to the age of 2 years of age and only 1.15% were kept after 3½ years of age for breeding purpose which was in accordance to the proportions of males retained by the farmers after 3 years of age in southern part of Rajasthan at 2.56% and whole Rajasthan at 1.89% as per 1997 census of the state.

Generally the males retained after 3½ year of age, were being used only for breeding purpose in this area. The ratio of such males to breedable females (above 3 years of age) was 1:477 in comparison to 1:84 in the state. Such huge difference might be due to fact that the survey was concluded in the command area of Vallabhnagar Farm and most of the farmers of these villages are using AI for serving their buffaloes and hence males were not retained for natural breeding. Among the adult buffaloes 54.6% (3915) were in milk and 45.4% dry (3256) in comparison to 78.9 and 21.1% in southern part of Rajasthan, 67.1 and 32.9% in other part of Rajasthan as per 1997 census data. The variation might be due to difference in month of survey because buffalo is seasonal breeder.

Feeding Management

Grazing

Most of the farmers (89.4%) let loose their buffaloes for grazing and remaining 10.6% farmers do not send their buffaloes for grazing. Among these, 25% farmers send their animal in own land, 60% in community land and 4.4% partially in their own as well as community land for grazing. Among the surveyed none of the farmer kept their buffalo purely on grazing basis instead they kept their animals on semi stall feeding system. The grazing time generally observed from 8.30 am to 5.00 pm during summer and 9.30 am to 4.00 pm during winter seasons, with an average time spent for grazing was 6.10 hrs/day.

Feeding

Green Fodder

More than 85% of buffalo breeders offered lucerne as green fodder during summer and winter seasons to their milch animals. Less then 4% of the farmers offered berseem as green fodder to their animals during winter season indicating berseem as green fodder in not popular in the area. The lucerne is more popular green fodder over berseem because of perineal in nature. Very few farmers (<2%) sowed jowar and maize during rabi season for feeding of green fodder to their milch animals. However, jowar and maize also frequently sown during rainy season for karbi and grain production but they are not used as green fodder. Less then 10% of the farmers do not feed green fodder to their milch buffalo during summer and winter seasons as they have no facility of green fodder production. Ad-lib green fodder was fed to all kind of animals during rainy season which is mostly collected from fields. Chaffing of green fodder is not in practice in the area.

Dry Fodder

In dry fodder, pure jowar karbi was used by 24% of farmers while 57% farmers used jowar karbi+ local grasses as dry fodder during winter season. Among the remaining 11% farmers used maize karbi,

4% wheat straw and 4% local grasses as dry fodder. During summer season, 96% farmers used wheat straw as dry fodder and remaining 4% used Jowar karbi, local grasses etc. Chaffing of dry fodder is not much popular in the area but it increased over the years due to frequent drought resulting significantly higher price of dry fodder.

Concentrate

No concentrate was fed to unproductive i.e. dry and growing stock except during last months pregnancy. Among feed ingredients 54% farmers do not mix any grain, 81% any cakes, 20% any oil seed and 73% any ready made feed in their concentrate ration. Barley among grains, Til and GNC among cakes and cotton seed among oil seed were mixed in concentrate rations for buffaloes.

Although, the proportion of ingredients in concentrate ration varies from farmers to farmers but on an average 54% was oil seed, 26% grain, 14% ready made feed and 6% cakes. About 60% of concentrate ration consist of cotton seed as well as oil seed cakes. This is traditional practice followed by farmers of this region since long because it serves as a good source of energy especially when supply of green fodder is not ad-lib and wheat straw being very low in energy value. Further, feeding of cotton seed and cakes to buffaloes is the best approach to match the ruminant production system with available resources in this region of the state. Moreover, feeding of cotton seed increases the size of fat globules which resulted increase collection of fat during churning. Cotton seed was fed to animals after boiling for 3-4 hrs resulted improve digestibility.

Pre and Post-Partum Feeding

Libral pre-partum feeding of animals from 8 months pregnancy or 6 week before calving with the object of securing full development of mammary glands for optimum milk production as well as foetus. In addition to normal diets, the farmers generally feed on an average 1.0-1.5 kg grinded barley per animal during last month of pregnancy. Moreover, some farmers add about 100-150 gm Til oil in soaked barley during last 7-10 days of pregnancy as high energy source as well increases the luxativeness of ration.

Just after parturition, the farmers provide luke warm water to animal and after one hour, animal is offered about 1.5-2.0 kg boiled whole barley containing 250-500 gm Til oil and 0.5-1.0 kg gur. This facilitates quick expulsion of placenta.

Feeding of boiled crushed wheat about 1-2 kg and gur about 0.5-1 kg daily for about 10-15 days after parturition is common. After 15 days following mixture of concentrate ratio was the part of traditional feeding for about one month:

Commodity	Per day range	Average total quantity / animal (kg.)
Ajwain	250-400 gm.	10.5
Gur	0.5-1.0 kg	20
Grind Barley	1.0-1.5 kg	30
Methi	250-500 gm	10
Til Oil	150-400 gm	10
Crussed Til	400-700 gm	15.5
Deshi Ghee	100-150 gm	3.5
Khopra (Coconut)	150-350 gm	8.0

Breeding Management

It is well known that Indian water buffalo are irregular breeders and display distinct sexual rhythm. This peculiar breeding behaviour has been duly recognized as one of the major problems of practical importance confronted by buffalo breeders owing to its direct bearing on reproductive and productive efficiency of buffaloes. So far as the influence of environmental factors on physiological and reproductive changes or rhythm of animals are concerned they may be due to atmospheric temperature, humidity and intensity of light.

Based on breeding records of 31, 207 Surti and Surti type buffaloes maintained in semi arid region of Rajasthan, the combined effect of temperature and humidity on percent buffalo exhibiting heat and conceived were found to be significant. However, its effect on conception rate and number of services per conception was non-

Table-5: Effect of temperature and humidity on breeding behaviour of Surti buffaloes

Temperature and humidity group	Traits			
	CR (%)	NSPC	PBEH	PBCON
Low temp. and moderate humidity (<25°C and 60-80%)	35.58	2.81	4.80[ab]	5.07[ab]
Low temp. and high humidity (<25°C and >80%)	35.19	2.84	12.85[d]	13.40[d]
Moderate temp. and low humidity (<25-35°C and <60%)	30.54	3.27	7.91[ac]	7.16[ac]
Moderate temp. and moderate humidity (<25-35°C and <60-80%)	34.19	2.92	39.72[f]	40.24[f]
Moderate temp. and high humidity 25-35°C and <80%)	33.52	2.98	27.73[c]	27.54[e]
High temp. and low humidity (>35°C and <60%)	33.11	3.02	3.91[ab]	3.83[ab]
High temp. and moderate humidity (>35° C and 60-80%)	31.02	3.22	2.63[ab]	2.42[ab]
High temp. and high humidity (>35° C and >80%)	27.07	3.69	0.42[a]	0.34[a]
Overall	33.76	2.96	100	100

Values of particular class bearing different superscript differed significantly (P<0.05)

significant. Significantly higher frequency of buffaloes exhibiting heat (39.72%) and getting conceived (40.24%) was observed during moderate temperature and moderate humidity period (Table-5). It is reasonably concluded that high ambient temperature (>35°c) with low (<60%), moderate (60-80%) and high (>80%) humidity; low temperature (<25°c) with high (>80%), moderate (60-80%) and low (<60%) humidity and moderate temperature with low humidity had an adverse effect on buffalo reproduction. The results also indicated that moderate temperatures (25-350c) with maderate humidity (60-80%) were found to be optimum physiological norms for optimum buffalo breeding.

Table 6. Effect of day length on breeding behaviour of Surti buffaloes

Day Length	Traits			
	CR (%)	NSPC	PBEH	PBCON
Longer	29.93 a	3.34 b	24.15 a	21.41 a
Shorter	34.98 b	2.86 a	75.84 b	78.58 b
Overall	33.76	2.96	100	100

The per cent buffaloes bred (75.84%) conceived (78.58%) and conception rate (34.98%) were significantly higher during shorter day length period as compared longer day length period (Table-6). The results thus indicated that during April to September when duration of day length was longer with high intensity of solar radiation, mechanism of oestrus regulation in buffaloes and other factors controlling conceptions seems to be adversely affected. Therefore, it may be concluded that photo-periodism plays a major role in breeding pattern in Surti buffaloes in sub-humid climate of southern Rajasthan.

For efficient and uniform breeding the animals should be protected from direct sunlight during longer day length period (April-September), housed in shed, allowed wallowing or splashing water during hot period of day. In addition to this main attention should be given on heat detection during this part of the year.

Buffalo of Andaman and Nicobar Islands

S. Senani

Central Agricultural Research Institute, Port Blair - 744 101 (A&N)

The Andaman and Nicobar islands union territory extends from 92-94 N latitude and 6-14 E longitude in a chain spread from North to South. The islands are hilly and are heavily forested. The Andaman and Nicobar island union territory comprises of 572 islands out of which only 36 islands are inhabited (Fig 1). Area of the Andaman is 6475 sq km (2500 sq mi). Total population of the islands according to 1999 census was 326265 persons. Geographically this chain of islands is a part of Yoma and Arrakkan mountain range of Burma. The Andaman group of islands is separated from the Nicobar group of islands by 10 ? channel. The total length of the road is about 720 km. The islands, formed by the peaks of a submerged mountain range, extend some 322 km (some 200 mi) in a northwestern to southeastern direction. Great Nicobar is the largest and southernmost of the islands. The chief occupations are fishing, woodworking, handicrafts, and the chief products are coconut, coffee, rice, and rubber. Area of the Nicobar Islands is about 1841 sq km (about 711 sq mi). The islands experience rain from both North-east and south-west monsoons to the tune of about 3000mm per annum equally distributed over eight months from April-Nov. This is followed by a dry period from Dec-April. During rains it is velvet green in the fields and hilly slopes but in the dry period the grass cover withers and very little grass is available for grazing.

Land Use

Major land use in the islands is forest which range from 19.83 to 100 per cent in different islands. Car Nicobar island has 19.83 % area and Narcondum, North Sentinell, Strait island, Peel island, Flat bay, Stewart island, John Lawrence, Aves, East and Interview island have 100% area under forest. On the other hand cultivable area ranges from 0.0-80.16 per cent. North Andaman, Car Nicobar and Katchal have over 40.05 % area under crops and rest of the islands has cultivable area ranging from 0-19 per cent. Out of 36 inhabited islands 11 islands have no cultivable land and 14 have less than 10 per cent area under cultivation. As such there is hardly any land earmarked for fodder production.

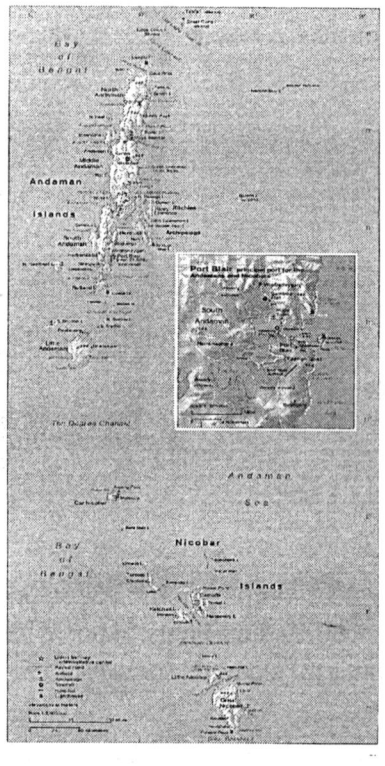

Buffalo

The islands support about 188 thousand of livestock, which includes 14,204 buffaloes as per 1997 census (Table-1). Over the last few decades the buffalo population is more or less stabilized due to the fact that the buffalo are reared for milk, meat and drought purposes. Buffalo were brought from the mainland India and over a period of time different breeds have so intermingled that it now cannot be defined as a distinctive breed. Rather characters of Murrah, Nagpuri, Bhadawari and Marathawadi could be traced (Chatterjee *et al.*, 2003). Buffalo are generally kept for the prime reason that their milk production is better as compared to non-descript cattle in this region. The average milk production is 4-5 liters/ day. In the earlier days when lumbering was done on a big scale in the forests, buffalo bulls were used for traction of big logs of wood from one place to other buffaloes.

Table-1: Population dynamics over last two decades

S. No	Species	1982	1987	% Increase	1992	% Increase	1997	% Increase
1.	Cattle	36560	47322	29.43	52941	11.87	60180	13.67
2.	Buffalo	11869	14400	21.32	14531	00.9	14204	-2.25
3.	Goat	33581	45062	34.18	56160	24.62	70923	26.28
4.	Pig	96029	29640	-6913	36464	23.02	42836	17.47
5.	Poultry	319916	442794	38.40	614242	38.71	800950	30.39

Basic Statistics: As per 1997 census

Distribution of Buffalo

In comparison to cow, buffalo are much preferred by commercial small to big dairy units. This is probably due to the fact that buffalo are a multi-purpose animal and their average milk yield is much more as compared to local nondescript cattle. Looking at the decadal variation of buffalo population it is evident that buffalo population from 1982 to 2001 has remained more or less steady or grown at a slower rate (Table-1). Out of 36 inhabited islands buffalo are reared in 10 islands. Majority of buffalo population is concentrated in Andaman districts and in Nicobar district Kamorta and great Nicobar has buffalo population (Table-2). A GIS based distribution of buffalo is shown in Fig 2 (Senani, 2004).

Table-2: Buffalo population in different islands (1997

Island	Island Id	Buffalo
Aves	4	0
Bambooka	28	0
Baratang	8	1257
Carnicobar	25	0

Buffalo Production under Different Climatic Regions

Chawra	26	0
Curlew	16	0
East island	1	0
Flat Bay Island	12	0
Great Nicobar	36	42
Havelock	10	186
Interview	3	0
John Lowrance	9	0
Kamorta	29	7
Katchal	32	0
Kondul	35	0
Little Andaman	24	0
Little Nicobar	34	0
Long island	6	88
Middle Andaman	5	2274
Nancowrie	31	0
Narcondum island	40	0
Neel Island	15	0
North Andaman	2	6378
North Passsage	7	0
North Sentinel Island	23	0
Parlob	17	0
Peel Island	14	0
Pilomilo	33	0
Rutland	22	2
Smith Island	13	0

South Andaman	20	3970
Stewart Island	11	0
Strait island	18	0
Teressa	27	0
Trinket	30	0
Viper Island	21	4

Morphological Characters

Local buffalo is an admixture of breeds and hence morphological characters are highly variable (Fig 3&4). Some buffalo shows deep massive body structure with short, broad back, light neck and head similar to Murrah buffalo. Some are medium sized, wedge shaped body with short and stout legs, copper coloured body coat with scanty hair skin to Bhadawari buffalo. While others are seen with horns tightly curved horns and others have long horns that are flat and curved and reach over the shoulder towards the back or are curved over the shoulder (Chatterjee *et al.*, 2003)

Production Performance

As the local buffaloes, not being a pure breed, growth, production and reproduction performances are highly variable. Adult male bodyweight was 400-450 kg and that of female 350-400 kg. Age at first calving varies from 42-45 month. Daily milk yield varies from 2-7 liters in a lactation length of 7-10 month. The milking is practiced both in the morning and evening but during late pregnancy it is restricted only in the morning (Chatterjee *et al.*, 2003).

The herd size is variable from 4-21, which includes animals in milk production, dry and calves. As such, buffalo was neglected till a few years back and was not considered for any upgradation programme. Of late genetic upgradation of local buffalo with Murrah semen has been initiated. Both natural insemination and AI with frozen semen are used for breeding. For AI Veterinary dispensaries, AI centres and sub-centres are being used. Some farmers have their own breeding bulls for the purpose (Chatterjee *et al.*, 2003).

Feeds And Feeding Practices

In general buffalo rearing heavily rely on grazing in the grasslands, wasteland, and fallow land and with no or a little of supplemental concentrate offered as stall-feeding. Due to over exploitation the productivity of the grazing land is deteriorating. Under rain fed conditions exotic perennial fodder grasses could be grown to yield 60-150 tons/ha/annum fresh fodders (Sharma *et al.*, 1990) and the pressure from the grazing lands could be reduced. Also if the grazing areas could be fenced, production augmented, grazing controlled and nutrients replenished high quality sustainable fodder and livestock production could be achieved from the grazing resources.

Cattle and buffalo are maintained mainly on roughage, grazing or paddy straw and self-made concentrate. Stall-feeding of concentrate (100%) followed by free grazing (81.39%) or partial grazing (11.63%) was the common method of feeding. Lactating buffaloes were maintained on straw and grazing or straw and self-made concentrate in urban areas whereas 21% cases lactating animals received straw, self-made concentrate and compound feed. Feeding of green fodders was a major problem faced by the buffalo farmers in the urban areas. The owners had to employ a labour for carrying green fodder for the animal and thus leading to extra cost on feeding.

A feed supplement, which is supposed to have concentrate feed, mineral, slat and vitamin supplement was highly variable at different farms. Maize was the most preferred concentrate (48.84%) followed by GNC (41.86%) and wheat brawn (13.95%). Wherever the compound feed was used it was mostly from a local manufacturer (13.95%). None of the farmers used standard concentrate feed. Soaking of concentrate feed was the common practice (90.7%) otherwise it was used as such. Concentrate feed was offered in a wooden tub or iron bucket. In no case the farmers for feeding lactating buffaloes used cultivated fodder. As far as the use of mineral and salt supplement was concerned common salt was the only routinely followed supplement (86.05%) and administered along with concentrate feed. However, in 86.72% cases no mineral supplement was used and in 6.98% cases it was not routinely followed and only in 9.30% cases it was routinely followed in urban areas. Source of drinking water was pond, streams or water supplied by the PBMC. Frequency of watering was thrice a day (87.72% cases) and in rest it was twice a day.

Requirement of Feed And Fodders

On the basis of population estimates of 1999 the human density is 371 per km^2 and that of livestock 174.5 per km^2. Senani (2003) worked out the requirements of total DM, roughage, green forage and concentrate as 276775, 13853, 418525 and 137463 tonnes per annum respectively (Table-3). Major chunk of concentrate feed is imported from the mainland. Considering the fodder productivity of the coconut orchards, fellow and wastelands for fodder production, these islands can supply the requirement of green forage but the requirement of concentrate remains unfulfilled.

Table-3: Annual requirement of roughage and concentrate feeds for the total

Sl. No.	Species	Number	DM (t)	Roughage (t)	Concentrate (t)	Green forage (t)
1.	Cattle	60180	141699.0	94142.0	47557.0	282426.0
2.	Buffalo	14204	41734.5	27840.5	13894.0	83521.5
3.	Goats	70923	119415.5	12943.5	6472.0	38830.5
4.	Pig	42836	46905.0	3127.0	46905.0	9381.0
5.	Poultry	800950	27021	-	22635.0	4386.0
		Total	276775.0	13853.0	137463.0	418545.0

Senani (2003). Unpublished data

Status of Productivity

Due to poor germplasm and indiscriminate breeding cattle population has further degraded and completely lost the higher milk production trait. As a result, the average milk production of cow is peltry 1.5-2.0 liters / day. Local buffaloes over the last one-decade have sustained in population size and largely kept for commercial milk production purposes at small dairy farms. Average milk production of local buffaloes lies between 4-5 liter/day (Senani, 2004).

Role of Buffalo in Island Development

In view of the milk requirement of local population and tourist inflow, buffalo offers possibility of becoming the major contributor as far as milk production in the islands is concerned. Buffalo keeping in periurban areas also offer lively hood options and job opportunities for people. Recently it has been noticed that green fodder is being transported in bundles on scooter or in the boot of taxies for feeding dairy animals. There is a need to sustain efforts on the improvement of buffalo by selection and artificial insemination, providing better management, critical feed supplements so that optimum production could be attained on a sustainable basis. Besides, buffalo could also promote rural tourism in the island villages, if it becomes part of some fun games and other activities such as buffalo race, bull fighting, tying and weighing of buffalo calves etc.

Research Needs

Buffalo is the least studied livestock species in this region. But being a multi-purpose animal used for milk, drought and meat purposes, it has great potential and scope for further improving its value in agriculture. Male buffalo bulls and calves could be used for draoght purposes, pulling bullock cart and ploughing the agriculture field for crop production. There is a need to standardize their use and estimate their power for these specific purposes. There is a need to study their growth, production and reproduction performance in depth and to find out ways to reduce inter calving period and also to improve milk and meat production. Carrying capacity of grazing lands, coconut orchards and all community lands vis-à-vis livestock needs to be assessed so that a decision may be taken which livestock and what population size should be reared in view of the demand of livestock product. As the land resources are limited and cannot be stretched, land productivity needs to be optimized along with integration of crops and livestock production systems. Buffalo farming may have a greater role in this integration as a multi-purpose animal. In hot and humid climate of Andaman and Nicobar, animals continuously suffer from the heat stress so the nutrient requirements of animals should also be determined for this climate zone.

Conclusion

In rural and periurban areas buffalo farming is preferred by farmers for the purpose of getting milk, drought power and meat. Most of the farmers had smallholdings, which were managed without labourers. Buffalo holds the key for increasing milk production in this region and serve as a multi-purpose animal for these islands.

References

Ahlawat, S.P.S, Chatterjee, R.N. and Senani, S.(2000). Sustainable cattle production in A&N Islands: future strategy. In proceedings of " Strategies for dairy development in eastern and northern India- vision 2000 on problems and prospects on dairying in eastern and northeastern India, 17-18Nov 2000, p46-51

Chatterjee, R.N., Ahlawat, S.P.S., Rai, R.B.(2003) Buffalo Production in the Bay Islands. Livestock International. **7(5):** 2,22-23

Sharma, A.K., Dagar, J.C and Bandyopadhyaya, A.K. (1990). Fodder farming in Bay Islands, Improved cultivars and agro-techniques. CARI Bulletin No. 4, CARI, Port Blair.

Senani, S.(2003). Status and prospects of increasing livestock production in Andaman and Nicobar Islands.

Senani, S. (2004). GIS based decision support for increasing livestock production. unpublished data.

Senani, S., Rai, R.B., Jai Sunder, Chatterjee, R. N. and Kundu, A.(2004). Resource mapping and development of Livestock intensification index for Andaman and Nicobar islands. In "Map India 2004" Extended Abs.120.

CHAPTER 9

Buffaloes of Orissa and Adjoining Regions Present Status and Perspectives

S. K. Dash

Department of Animal Breeding and Genetics, Orissa University of Agriculture and Technology, Bhubaneswar (Orissa)

Orissa possesses some indigenous breeds of buffaloes in different regions of the state. These are Paralakhemundi, Jirangi, Manda, Kalahandi, Sambalpuri (Cockrill, 1984, Mason, 1974). Many other strains like Chilika, Paradip and Boudh buffaloes are common in the state. Besides these, there is a large population of indigenous nondescript buffaloes found in different parts of Orissa. The buffaloes found in the coastal part of Orissa are swamp type having 2n=48 chromosomes (Rao, 1981). All these buffaloes are low milk producers but are very good for cultivation and transportation in both plain as well as in jungle and hilly tracts. Out of 13.88 lac buffaloes population in Orissa (Directorate of Economics and Statistics, Orissa) 20 to 30 % belong to indigenous breeds i.e. Kalahandi, Chilika, Paralakhemundi, Sambalpuri and Paradeep buffaloes.

Present Status

Usually the buffaloes in Orissa are reared under range system. Unlike others, Chilika buffaloes possess special quality of entering deep into the salt lake of Chilika and feeding on the vegetations that grow there. Other types thrive entirely on the natural vegetation available in the nearby jungle, hillocks and paddy fields. In most of the areas feeding of straw and rice husk as a routine practice is confined to the lactating animals. Chilika buffaloes are never provided with shed for shelter, they are let loose throughout the year. All other buffaloes are provided with thatched houses with kachha floor. The usual practice is to milk the buffaloes once in a day. The milk yield ranges from 2 to 3 liters per day with about 9 to 10 months of lactation length. Mishra (1993) observed a higher lactation yield of 1048.85=9.73 liters in Paralakhemundi buffaloes.

Table-1: Average Value and standard error of conformation traits of Buffaloes in Orissa.

Sl. No.	Traits	Kalahandi (Dash, 2002)	Chilika (Mishra, 2001	Paralakhemundi (Patro, et al. 1987)
1.	Height at withers (cm)	118.53 ±0.31	123.85 ± 2.75	122.90 ± 1.47
2.	Body length (cm)	116.51 ± 0.40*	122.33 ± 0.20	114.68 ± 2.16
3.	Chest girth (cm)	174.55 ± 0.65	169.65 ± 0.28	184.48 ± 3.95
4.	Length of forelimb (cm)	67.42 ± 0.47	67.33 ± 0.53	—
5.	Length of hindlimb (cm)	72.57 ± 0.47	72.42 ± 0.33	—
6.	Body weight (kg)	330.48 ± 6.70	321.98 ± 0.67	325.19 ± 3.79
7.	Horn Length (cm)	51.77 ± 0.42	49.05 ± 0.23	53.37 ± 1.94
8.	Tail Length (cm)	75.63 ± 0..37	67.59 ± 0.15	—

Majority of buffaloes are excellent draught breeds. In many cases the female buffaloes are also used for the purpose. Dash (2002) studied this in Kalahandi buffaloes and reported that these animals can cover a distance of 20 to 25 km. with a medium load and can cover 14 to 15 km. continuously with a full load of 10 to 15 quintals at an average speed of 4 to 5 km. per hour. The buffalo bullocks are used for ploughing continuously for 5 to 6 hours in morning and 2 hrs. in afternoon with a rest of 2 to 3 hrs. in noon time. The animals in general and Chilika buffaloes in particular appear to be very much disease resistant and acclimatized well to the extreme climatic conditions of respective regions. The breeding of buffaloes performed by natural service using bulls of same type that remain along with the females when left loose for grazing. Artificial insemination or breeding with other type bulls is not preferred by the people of the locality, because they fear for using draught quality in next generations and acquiring more attention in management of crossbred buffaloes.

Table-2: Average Value and standard error for Reproduction and Production Traits of Buffaloes in Orissa.

Sl. No.	Traits	Kalahandi (Dash, 2002)	Chilika (Mishra, 2001	Paralakhe-mundi Mishra, 1993)	Paradip (Mishra, 1991)
1.	Age at Sexual Maturity (Days)	1148.24± 7.25	1013.41± 4.08	1171.16±6.13	1383.04± 4.29
2.	Age at first Calving (Days)	1528.05± 7.40	1331.42± 6.82	1499.83±5.67	1704.89±4.65
3.	Calving Interval (Days)	526.89± 3.35	431.74±3.56	691.00±2.65	583.67±2.02
4.	Daily Milk Yield (Ltrs)	2.64± 0.07	2.59± 0.03	2.48±0.12	3.08±0.08
5.	Lactation Length (Days)	308.30± 2.21	238.72±2.1	429.33±3.86	297.7±0.95
6.	Dry Period (Days)	218.593.97	192.02± 1.78	260.50±2.32	286.15±2.32
7.	Service Period (Days)	194.26± 2.23	134.00± 2.53	386.00±2.55	193.28±2.05
8.	Lactation Yield (Ltrs)	813.12± 8.76	622.31± 6.72	1048.85±9.73	895.42±9.00

All these breeds are medium sized, compared to other types found in India. The average values and standard error of conformation traits are shown in Table-1. The colour of these animals ranges from black, grey to ash grey and brownish grey. The photographs of different types of buffaloes are presented in Fig. 1 to 6. The horns are long. In Kalahandi buffaloes horns grow laterally and horizontally but in Paralakhemundi buffaloes horns grow backwardly and inwardly. Forehead is slightly protuberant in Kalahandi buffaloes, but flat in Paralakhemundi and Chilika buffaloes. The legs are short, stout and thin in Paralakhemundi but muscular in Kalahandi buffaloes. Body length of Chilika buffaloes is more than in other types. The chest girth of Paralakhemundi type is more than others. The average body weight of all these breeds ranges from 320 to 335 kg.

Buffaloes in coastal region viz. Paradeep and Chilika are reared mainly for milk purpose with zero input basis. Very few males are used for draft purpose in Chilika region, but no males are used for this purpose in Paradeep region. But, Kalahandi and Paralakhemundi buffaloes are kept as dual purpose breed, primarily for cultivation and transportation. Both the breeds prove their excellency in undulated land of hilly tracts. The bullocks are docile and slow but can carry heavy loads and plough in hot sun. Kalahandi buffaloes are very hardy and can tolerate heavy rainfall, icy-winter, extreme climatic conditions and can climb steep hills easily. They can withstand drought conditions and live on sparse feed with least wallowing requirements and are famous for its long years of service and longevity.

Average milk yield of these buffaloes ranges from 2 to 3 litres per day on single milking. Animals yielding upto 5 kg milk per day are also found in all types found in Orissa. The average values and standard errors for production and reproduction traits of breeds of buffaloes in Orissa are presented in Table-2. The farmers in respective habitats are satisfied with the performance of their animals because they thrive well and multiply efficiently with zero input and with least botherations of medication, thereby conferring good net economic returns.

Cytological investigation conducted in the Deptt. of Animal

Fig. 1. Male Kalahandi Buffalo

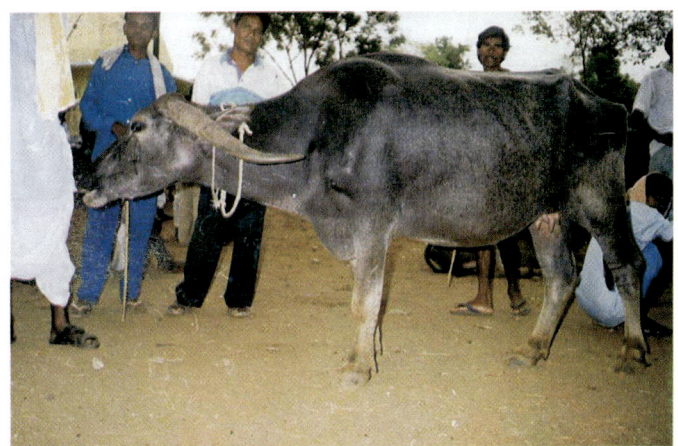
FIg. 2. Female Kalahandi Buffalo

Fig. 3. Kalahandi Bullocks at work

Fig. 4. Paralakhemundi Buffalo

FIg. 5. Female Paradip Buffalo

Fig. 6. Murrah X Paradip Buffalo (Female)

Breeding and Genetics, Orissa Veterinary College, OUAT, Bhubaneswar revealed that buffaloes in the coastal districts of Orissa had 2n=48 chromosomes and buffaloes in hilly and jungle areas had 2n=50 chromosomes. Inheritance of the major conformation and economic traits in above said breeds/ types of buffaloes indicates that these buffaloes can be improved through individual selection. So in habitats of such exclusive breeds of buffaloes with unique characteristics, selective breeding through exploitation of existing genetic variation should be tried for conservation of valuable germplasm. However, farmers desirous of commercial diary farming with proper demand for buffalo milk may opt for rearing Murrah and Murrah crosses for higher milk productions.

Summary

The buffaloes in Orissa and adjoining areas are well adapted to extreme climatic conditions of heat, rain and cold, prevailing in different regions. Out of 13.88 lac buffaloes population in Orissa (Directorate of Economics And Statistics, Orissa) 20 to 30 % belong to indigenous breeds i.e. Kalahandi, Chilika, Paralakhemundi, Sambalpuri and Paradeep buffaloes. Chilika buffaloes in particular have special quality of feeding on the weeds, aquatic plants and other vegetations submerged in water. These animals are poor milk yielder, but an average of about 2.5 litres of milk is produced in a day with high fat percentage. In western part of Orissa the buffalo rearing is mainly envisaged at using those as draught animals. They do play a vital role in crop cultivation, transportation, and supply of food nutrients particularly in their respective regions. Almost all the types of buffaloes are raised under extensive system with almost zero input. So the milk production and draught service is considered as net income from buffalo rearing.

References

Cockrill, W. Ross (1984) Water buffalo. In the evolution of domesticated animals, published by Longman Group, New York, USA, pp 52-63.

Dash, S.K. (2002) Studies on Kalahandi buffaloes in Orissa. M.V.Sc thesis submitted to Orissa University of Agriculture and Technology, Orissa.

Mason, I.L. (1974). Species, type and breeds : In the husbandry and health of domestic buffalo. Published by FAO of United Nations, Rome, Italy, pp 1-47.

Mishra, K. P (1993). Studies on the productive and reproductive performance of Paralakhemundi buffaloes and its crosses. M.V.Sc thesis submitted to Orissa University of Agriculture and Technology, Orissa.

Mishra, P.K. (2001). Study of production and reproduction potentialities in Chilika buffaloes found in Orissa. M.V.Sc thesis submitted to Orissa University of Agriculture and Technology, Orissa.

Mishra, S. (1991). Reproductive status of crossbred buffaloes (indigenous, Murrah) in Orissa. M.V.Sc thesis submitted to Orissa University of Agriculture and Technology, Orissa.

Patro, B.N. and Das, K. (1987) Paralakhemundi and Manda buffaloes of Orissa, Buffalo Bulletin, **6(3):** 60-64.

Rao, P.K. (1981) The Chromosomes of local and Murrah, buffaloes M.V.Sc thesis submitted to Orissa University of Agriculture and Technology, Orissa (India).

CHAPTER 10

Strategies for Enhancing Productivity of Kuttanad Buffaloes

K. Anilkumar

Centre for Advanced Studies in Animal Genetics and Breeding,
College of Veterinary and Animal Sciences,
Kerala Agricultural University, Mannuthy, Thrissur- 680 651 (Kerala)

Cattle is the major domestic animal for the state of Kerala. The presence of buffaloes in the state is limited to some of the commercial dairy herds and work animals in rice growing areas like Kuttanad and Palghat. Three groups of buffaloes namely river type, swamp type and their crosses were established in Kerala (Anilkumar, 1991). The current breeding policy of the state envisages use of Murrah semen to upgrade the local buffaloes. (Anon.1998).

The buffalo population was reduced from 4.72 lakhs in 1972 to 1.65 lakhs in 1996 and 1.11 lakhs in 2000 (Annon.2000) in Kerala. Contribution of buffaloes in meat production of the state is 24.43% of total meat production from large animals. It is estimated that 13.74% of animals slaughtered in the state for meat are buffaloes.

Breeding Tract

Kuttanad buffaloes are found only in Kuttanad area of Kerala. It is comprising two districts of the state namely Kottayam and Alappuzha. Major peculiarity of the area is that paddy fields of the Kuttanad are under water for most of the time. At the time of rice cultivation the water is pumped out to facilitate the seeding. Even though the rice cultivation system is unique and expensive, it is being practiced by the farmers as no other option is available.

The paddy cultivation is a seasonal work. Hence the use of the bullocks for the farmers is also seasonal. It is limited to three to four months a year. That too spread over two different periods.

Management Practices

Kuttanad buffaloes are work animals. Males are exclusively used for working in paddy fields. The paddy fields are below water and are having thicker muddy layer so that the farm operations with tractors and tillers are difficult. The Kuttanad buffaloes are best suited for the operations.

The use of female animals are as producers of male calves. The milk yield from these animals is very less and is usually fed to the male calves. During the working season, the bullocks are fed with concentrates. It is mainly oilcakes like coconut oil cake, gingili oil cake or groundnut oilcake. The quantity fed to the bullocks usually varies from 500 grams a day to 1.5 kg. The major source of nutrition for these animals is the abundant supply of grass available in the area.

In Kottayam district more than 80% of the owners of Kuttanad buffaloes are labours and the other 20% are agriculturists by profession. Whereas in Kuttanad agriculturists amount to only 66.66% and the rest of the owners are labours. Most of the owners of Kuttanad buffaloes keep only buffaloes and not cows.

Milk Production

Milk production of these animals is meager and amounts 1 to 2 liters per day. This is against the average yield of 5.79 kg / day for milch buffaloes in the state. The Kuttanad buffaloes are usually milked only once a day. In case of male calves the calf is given major part of the milk of the mother. Hence proper recording of the milk production is not practical.

Fodder

The abundant supply of fresh grass available in the breeding tract of these animals is the major fodder source for Kuttanad buffaloes. The fodder is available throughout the year in one or other areas and the buffaloes are transported to different areas based on availability of the fodder. Feeding of paddy straw is very limited due to availability of green fodder.

Feeds

Major type of concentrate fed to these animals is oilcakes. The oilcakes fed to the animals are limited to milking buffaloes and working bullocks. The supplementation of concentrates in other forms is not practiced in the area. Hence the nutrient requirement of the animals is met mainly through grass and little quantity of oilcakes.

Improvements

Improvement in the animal can be through improving the performance of the bullocks of this genetic group. One of the major constraints faced by the farmers is the availability of good quality bulls. The male calves are castrated and sold as bullocks, since it fetches good money for the farmers. The keeping of bulls for breeding is very limited and it results in crossbreeding of the buffalo cows with the semen from Government agencies which is invariably Murrah semen.

One of the important measures to be taken is to make facilities in the breeding tract of the animal for supply of semen of Kuttanad buffaloes. This can be done through the Department of Animal Husbandry centres of the state.

Selection

The selection of bulls is not practiced for Kuttanad buffaloes. Hence it is high time to establish a milk recording system for the buffaloes. Based on the information collected good animals can be identified and used as bull mothers. Individual animals with reasonable milk production are available in Kuttanad type buffaloes. These animals can be identified and used as bull mothers. Further improvements can be made through selection of breeding animals from the population after analysis of records. A progeny testing programme for the breed is to be started to identify good bulls.

Conservation

The population of Kuttanad type buffaloes is estimated to be around 500 (Anilkumar *et al.* 2003). Hence the efforts to conserve these animals have to be undertaken on war footing. Two strategies should be employed for the conservation programme.

In situ conservation

The animals are being kept by the agriculturists and workers connected with rice growing operations. Hence the efforts to popularize this breed among these groups need be considered. The major hurdle for keeping these animals is the seasonal nature of their use. These animals are required only at the time of paddy cultivation. Hence they are usually sold after the farming operations to limit the maintenance expenditure. Many times these animals find their way to butchers because of the high demand for carabeef in the area.

In situ conservation units of Kuttanad buffalo can be established in farmers premises. Support for the farmers in form of meeting part of the maintenance cost can be thought of. At least fifty such *in situ* units of a couple of animals each should be established. The inputs like semen, veterinary aid, insurance coverage and periodic health checkups should be arranged for these animals. Their performance should be closely monitored and recorded.

Ex situ conservation

Conservation of the genetic group in form of embryos and semen need be employed to check their erosion. Facilities for this need be established. Some steps are already taken to collect and store semen of Kuttanad buffalo bulls. But this should be strengthened with more facilities.

Establishment of a bull mother farm of Kuttanad buffaloes which can also act as a ex *situ* conservation unit is highly essential. A small unit with around 50 animals including 10 bulls need be established first. This unit will be the focal point of developments for the Kuttanad buffaloes. Interested Non-Government Agencies should be given

option to start conservation
units. The government farms
available should be provided
with few heads of the
buffaloes to augment the
conservation efforts. A draft
plan for conservation of the
Kuttanad buffaloes are
presented. Participation of

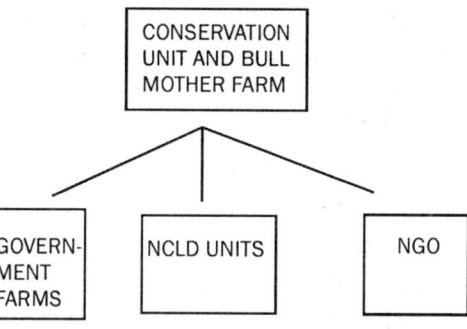

government agencies and NGOs will ensure successful results. The
success story of Vechur conservation by Kerala Agricultural University
may be considered as guideline for conservation efforts of Kuttanad
buffaloes.

Genetic Characterization

Genetic characterization of the buffaloes need be undertaken
immediately. Comparison of this group of animals with established
breeds of the country need be done to establish the unique nature of
the animals. Molecular genetic tools like microsatellite markers need
be used for such studies. At least 50 unrelated Kuttand buffaloes
need be selected for the study. 10-20 microsatellite markers can be
used for genetic characteriation and comparison with other groups.
Surti, Murrah and South Kannara buffaloes need be compared with
Kuttanad type as they are available in the area.

Molecular genetic studies on different gene loci like DRB3,
Prolactin, Lactalbumin, Lactoglobulin and similar others need be
undertaken. The association of alleles in these loci with the beneficial
characters like disease resistance, feed conservation efficiency and
milk composition need be established. Once this is possible application
of Marker Assisted Selection (MAS) can be employed for breeding
bulls.

Biotechnology Tools for Multiplication

Due to the highly endangered nature of the genetic group, use
of molecular technologies for multiplication of these animals can be
tried. MOET and use of foster mothers from other breeds is one option

for speedy multiplication. The number of animals can hence be increased at a higher pace compared to natural way.

Conclusion

The report of cattle commission suggests ban of slaughter of cows and calves. This will surely increase the importance and viability of buffalo production systems. Kerala being a state with majority non-vegetarian population is required to concentrate more on developments in buffalo based enterprises. Kuttanad buffaloes being the local breed has a lot of importance in such systems. It is the need of the hour to conserve and characterize this valuable germplasm of Kuttanad buffaloes.

References

Anilkumar,K. (1991) Genetic charecterisation of buffaloes in Kerala using cytogenetic technique. M.V.Sc thesis submitted to Kerala Agricultural University, Vellanikkara.

Anilkumar, K and Raghunandanan.K.V. (2003) The dwarf cattle and buffalo of Kerala. Kerala Agricultural University, Thrissur, Kerala.

Anilkumar, K., Raghunandanan, K.V and Anil, K.S. (2003) Physical profile of Kuttanad buffaloes.

Anon. 1998. Report of the committee to evaluate and formulate livestock breeding programme and policies in the state of Kerala. Government of Kerala, Thiruvananthapuram. p 18.

Anon. 2000. 16th quinquinal census report . Government of Kerala Thiruvananthapuram.

CHAPTER 11

Present Status and Future Perspective of Bhadawari Buffaloes

B P Kushwaha, S B Maity, and S S Kundu

Network Project on Bhadawari Buffalo
Indian Grassland and Fodder Research Institute,
Jhansi- 284 003 (UP)

Bhadawari, one of the recognised buffalo breeds of India is apprehended to be on the verge of extinction. The force of development driven by short-term tangible economic consideration has swept aside the Bhadawari buffaloes. Bhadawari, the brown beauty of ravines and predominantly the only breed to have adapted to the harsh conditions of the ravines with undulating topography, thorny and scanty bushes, climatic stress and draught conditions. The buffaloes are of medium size with medium milk yield but the fat content may go as high as 13 per cent. This breed received lesser attention of animal breeders and the planners. With the pace of development bringing canals for irrigation thereby increasing the agricultural production has seen large influx of high yielding Murrah into the breeding tract of the Bhadawari. The indiscriminate crossbreeding with Murrah buffaloes during the last 3 to 4 decades has almost wiped out the Bhadawari buffaloes in the region where agriculture has kept pace with development. The Bhadawari buffaloes are found in the ravines of Yamuna and Chambal rivers spred over in Uttar Pradesh and Madhya Pradesh. The breeding tract is Agra and Etawah districts of Uttar Pradesh and Bhind and Morena districts of Madhya Pradesh. This tract was a part of the erstwhile Bhadawar estate from where the name of these animals originated. Immediate focus should be given for its conservation and improvement lest this genetic resource is lost forever.

Population and Distribution

Population of Bhadawari buffaloes was 1.139 lakhs, constituting 0.82 per cent of total buffalo population (619 lakh) in Uttar Pradesh in year 1977. It was reduced to 0.982 lakh (constituting 0.54 per cent of buffalo population of UP) in year 1991 showing overall decline of 13.78 per cent during the period (Table-1). During the same period, the buffalo population in India increased by 24.33 per cent and that of UP by 30.9 per cent.

Year	Total buffaloes (lacs)		Bhadawari buffaloes (UP) (lacs)
	India	Uttar Pradesh	
1977	619	139	1.039
1991	770	182	0.982
% change (1977-91)	+24.33	+30.9	-13.78

Source: Report UP Govt. 1991

Table-1: Change in Bhadawari population over the years

A similar trend was also reported by Singh *et al.*, (1993) who made a preliminary survey covering three blocks of Agra and one block each of Etawah and Bhind district. Of the 40 villages surveyed none of the villages had even 20 Bhadawari buffaloes except the Pachhay village in Badhpura block of Etawah district. It was reported that only 11 per cent of buffaloes reared by the farmers were Bhadawari. Estimated population of this breed in the whole breeding tract was reported to be 37,706.

Kundu *et al.* (2002) conducted the survey of Etawah, Jalaun, Lalitpur districts of U.P. and Morena district of M.P. and reported presence of 1- 6 Bhadawari buffaloes per village. Sharma *et al.* (2003) conducted the survey of Bhadawari buffaloes in Etawah and Agra districts and reported that the Bhadawari population in most of the villages ranged from 1-3 buffaloes per household. After surveying 241 villages they could find only 1337 Bhadawari buffaloes. Majority of the person keeping Bhadawari buffaloes were either landless labourers or small/marginal farmers.

A survey was conducted by IGFRI, Jhansi (Kushwaha *et al.*, 2004) under network project on Bhadawari buffaloes and reported that the number of Bhadawari animals has drastically reduced. A total of 93 villages were surveyed. Bhadawari buffaloes true to its type were found in Etawah, Bhind and Jalaun districts though the number was quite low. In Lalitpur, and Jhansi districts few buffaloes having Bhadawari type character were found. In Morena there were hardly any Bhadawari animals.

Bhadawari animals were not found in all the villages surveyed and in most of the villages where Bhadawari animals were present the number varied from 1 to 5 animals per village. Looking into the present status of the Bhadawari animals averaging 3.2 per village and considering Agra, Etawah, Bhind, Morena and Jalaun as the breeding tract. The estimated population of Bhadawari animals, in 4,214 villages of the 5 districts, would be around 13,500. In Datia district there were no Bhadawari animals. In Lalitpur district some animals of Bhadawari type could be seen in the villages around the Saidpur farm. The total population of Bhadawari buffaloes is expected to be 13,500 to 15,000 including Bhadawari type animals from Jhansi and Lalitpur districts.

Increase in the population of Murrah in U.P. and decrease in Bhadawari population indicates a shift towards high milk production rather than for high fat producing buffaloes in the native tract of Bhadawari breed–a trend which may be dangerous for the existence of this important germ plasm.

Reason for Decline in Population

Following observation were recorded, during the surveys conducted by various workers (referred above), which may be responsible for decline in population of Bhadawari buffaloes:

◉ It has been observed that the farmers are more interested in keeping high yielding buffaloes. Because of non-availability of grazing land/pasture farmers prefer to keep high yielders viz.

Murrah buffaloes as stall-fed rather than Bhadawari buffaloes which have traditionally been kept on grazing.

- Ghee making was limited to the extent that they can meet their family demand and most of the farmers showed their interest on sale of milk rather than ghee.

- Milk yield of Bhadawari buffalo was reported to be poor than other breeds prevalent in the area and this was reported to be one of the reasons for decline in their numbers.

- There is an acute shortage of Bhadawari bulls in the field, which forces the farmers to go for the bulls available in the vicinity.

- The farmers reported that the lactation length of Bhadawari buffaloes is shorter than other breeds viz. Murrah.

- Farmers were of the opinion that Bhadawari buffaloes generally produce milk up to 4-5 months, which declines sharply as soon as it gets pregnant. However, this trend was not true in Murrah buffaloes.

- Nuts (a Banjara community from Rajasthan, who keep on moving along with their animals) keep one to two Murrah bulls with them and charge Rs. 50 to 100 per service from the farmers. For breeding of buffaloes farmers either take their animal to them (Nut) or on demand the bull is taken at farmers' door. It was also reported that the nuts are not allowing the other bulls of the village to survive. Very often poisoning of bulls owned by the villagers was also reported.

- It was reported (in Morena and Bhind districts) that keeping of breeding bull is not considered good and not acceptable in the society. If any farmers dared to keep the bull, he faced the risk of being expelled from the society. Because of this social problem livestock owners have to depend on Nuts for breeding their buffaloes, who were keeping only Murrah bulls.

- Madhya Pradesh Govt. has adopted the policy of upgrading local buffaloes with Murrah by providing Murrah bulls and semen. As a result the Bhadawari animals from Bhind and Morena districts of M.P. have almost vanished.

- It was informed in Morena district that during 1985-92 around

2,500-3,000 Bhadawari buffaloes were purchased every year from this region and were sold to MP Govt in Mandla, Sahdol, Sewni, Narsinghpur and Chhtarpur districts from where these buffaloes were distributed to different farmers by the MP Govt. under some scheme. Due to regular migration through purchase the number declined. At the same time infiltration of Murrah buffaloes started in this region in urban and periurban areas.

⊙   Bhadawari buffaloes are mainly concentrated in the villages, which are remotely located and facilities for milk transport/sale are not available.

Characteristics of Bhadawari Buffaloes

Bhadawari are usually copper coloured with scanty hair, which is black at the roots, and reddish-brown at the tips, sometimes the hair are completely brown (Ranjhan and Pathak 1979). Kaura (1950) has described the Bhadawari animals as medium sized with wedge-shaped bodies. Generally all four legs from hooves to knees and hocks are found to be partly brownish-white. The head is comparatively small, light and bulging out between the horns but sloping down slightly towards the forehead. In the males it is a bit heavier and coarse. The horns are characteristically placed, flat, compact and of average thickness, growing backwards and than upward turning inwards with slightly pointed tips. Forehead is slightly broad and deep in the middle. Ears are of average size, rough and pendulous. Dewlap is non-existent. The tail is long, thin and flexible running down the hocks with black and white or pure white switch. The vulval opening is situated not so close to the anus as is found in the other breeds of buffaloes. The udder is not so well developed as in the case of Murrah buffaloes, but the milk veins are fairly prominent. The teats are of medium size though not of uniform length

Observations on physical characteristics of Bhadawari animals recorded during the survey (survey report, 2003) indicate slight diversion/changes in the physical characteristics of Bhadawari animals in the field. Physical observation on 301 buffaloes was recorded during the survey. About 70 per cent of Bhadawari buffaloes were of medium size, 65 percent were having copper colour skin and rest about 35 per cent of the buffaloes were having blackish skin colour and comparatively larger body size. Deviation in the horn pattern was also observed and about 20 per cent of the buffaloes had curved and twisted horns. Colour of udder was pinkish in about 80 per cent buffaloes. Some buffaloes were having pinkish udder and curved and twisted horn. Presence of twisting of horn, like Murrah buffaloes, indicates the mixing of Murrah blood. About 50 per cent of Bhadawari buffaloes had two white lines (chevron) at lower side of the neck, 30 per cent of them having one white line and rest about 20 per cent were not having the chevron though having all other characteristics of Bhadawari type. The average body measurement of Bhadawari

Body Length	Height	Heart girth	Pounch girth	Tail length	Reference
140.0	130.0	180.0	-	-	Bhat 1981
142.6	127.3	198.0	211.6	-	Pundir *et al.*, 1996
115.78	125.82	185.39	204.52	-	Singh *et al.*, 2003
137.0	-	188.7	200.53	-	Singh and Desai, 1962
133.38	122.90	185.21	195.64	89.66	Survey report, 2003
138.78	126.33	185.02	199.48	-	Goel *et. al.*, 2003

buffaloes reported by various workers are presented in table below:

Table-2: Average body measurement of Bhadawari buffaloes (in cm)

Singh and Desai (1962) and Bhat (1981) reported body weight of adult Bhadawari buffaloes as 425.7±7.7 and 385.5 kg, respectively. Average adult body weight of buffaloes maintained at IGFRI was reported to be 392.8 kg. The average milk yield of these buffaloes was reported to be 3.75 kg/day (Singh and Singh 1977), while the well maintained good animal may produce upto ten kg. milk a day. Age at first calving in Bhadawari buffaloes at Etawah farm was 58.3

months (Singh and Nivsarkar 1997).

Due to high butterfat per cent the milk of this breed is highly suited for ghee making, which is the common village industry. Male are frequently used for draught and probably can stand better than the other breeds of buffaloes, e.g. Murrah. Due to copper colour,

Breed	AFC (m)	CI (Days)	Lact. yield (kg)	Lact. Period (days)
Surti	44.5 ± 2.0	461.1± 15.3	1772.0 ± 10.3	350.1 ± 5.4
Bhadawari	50.7 ± 0.8	453.6 ± 10.2	1111.0 ± 12.9	276.0 ± 2.2
Nili	53.2 ± 0.6	461.6 ± 18.8	1855.2 ± 10.6	316.0 ± 8.5
Murrah	41.3 ± 1.4	495.1 ± 12.2	1744.0 ± 10.5	279.0 ± 6.5
Non-descript	49.5 ± 0.8	481.0 ± 18.4	5541.0 ± 39.9	272.3 ± 8.5

they thrive well under hot climate. This breed therefore, appears to be more suited to need of the villages (Kaura, 1950).

Table-3: Productive and reproductive traits of different Indian breeds of buffaloes

Source: Bhat (1981)

Among all breeds, the lowest calving interval was observed in Bhadawari buffaloes. However the poor yield is suggested to be mainly due to poor nutrition and management.

Production Performance

The studies carried out on Bhadawari buffaloes are scanty. Based on the reports of farms at Kanpur, Mathura and Bharari (Jhansi) it has been reported that the birth weight of male and female Bhadawari calves were 25.26±024 (222) and 24.86±0.21 (198) kg respectively (Katiyar, 1987) and the overall average for the breed was 25.07±0.23 kg. Birth weight of Bhadawari buffaloes calves kept at CSA University of Agriculture and Technology was 25.1±0.23 kg which ranged from 23.50 to 26.56 kg (Report 1988). Average birth weight of Bhadawari calf maintained at IGFRI was reported to be 25.10 kg.

Milk Production

Bhadawari animals are moderate to low yielder, but the feed they consume are of poor quality. Sharma and Singh (1978b) observed that lactation length significantly influenced first lactation total milk yield but had no influence on milk producing ability. Buffaloes with more than 334 days lactation length produced significantly higher milk yield. Service period and breeding efficiency had no influence on the first lactation milk yield or milk producing ability. The average milk producing ability was 1102.2±31.6 kg.

Singh *et al.* (1993) recorded very low estimates of first lactation total milk yield of Bhadawari buffaloes (699.3±48.2 kg) based on 99 observations and observed that only period of calving significantly influenced first lactation total milk yield. Milk yield was highest in the first period and started declining is subsequent periods.

The overall weighted least square mean of first lactation milk yield of 300 days, based on 215 observations was 711.2±28.2 kg and varied between 660.4±44.4 kg (Singh *et al.*, 1993) and 753.8±36.1 kg (Anonymous 1988). Sharma and Singh (1981) observed that if first lactation milk yield is increased in Bhadawari buffaloes, than the total yield at fourth lactation decreases, while total wet period and lifetime milk production increases.

The period of calving had significant (P<0.01) influence on first lactation milk yield (FLMY) of 300 days. The farm type and season of calving had no significant influence on milk yield. There was a decrease in FLMY of 300 days along with period of calving.

Milk yield was maximum in the second lactation and remained static upto parity 4 and then gradually declined upto parity 7. A very wide variation was recorded in the total milk yield with maximum milk yield being recorded by Singh and Singh (1977) for animals of Saidpur farm (1098 to 1165 kg) and minimum yield (554 to 734 kg) recorded by Singh *et al.* (1993) for animals of the farm of CSA University of Agril. and Technology, Kanpur.

Singh and Desai (1962) recorded very high estimate of 1111.0±12.36 kg for the pooled, 300 days lactation milk yield for the Bhadawari buffaloes at Bharari farm (ranges from 471.8 to 2071 kg). Relatively lower milk yield (751.3±45.3 kg) was recorded for the Bhadawari buffaloes maintained at the CSA University of Agriculture and Technology, Kanpur. Range in lactation milk yield reported was 472.4±90.9 kg to 884.4±39.5 kg (Anonymous, 1988). Average lactation milk yield, lactation length of Bhadawari buffaloes, of IGFRI herd, were 1064.53 kg and 296 days respectively. Average peak yield was 6.62 kg ranging from 5.1 to 9.0 kg. Average fat in their milk was 8.71 per cent with the range of 5 to 12.00 per ecnt.

The pooled lactation milk yield for 300 days recorded by Singh *et al.* (1993) were lowest (650.4±44.8 kg) among all the reports on this breed. This was recorded in animals maintained at Kanpur and Dalipnagar farms. However the milk yield differed significantly between the two farms. Animals at Kanpur farm gave higher milk yield (690.1±44.1 kg) than those at Dalipnagar farm (610.8±44.1 kg). Singh *et al.* (1993) further observed that order of parity, period and season of calving did not influence 300 days milk yield, however milk yield was highest for parity 2, first period of calving and spring season of calving. There was continuous decline in the pooled lactation milk yield for 300 days along with period of calving.

Reproductive Performance

The age at first fertile service was 38.64±0.48 (135) months for the Bhadawari animals maintained at CSA University of Agril. and Tech. Kanpur (Anonymous, 1988). However, Singh and Nivsarkar (1997) reported comparatively higher age at first conception (48.1 months), but their data was based only on 25 observations.

Seasonal breeding has been recognized as one of the problems of practical importance owing to its direct impact on the reproduction and as well as production efficiency of buffaloes. Sharma *et al.* (2003) have reported that 50 per cent of the Bhadawari buffaloes expressed heat during winter season while only 2.1 per cent buffaloes expressed estrus in summer months. Dwivedi (1991) observed that Bhadawari buffaloes are seasonal breeder, maximum number of animals were bred between September and December with highest percentage (26%) in October and lowest in May (0.67%). Mature animals were more fertile than old or young animals. The weighted means for age at first calving for Bhadawari buffaloes on the basis of 596 records was 48.6±0.58 months with a variation from 44.4±0.67 (Singh and Singh 1977) to 51.3±1.55 months (Singh *et al.* 1993) for animals of Saidpur and Kanpur farms. Higher age at First calving (58.3 months) was recorded at Etawah farm (Singh and Nivsarkar 1997).

Singh and Desai (1962) reported average age at first calving for Bhadawari buffaloes maintained at state livestock-cum-agriculture farm, Bharari as 50.7±0.83 months. The year of calving significantly influenced age at first calving while season of calving did not have any significant influence on age at first calving (Sharma and Singh 1978a). Mishra *et al.* (1986) found that period and season of calving significantly influenced age at first calving of Bhadawari buffaloes. The influence of farm, season and period of calving found to be non-significant on age at first calving in Bhadawari buffaloes (Singh *et al.* 1993). The av. breeding efficiency of Bhadawari animals was 83.1±2.6%. Sharma and Singh (1978a) recorded significant effect of dry period. Katiyar (1987) reported significant effect of parity and Dwivedi (1991) reported significant effects of year and month of calving on breeding efficiency. Sharma and Singh (1978c) observed that age at first calving was negatively and significantly related with milk producing ability and breeding efficiency in Bhadawari. The selection of early calvers is therefore, recommended for improvement in milk production and breeding efficiency.

Feeding and Management Practices

The animals are usually housed in katchha houses with floor plastered with mud and dung and roof thatched with straws, kadbi and other locally available material. These thatched shelters lean along with the sidewalls of houses. Mangers are constructed in open, which may be round; square shaped and permit only individual feeding. Wherever, tree is near the house, mangers are built beside it. Buffaloes are tied near the manger in the morning and evening. During hot sun and rainy days the buffaloes are kept inside (Survey report, 2003).

Buffaloes in the past used to be grazed extensively in the ravines. Their compact small body and smaller legs (Bhadawari have also been described as "Suar Gori" meaning having smaller legs like pig) help in making balance while grazing in undulated ravine topography. Some 20-30 years ago, these ravines were seeded with babul, which have taken roots and proliferated extensively. Farmers complained that the babul tree has destroyed the natural grassland and it does not allow grasses to grow under its canopy. Moreover, the prickly thorns prevent its use as fodder besides restricting the movement in ravines. Hence buffaloes are mostly stall fed these days, with locally available roughage like sorgham kadbi and wheat straw. The chopped fodder/ kadbi is liberally mixed with water and mixing of concentrate at regular interval is done to induce feeding. The concentrate mixture is soaked in the evening left overnight and then boiled in the morning and fed to the animals after cooling. The practice of boiling concentrate is discontinued once the buffalo successfully conceives. Farmers have a belief that boiling of concentrate helps keep buffalo warm and conception is easier. Buffaloes are offered water twice a day in general. Bhadawari buffaloes are hardy animal. They easily tolerate the extreme weather condition in the ravine where the maximum temperature goes up to 48⁰C. Unlike Murrah buffaloes, they do not demand wallowing and frequent bathing.

Bhadawari are reported to be a regular breeder giving a calf per year in field condition. They have a short lactation length of about 7-8 months but the milk tastes sweet with high fat content and a flavour that is unmatched.

Yadav *et al.* (1999) studied Bhadawari buffaloes maintained at livestock farm J.N.K.V.V. College of Agriculture, Gwalior under NARP-II project produced on an average 3.7 litres of milk/day containing percent fat, S.N.F, T.S. and sp. gravity as 7.3, 9.41, 16.64 and 1.031, respectively. In a trial from July 92 to June 93 group feeding of buffaloes indicated that if the leguminous green fodder like berseem is available *ad libitum* (about 50 kg/day/animal) the concentrate feeding may be avoided without any adverse effect on production of milk and its composition for production up to 7 litres of milk a day. Trials on soaked and unsoaked concentrate and wheat straw feeding showed that there was no significant effect of soaking of feeds on the production and composition of milk.

Sachan *et al.* (2003) studied the dry matter intake and eating pattern of Bhadawari buffaloes maintained at IGFRI Jhansi. Average total dry matter intake was reported to be 9.54, 8.90 and 9.69 kg respectively in three groups. Dry matter intake was more in stylo than jowar and grass hay roughage based groups.

Need for Conservation of Bhadawari Breed

The Bhadawari need to be conserved for being a unique buffalo breed and a valuable genetic resource. Many characterstics of this breed have not been fully studied and therefore it needs to be conserved for posterity. These buffaloes are well adapted to the climatic conditions of the area with medium sized body, less water requirement, thriving well on low quality forages, better draught ability, higher reproductive efficiency and high milk fat content. This fits very well to the requirements of small and medium farmers provided the milk yield is enhanced through concerted efforts by selection and progeny testing programmes.

These buffaloes are most suited to far off inaccessible areas and ravines where milk procurement is difficult. The Bhadawari buffaloes in these areas could very well sustain ghee making industry having longer shelf life. Moreover, the meat production from these buffaloes has remained unexplored which needs to be understood.

Conservation Efforts Made so far

The earliest reference about Bhadawari buffalo appeared in 1941 when Zachariah described this as 'Bhadavan' buffalo– the best breed of buffaloes in Uttar Pradesh found in the neighbouring districts of Agra and Etah. But the popularity of this breed was known to the world by the detailed description of this breed by Kaura (1950).

The conservation efforts for Bhadawari started immediately after independence when a Bhadawari herd was maintained at State Govt. Farm, Bharari, Jhansi. This breed remained here up to 1965. These were then transferred to the Saidpur farm in Lalitpur district. These animals were thereafter shifted to Bhadawari buffalo and Jamunapari Goat breeding farm, Etawah in 1989-90. The Bhadawari buffaloes were also maintained at Kanpur and Dalipnagar farms in the past. In between, an ICAR project under NARP for conservation of Bhadawari buffaloes was completed at Agricultural College, Gwalior. During last 50 years the conservation efforts were limited to maintaining a small herd at a farm without breed improvement programmes probably due to lack of resources and technology. Moreover, conservation of bio-diversity was an unknown term during those days especially in case of large ruminants. However, during last few years the nation has awaken to protect its bio-diversity for future generations. The Bhadawari buffalo farm at Etawah has reoriented its efforts under UP diversified Agricultural Support Programme. Indian Council of Agricultural Research has also launched its programme for conservation and improvement of Bhadawari buffaloes. A project under Network has been started at Indian Grassland and Fodder Research Institute, Jhansi (UP) under which a herd of Bhadawari buffaloes is being maintained with the objective to provide improved bulls and semen for breeding purposes and Cryopreservation of semen for future use. A project on Cryopreservation of Bhadawari buffalo semen is in progress at Pandit Deen Dayal Upadhya University of Veterinary Sciences, Mathura (UP) under NATP programme. National Bureau of Animal Genetic Resources, Karnal (Haryana) have also started *in-vivo* conservation of this breed.

Recommendations

In spite of all the efforts for conservation of this breed during last 50 years, the number of Bhadawari animals continued to decline at a faster rate (according to various reports the total population of this breed in the whole breeding tract is estimated to be around bout 15 to 30 thousand).

Bhadawari buffaloes present in the field are probably the last representatives of their breed, which the humanity will lose, are not so too distant future. Because most of the presently available buffaloes in the field have completed three lactations. After completing another 3-4 lactation they will find their way to slaughter house. With virtually no Bhadawari bulls present in the field, the next generation of this breed is ruled out. Hence, there is an urgent need to conserve the existing Bhadawari buffaloes in the field.

The first and foremost step should be to make slaughtering of Bhadawari and Bhadawari type buffaloes illegal and heavy penalty be imposed on those violating it. This ban should be enforced in all the districts where Bhadawari animals are found.

Secondly the upgardation of desi buffaloes including Bhadawari through Murrah, as MP Govt. in Bhadawari breeding tract in Madhya Pradesh is doing it, should be stopped immediately.

For *in-situ* conservation of Bhadawari animals in their native environment nucleus herd of elite Bhadawari buffaloes should be established (already in operation at IGFRI Jhansi and at Etawah) which would act as bull mother farm. Three or more such herds should be raised at different locations, which will exchange their bulls at periodic interval.

In the field it has been reported that some non-descript/desi buffaloes are giving birth to Bhadawari type calves implying that genes of Bhadawari genotype are present in the population. Supplying of promising purebred Bhadawari bull calves/bulls in the field will spread the genes in a short span of time. But a coordinated effort is required wherein the bull should be rotated to another place after 3-4 years, to avoid inbreeding.

The performance of the artificial insemination in the field is quite poor due to obvious reasons. Farmers are also not coming forward for AI and they prefer to breed their animals by bulls, which may or may not be of the desired breed. Looking into scanty population (per village 2-3) progeny testing may not be practical. There is an urgent need to increase the number of Bhadawari animals. This can be done by distributing bulls in the field, which will always be available for breeding and will cover maximum population and thereby increase the number at a faster rate.

Since the Bhadawari buffaloes are more suited to the needs of small and medium farmers and for remote/inaccessible areas, vigorous campaign to create awareness among farmers should be undertaken.

It would also be important to maintain the community grassland and feed fodder resources optimally, so that the land less and small farmers (who own the Bhadawari buffaloes) can graze their animals.

Under *in-vivo* conservation sperm, oocytes and embryos can be cryopreserved. Storage of DNA, establishment of embryonic cell lines and somatic cells can be stored for longer period and can be used to produce a normal animal.

Sustained efforts over few decades will restore the Bhadawari buffaloes in the field. Bhadawari buffaloes are an unique breed evolved in this region over thousands of years adopting to the harsh climate and undulating ravine topography surviving on feed resources on which other animal like Murrah would normally starve. Having such excellent characteristics/qualities besides milk fat (13%), which is unmatched among buffalo breeds, it is high time to conserve and propagate this breed for posterity.

References

Basic Agricultural Statistics: Madhya Pradesh,2000. Govt. of Madhya Pradesh, Gwalior.

Bhat, P. N (1981). Cattle and buffaloes: Animal Genetic Resources in India. NDRI Publication No. 192, Karnal.

Dwevedi, A.D. (1991). The effect of month and season upon breeding efficiency of Bhadawari buffaloes. M.Sc. Thesis submitted to the CSA University of Agril. & Tech., Kanpur.

Goel, R., Sharma, K.C. and Singh H.N. (2003). A study on general description of Bhadawari buffaloes. Abstract in proceedings of 4th Asian Buffalo congress for food security and rural employment, 25-28 Feb. 2003, New Delhi

Katiyar, S.S. (1987). Effect of parity on economic traits in Bhadawari buffaloes. M.Sc. Thesis submitted to the CSA University of Agril. & Tech., Kanpur.

Kaura, R.L. (1950). Cattle Development in the Uttar Pradesh. pp 21-22, Depertment of Animal Husbandry, Lucknow, U. P.

Kundu, S.S., Anil Kumar, N.P. Singh, S.B. Maity, C.B. Sachan, P. Sharma and Sultan Singh (2002). A survey of Etawah, Jaluan, Lalitpur and Morena district for Bhadawari Buffaloes. Abstract in Xth international Congress- September 23-27, 2002- New Dehi, INDIA

Kushwaha, B. P.; Anil Kumar; Sultan Singh; S.S. Kundu; S. B. Maity and P. Sharma (2004). Current status of Bhadawari Buffaloes in the country. Abstract in Proceeding of National Symposium on " Livestock Biodiversity via-a-vis Resource Exploitation: an Introspection" held at NBAGR, Kranal 11-12 Feb, 2004.

Mishra, R.C., Kushwaha, N.S. and Singh, R. (1986). Study on age at first calving in Murrah and Bhadawari buffaloes. Indian. vety. Med. Journal **10**: 13-15

Pundir, R.K. Singh, R.V., Vij, P.K., Vish, R.K. and Nivsarkar, A.E. (1997). Characterization of Bhadawari buffaloes NBAGR, Karnal Research Bulletin No. 7.

Pundir, R.K., Viz, P.K., Singh Ran Vir and Nivsarkar, A.E. (1996). Bhadawari Buffaloes in India. Animal Genetic Resources Infromation, No. **17**: 109- 122.

Ranjhan, S. K and Pathak, N. N (1979). Management and feeding of buffaloes, Vikash Publishing House Pvt. Ltd., New Delhi.

Report (1988). Terminal Report of Studies on Bhadawari Buffaloes as a Genetic resource in its breeding tract. C.S.A. University of Agril. & Tech., Kanpur.

Sharma, R.C. and Singh B.P. (1978a). Evaluation of genetic potential of Bhadawari buffaloes (Reproductive characters). Indian Vet. J. **55**: 595-600

Sharma R.C. and Singh B.P. (1978b). Evaluation of genetic potential of Bhadawari buffaloes (Productive characters). Indian. Vet. J. **55**: 751-55.

Sharma, R. C and Singh, B. P (1978c). Evaluation of genetic potential of Bhadawari buffaloes (Genotypic and Phenotypic correlations). Indian. Vet. J. **55**: 943-946.

Sharma, R. C and Singh, B. P (1981). Factors affecting the lifetime milk production in Bhadawari buffaloes. Indian. Vet. J. **58**: 47-52.

Sharma, R.C., Shukla, S.N. and Singh, H.N. (2003). Breeding behaviour of Bhadawari buffaloes. Abstract in proceedings of 4th Asian Buffalo congress for food security and rural employment, 25-28 Feb. 2003, New Delhi.

Sharma, K.C., Goel, R. and Singh, H.N. (2003). Socio-economic condition of the farmers in native tract of Bhadawari buffaloes. Abstract in proceedings of 4th Asian Buffalo congress for food security and rural employment, 25-28 Feb. 2003, New Delhi.

Sachan, C.B., Kundu, S.S., Sultan Singh, Singh, N.P. and V. P. Singh (2003). Dry matter intake and eating pattern of Bhadawari buffaloes on different sources of roughage. Abstract in National symposium on Sustainable Livestock production A National Thrust, held at JNKVV Jabalpur 6-8 March, 2003.

Singh, H.P. and Singh, R.C. (1977). Statistical studies of some economics traits of Indian buffaloes. Indian. Vet. J. **54**: 823-33.

Singh, Ranvir and Nivsarkar, A.E. (1997). Bhadawari buffaloes as an endangered genetic resource Buffalo J. **13**(2): 115-125.

Singh, Ranvir and Pundir, R.K., Nivsarkar, A.E., Sahai, R. and Singh C.D. (1993). Production and reproduction performance of Bhadawari buffaloes. Proceedings of National Seminar on Animal Genetic Resources and their conservation. April 22-23. NBAGR/ NIAG, Karnal.

Singh, S.B. and Desai, R.N. (1962). Production characters of Bhadawari buffaloes. Indian Vet. J. **39**: 332 – 43.

Singh, V.K., Singh, B.P. and Singh, H.N. (2003). A study on body measurements and production traits in Bhadawari buffaloes. Abstract in proceedings of 4th Asian Buffalo congress for food security and rural employment, 25-28 Feb. 2003, New Delhi.

Survey report (2003). Survey report of the survey conducted under network project on Bhadawari buffaloes, IGFRI, Jhansi.

Yadav, R.S., Yadav, M.S. and Mehla, O.P. (1999). Effect of three rearing systems on growth performance of young buffalo calves. Indian J. Anim. Prod. Mgmt. **15(1)**: 26-28.

Yadav, S.S., Sachan, C.B. and Gupta, R.D. (1999). Studies on milk production performance of Bhadawari buffaloes as influenced by feeding management. *In:* Proceedings of international conference on sustainable animal production health and environment future challenges, November 24-27, C.C.S. Haryana Agril. University, Hisar.

CHAPTER 12

Export Potential of Buffalo Meat and other Products

Nagendra Sharma
Director

National Dairy Research Institute, Karnal - 132 001 (Haryana)

India has the world's largest cattle population with 209 million cattle and 94 million buffaloes. There are 27 indigenous breeds of cattle and seven breeds of buffaloes. The gross value output of the livestock sector is approximately worth $ 27 billion.

According to statistics compiled by the Food and Agriculture Organization of the United Nations (FAO), the total number of animals slaughtered for meat in India rose from 66,299,600 head in 1980 to 106,239,000 head in 2000, nearly doubling. There was a rise in the total number of cows and buffaloes slaughtered for meat in India from 15,644,000 head in 1980 to 24,300,000 head in 2000, an increase in total beef meat and buffalo meat production of 1,673,972 in 1980 to 2,863,400 in 2000. Currently, about 11 million buffaloes are being slaughtered annually contributing 1.42 million tons of meat accounting for 29.1% of total Indian meat production (FAO, 2001). It is expected that buffalo will ultimately emerge as the future animal of Indian dairy-cum-meat industry. As the demand for lean red meat has been increasing consistently worldwide, buffalo meat is expected to get customer preference due to its leanness and lower cost.

Features of Buffalo Meat

Buffalo meat is 25 per cent richer in protein content than beef and 50 per cent lower in cholesterol. The amino acid composition especially of critical amino acids of skeletal muscles is higher than of beef. The salient features of buffalo meat include lower proportion of fat in meat and large muscle fiber. The meat form young buffaloes reared for early slaughter (termed as buffalo broilers) are lean, tender and palatable. Such meat is of excellent quality and is considered a delicacy.

Almost 93 per cent of India's meat exports consists of deboned and deglanded frozen buffalo meat. This is a risk-free product having pH status below 6.0, which is brought about by compulsory chilling of the carcasses for a minimum of 24 hours monitored between two to four degrees Celsius. Even the OIE in Paris has confirmed that international trade in deboned and degalnded frozen meat prepared in accordance with guidelines formulated under the Zoo Sanitary Code (International Animal Health Code) will ensure that there is no risk of transmission of FMD virus.

Potential of Buffalo Meat and other Products

Buffalo has an outstanding potential as a source of quality meat for human consumption 1.4 million tonnes of buffalo meat in 1997-98 and exported 15,7000 tonnes valued at Rs 110 crores. It is estimated that India has a potential of producing 4 million tonnes of buffalo meat annually with a slaughter rate of 30 per cent of total buffalo population and carcass yield of about 138 kg per animal slaughtered. The ban on cow slaughter is not applicable on buffaloes and, therefore, religious or social ban on good quality food does not come into picture.

The value of total output of meat products, including poultry meat and eggs, is Rs 3,208.00 crores which is about 19.5% of total output of livestock. The major portion i.e. 54% of meat and meat products is contributed by sheep and goats, cattle and buffalo meat constitutes 26% of the total output.

Domestic Consumption

The major share of domestic consumption is that of meat obtained from sheep and goats. Meat from buffaloes and pig has restricted demand. The per capita consumption of beef/buffalo in India is 2.8 kg, about half that of fish, but more than twice the average intake of mutton, pork and poultry.

It should be kept in mind that due to social reasons, the consumption of buffalo meat is restricted in India. While over 70% of Indians are non-vegetarian by choice or religious belief, consumption of beef and pork is limited. The slaughter of cow / bull is prohibited in all but two states (Kerala and West Bengal). At the same time, meat products like sausages, bacon, kababs, ham, patties, hot dogs, pies are becoming popular in urban markets, as the annual production of 1,152 tonnes valued at Rs 190 lakhs.

Estimates of existing production details

Sectors	Figures in ('000' tones)
Mutton	675
Pork	420
Poultry	600
Cattle Meat	1,295
Buffalo Meat	1,210
Total Meat	4,500
Fish Production & Processing	5,26,000
Milk	81,000
Milk Products	307
Fruit and Vegetables Products	990
Soft Drinks (FY 2000-2001)	6,540 Million Bottles

(*Source:* Annual Report 2000-2001, Dept. of Food Processing, Ministry of Agriculture)

Export Potential

Buffalo meat constitutes the major item of Indian meat exports due to domestic availability, price differential between domestic and export markets and export policy (buffalo meat is permitted for export while beef is banned). The potential advantages of Indian meat exports include: large raw material resource, price competitiveness, cheap labour costs, proximity to importing countries, and preference for Indian lean meat produced on natural grazing and liberalized economic policies.

Out of the total meat exports, buffalo meat alone constitutes 72%. Buffalo meat is being developed as an export oriented commodity. Buffalo meat valued at more than Rs 90 crores was exported last year. Beef meat available in India is the meat of the water buffalo.

India is trying to build markets for animal meat. Indian buffalo meat is well accepted in Indonesia, South Africa, and the Commonwealth of Independent States (CIS). Exports of animal meat products continue to grow at a fast clip. The allocation for various programme in meat sector have been increased from Rs 1,384 crores to Rs 1,804 crores. There is also a proposal to establish National Meat Board to monitor and guide development programmes in Meat Sector. An estimated cost of Rs. 2 crores is proposed to be shared by GOI, Meat Exporters and Leather Exporters in the ratio of 50:25:25. It would also be important to review the State Animal Preservation Acts and suggest Amendments in view of the changing Animal Production and Utilization Scenario.

Indian meat exports started with as small a quantity as 2,000 tonnes in the year 1973-74 and increased to about 60,000 MT by 1987-88 valued at Rs. 88 crores. Meat and preparations have shown an annual compound growth rate (%) of 22.49 during 1991-98, and 11.65 during 1974-98.

Export performance of meat during the last decade indicates that buffalo meat export was 93% of total meat export meat in quantity terms and 88% in value terms. During the decade 1991-2000, buffalo meat exports increased by 163% registering an average annual growth rate of 16%. However, the growth rate was considerably higher during the first 5-year period 1991-95 as compared to the later 5-year period. In value terms buffalo meat exports increased by 560% during the decade as compared to 191% increase in sheep and goat meat exports.

Realization per ton of buffalo meat increased by 66% during first half of the decade while in the second half the increase was by 51%. And for sheep and goat meat increase in per ton realization was by 63% and 20% first and second 5-year periods of the decade.

The Indian meat and meat products are in great demand and their popularity is increasing as the livestock in India is reared naturally on green pastures and are not fed any growth promoters, hormones, antibiotics and chemicals, therefore, the meat is wholesome and safe for human consumption. Indian buffalo meat is exported frozen in boneless and de-glanded forms and is free from FMD. Thus, the Indian buffalo meat has established itself in the markets of South East Asia, Middle East and African countries.

The number of countries to which Indian meat is exported has increased from about 30 in 1994-95 to 50 in 1999-2000. During the year 1998-99 buffalo meat exports have declined by 13% in quantity and 5.2% in value terms compared to the previous year due to economic crisis in South Asian countries. In the following year buffalo meat exports have increased by 11% in quantity and 3.8% in value terms and in the current year (2000-01) the meat exports are estimated at Rs. 1,200 crores, an increase of 50% in value terms.

Impact of Lifting of Buffalo Meat Ban, by the Importing Nations

Meat exporters recovered lost grounds following the lifting of the ban on imports of Indian buffalo meat by Jordan, Egypt and Kuwait in 2002. The outbreak of foot and mouth disease (FMD) and Bovine Spongiform Encephoalopathy (BSE or mad cow disease) in Europe had forced these countries to review their protocols pertaining to imports of livestock products as a precautionary measure. As a result, a temporary suspension on issuance of permits for import of buffalo as well as frozen sheep meat from all countries, including India, was imposed towards end 2001.

According to AIMLEA, unlike in Europe or the US, there was no possibility of transmission of BSE from the country's meat exports because the livestock here are exclusively fed on natural pastures and crop residues. Feeding of meat and bone meal to ruminants is prohibited by Indian law and neither are the animals administered any hormones or growth promoters. As regards FMD, it is claimed that nearly almost 93 per cent of India's meat exports consisted of deboned and deglanded frozen buffalo meat. This is a risk free product with pH status below 6.0 brought about by compulsory chilling of the carcasses for a minimum of 24 hours at between two to four degrees Celsius. This reduce the risk of spread of FMD virus to insignificant levels.

Accordingly, as per a Report of APEDA, exports of frozen buffalo meat touched a record 2.88 lakh tonnes in 2000-01 (valued at Rs. 1,375.04 crore) against 1,67 lakh tonnes in the previous year. Of the 2.88 lakh tonnes exported last year, the major buyers included Malaysia (77,000 tonnes), Egypt (48,000 tonnes), Philippines (47,000 tonnes), the United Arab Emirates (41,000 tonnes), Iran (12,500 tonnes), Jordan (12,000 tonnes) and Kuwait (4,500 tonnes).

Constraints

The Indian meat export industry has not developed on pragmatic lines due to the controversies against meat sector development in the country.

Major constraints affecting Indian meat exports include: livestock disease situation (prevalence of foot and mouth disease), inadequate modern abattoir facilities, negative propaganda of some social groups against meat exports and lack of pragmatic slaughter policy for effective utilization of livestock resources.

A number of organizations are working for imposing total ban on meat exports. The consequences of ban on meat export would be more on buffalo production as buffalo meat export forms 92 per cent of total meat export. A marginal effect on goat production could be observed. However, sheep production gets adversely affected fetching decreased returns to sheep farmers. The demand for mutton (sheep), particularly in Northern India is limited and a ban on meat export would adversely affect sheep producers of Rajasthan, Uttar Pradesh, Haryana and Gujarat States. Consideration of ban on meat export is neither in the interest of the livestock producers nor in National interest. A study on the economics of meat export would be desirable to critically examine the benefits of meat exports in National interest and answer the frequent criticism-demanding ban on meat exports.

1. Limitation in Development of Livestock Markets

India has over 2,000 markets where livestock are traded. Livestock markets are under the jurisdiction of the state government although the direct operation and supervision would generally fall within the purview of the local bodies. There are a few privately owned markets. State Acts regulate marketing of agricultural produce and the marketing committees are responsible for implementing and enforcing the provisions of the Act.

The market for live animals in the country unfortunately has not developed on scientific lines. There are no separate markets for different species of animals. There are no separate enclosures for different species/animals. Brokers facilitate most of the trade. Vertical linkages between the processors/butchers and livestock producers are rare. Wholesale marketing margins amount to about 30% of the consumer price. Market facilities are generally inadequate and if available are poorly maintained. Weighbridges, ramp facilities for loading and unloading, feeding and watering and veterinary facilities are not available. Revenues generated under the act are supposed to be allocated to the markets for operations and improvement but not happening.

The development of live animal market information system is vital as data is a key input to informed planning and decision-making. Appropriate scheme should be formulated to strengthen the market facilities and introduce a scientifically managed market for conducting marketing operations as well as collecting proper data on livestock marketing.

2. Limitation in Slaughterhouses

There are 2702 slaughterhouses in the country, which are

recognized or authorized by local bodies. In addition a considerable number of animals are slaughtered in unauthorized places. A rough estimate indicates up to 50 per cent of animals slaughtered in any urban centre are from unauthorized slaughter.

Over the years, the facilities and hygienic conditions in most of the slaughterhouses have deteriorated. Compared to 1951, livestock population increased by about 62 per cent and human population increased by 134 per cent but the number of authorized slaughterhouses have not increased to meet the demand for meat production. The increased demand for meat is met either through overcrowding operations in the existing slaughterhouses operating at much higher capacity than feasible in the facilities or through unauthorized slaughter at many places. In both these situations not only meat hygiene is a casualty, increased pollution and adverse public reactions are observed. The existing slaughterhouses capacity in the country is unable to meet the growing public demand for clean and hygienic meat. This can be achieved by improving existing slaughterhouses to accommodate higher capacities and creating new slaughterhouses with modern facilities.

3. Shortcomings in Carcass Utilization Centres

During 8th Plan a Centrally sponsored scheme "Assistance to states for establishment of carcass utilization centres and hide flaying units" was implemented to effectively utilize dead animals and prevent environmental pollution through proper disposal of the animal waste materials. Continuation of the scheme on dead animal disposal in an appropriate manner is a necessity for recovery of hide and skins and prevents environmental pollution. A critical assessment of the established centres is necessary to evaluate viability and continuation of the scheme. A concerted effort is required for popularizing appropriate disposal of dead animals including their burial in the event of unsound economics of modern rendering for prevention of environmental pollution and livestock disease control.

Immediate Issues

There are a number of issues, which presently affect the

effectiveness of the meat processing industry in India. These are summarized below:

- Quality of animal herds, with the possible exception of milking herds, has received little attention resulting in low efficiency of food conversion and poor ratios of lean to fat.

- Animal feed, which is vitally important in tropical countries when dry season lack of rain reduces grass production, is very limited and distribution is poor.

- Lack of veterinary support reduces the level of both dairy and meat production. Fragmented herd ownership means farmers do not have sufficient funds to pay for services and drugs. Animal diseases are widespread with consequent effect upon meat and dairy quality.

- Abattoir management is poor and technologies employed are out of date. During the 1990s a number of initiatives have been introduced to improve the quality of abattoirs. However, overall there has been little improvement even in the licensed sector. In the unlicensed sector there has been none at all.

- Slaughter levels remain low, particularly among cattle and beef populations. This results in poor exploitation of the animal population.

- The use of byproducts is very limited which substantially raises the cost of the meat to the consumer. In particular, the leather industry is continually commenting upon a shortage of raw materials and very little use is made of the bones, blood or other byproducts.

- The lack of a chilled distribution system means that the majority of slaughterhouses are located close to the metropolitan centres.

Role That Rural Abattoirs Can Play

Condition of many of the urban slaughterhouses is far from satisfactory. Improvement of these slaughterhouses is facing more number of problems and efforts to establish new abattoirs or improving and expanding the existing abattoirs have not been successful. With large number of meat consumers living in urban areas, it has become necessary to critically examine the supply of hygienic meat to urban consumers. Production of meat in rural areas and transport of meat to cities has been viewed as an alternative with many added advantages.

For comparison of economics of rural and city abattoirs additional cost of meat in terms of chilling of meat and refrigeration transport need to be considered in case of supply of meat to city consumer from rural abattoir as compared to the transportation of animals to city abattoirs and re-transportation of hides and skins to tanneries and other waste to out side city limits. A preliminary estimate of the comparative economics has indicated that additional cost of sheep/goat meat due to chilling and transport from rural abattoir comes to Rs. 2.18 per kg as compared to the additional cost of Rs. 4.03 per kg when the same meat is produced in city abattoir. Similarly in case of buffalo meat additional from rural abattoir comes to Rs. 1.56 per kg as compared to Rs. 2.72 when produced from city abattoir. Thus, there is a possibility of reduction in consumer cost of meat and better prices to the meat animal producers. This would also be one of the major steps toward animal welfare, which save millions of animals from stress and pain while transportation from place of production to place of consumption.

Recommendations of the 4th Asian Buffalo Congress

Buffalo meat, branded for its high protein content with low fat, cholesterol and calories and best for processing of products like hotdog, sausage, corned beef and others, having excellent binding quality, needs to be developed further, especially Asian countries need to: -

⊙   Establish world-class abattoirs adopting international

standards such as HACCP and ISO 9002 systems.

⊙ Link meat buffalo farmers and meat processors.

⊙ Establish disease free zones recognized by the office International des Epizooties (OIE) around major meat export oriented units.

⊙ Establish a National Meat Development Board in India, which will ensure that meat for export is free of buffalo diseases.

⊙ Maintain close coordination between the Indian Ministry of Agriculture and the country's food processing industry to ensure that production, processing and marketing of milk and meat products follow international standards for food safety.

Suggested readings

Annual Report, 2000-01, Dept. of Food Processing, Ministry of Agriculture

Business Line, Oct. 20, 2001

FAO, 2001, Statistical database www.fao.org

Ministry of Food Processing Industries Web

Report of Agrifoods Exports Division

Review of past performance and assessment of future needs of the Report of the Working Group on Animal Husbandry, set up by the Planning Commission for the Tenth Plan Proposals

CHAPTER 13

Meat Production Potential of Rajasthan: An Overview

S. A. Karim

Central Sheep and Wool Research Institute,
Avikanagar- 304 501 (Rajasthan)

Livestock contributes sizably to agrarian economy and plays a vital role in livelihood security of small and marginal farmers and landless labours particularly in hot semiarid and arid zones and Himalayan foot hill region of the country. Consumption of beef/pork has associated religious taboos while meat from sheep and goat and to some extent buffalo is uniformly consumed by all segments of the society. Livestock have definite advantage over conventional agriculture in semiarid and arid environment as centuries of adaptation to perennial feed scarcity in the region have enabled them to survive under the harsh environment by developing efficient feed selection mechanism for meeting their DM and nutrient requirement. Although the average carcass yield of Indian sheep and goats is relatively low (10 kg) than the developed countries (25 kg) still higher population size and leaner carcass have definite commercial advantages.

The sheep, goat and buffalo contribute 0.16, 0.47 and 1.42 m MT meat to the Ag GDP amounting to 3.3, 9.6 and 28.9 % of total meat production (4.9 m MT meat) in the country (Ranjhan, 2003). Although cow slaughter with some riders is allowed in certain states of the country, still the contribution of beef to total meat production in the country is negligible more so in cow belt of north India where it is banned. All males surplus to the breeding requirement, spent ewes/ does, rams/bucks and buffalo are slaughtered for meat purpose.

Livestock sector is one of the most important component of agriculture in the country and contribute 27 % to Ag GDP: the importance of livestock in critical semiarid and arid region of the country is still higher. Conventional agriculture is always a gamble in critical zones due to prevailing harsh climatic conditions and recurrent drought hence the mobile (on hooves) livestock wealth of the farmers act as a mean of survival in the region (CSWRI, 2003). During summer, with depletion in surface vegetation, the farmers of the region with their total livestock migrate to neighbouring states seeking greener pasture and return back to their native tract in favourable season.

The cost of feeding in organized livestock sector accounts for more than 70 % of total inputs while the investment of the farmers in Rajasthan on this account is negligible (Karim, 2004). Livestock rearing in this part of the country is mostly under extensive range management on community rangeland although some resource full farmers supplement their productive stock with crop residues/concentrate that too mostly for dairy production. The surplus male buffalo calves, lambs and kids are solely maintained on extensive range management. Earlier sheep were maintained for wool as major produce while in the changed scenario of production, mutton has replaced the wool as the major produce. Likewise major source of income from goats is meat while milk, manure and hide are other produce of economic significance. Buffalo meat was earlier considered as low grade meat and the male surplus to the breeding requirement were not exploited for meat purpose rather starved to death without serving any purpose. However, off late, the situation in buffalo meat production has improved with adoption of organized feeding practices for quality meat production.

Population and Production Trends

The livestock population of the state as per the Basic Statistics of Rajasthan (2000) is presented in Table-1. The livestock species for meat production viz. buffalo, sheep, goat, pig and poultry have shown 25.9, 17.6, 12.4, 22.2 and 22.2 % increase in population during the period 1992-1995 indicating their multifaceted utility. The sheep and goat population is concentrated in arid region of the state particularly in the districts of Barmer, Bhilwara, Bikaner, Churu, Jaiselmer, Jalore, Jodhpur, Nagur while buffalo in Ajmer, Alwar, Bharatpur, Bhilwara, Chittorgarh, Dausa, Ganganagar, Jaipur, Jalore, Junjhunu, Nagur, Sikar, Udaipur, Hanumangarh and Karauli, the intensively cropped area or districts with large urban population (Table-2).

Sheep

Rajasthan has 14.3 million sheep population and 32 % off take rate hence 4.58 m sheep are slaughtered annually. Considering the average carcass yield of 10 kg, the state produces 45.8 m kg meat which selling @ Rs.100/kg contribute Rs.4580 m to Ag GDP. Although the state is richest in the country in sheep population and genetic resources, still their production figures do not match with the population trends primarily due to low individual production. The low production potential of sheep is ascribed to indiscriminate breeding and intermixing of breeds, under feeding and lower plane of nutrition for most part of the year, inadequate managemental support, poor health coverage and above all unorganized marketing structure for the sheep produce.

All sheep, whether males, spent ewes and rams surplus to the rearing requirement are marketed for meat purpose. The important sheep breeds of Rajasthan for mutton production are Marwari, Malpura, Jaiselmeri, Sonadi and Pugal. Earlier sheep were reared for wool as the main produce whereas in the changed scenario of sheep production mutton has replaced wool as the main produce. In the existing sheep production system 70 % of total income of the farmer is realized from sale of animals for meat purpose. Hence farmers have changed their sheep breeding plan by opting for large stature animals to get higher mutton production. Accordingly Kheri breed was evolved by mixing Marwari and Malpura to achieve higher weight and adapted animal to withstand the migration stress. Continued drought in the Rajasthan further changed the sheep breeding plan. Earlier surplus male lambs were disposed off for mutton purpose around 9-12 months of age weighing about 20-22 kg with average carcass weight of 10 kg. The sheep farmers preponed the migration in view of the drought condition and preferred to dispose off their surplus lambs just after weaning (at 3- 4 months of age) and accordingly introduced Patanwadi sires with proven record of higher milk yield and better pre- weaning gain.

The under *in vogue* production system of extensive range management the sheep has birth, weaning, 6-month, 9-month and 12-month weight of 3.0, 12.0, 18.0, 20.0, and 22.0 kg, respectively (Singh and Karim, 2004). Three decades of concentrated research in developing technologies for quality mutton production at Central Sheep and Wool Research Institute, Avikanagar, improved the growth rate of the lambs by 300 % achieving 35 kg finishing weight at 6 months of age (90 days weaning and 90 days feeding) under intensive feeding or 25 kg at age of 4 ½ months (60 days weaning followed by intensive feeding). The cost of feed input/kg gain in live weight under intensive feeding ranged from Rs.28-30 in different studies. Grazing with *ad lib.* concentrate supplementation of weaner lambs is another technology developed for commercial application wherein the lambs adopted from the farmers achieved 30 kg finishing weight at 6 months of age. The concentrate to roughage ratio in such feeding is stabilized at 75:25 and the cost of concentrate input/kg gain in live weight ranged from Rs.23- 25 (Singh et al. 2003)

The finisher lambs raised under intensive feeding or grazing with supplementation have dressing yield of 50 % with average carcass fat content ranging from 12- 13 % providing a carcass of choice quality fit for calorie conscious consumers. The spent ewes and breeding rams weigh on an average of 27 and 35 kg with dressing yield of 40%. Moreover the carcass from the spent animals is tougher in nature and has low preference by the consumers. The butchers however, sell the meat both from finisher lambs and spent sheep at uniform price thereby cheating the consumers. The low grade meat and offal from spent animals could be effectively utilized by converting to ready eat value added products (Nuggets, sausage, salami, kofta, kebab, pickle etc.). Moreover, the meat products developed had ready acceptability by the consumers. The concept of meat grading based on defined cuts with variable prices as prevalent in developed county is non-existent in India. However, for evaluation purpose the dressed carcass is split to leg, loin, rack, neck and shoulder and breast and fore shank cuts having cut yield of 32, 13, 13, 25, and 17 %, respectively.

Goat

The goat population in Rajasthan was recorded at 16.9 m in 1997 while in recent years because of the prevailing drought condition in the state their population has shown steady increase. Wider choices of vegetation in goat and their better adaptability to hot conditions were major factors contributing to the population trend. Considering that on an average 37 % goats are slaughtered annually having average carcass yield of 10 kg, the contribution of the species to meat production of the state is 62.53 m kg. The arid districts of the state viz. Barmer, Churu, Jaiselmer, Jodhpur, Nagur and Sikar have sizable goat population (Table-2). The goat breeds of Rajasthan are medium stature and leggy animals primarily raised for meat and milk production.

Goat rearing is traditional occupation of small and marginal farmers and landless labors in the state. The important goat breeds of Rajasthan for meat and milk production are Sirohi, Marwari, Parbatsari, Jhakrana. Like sheep, all goats surplus to the rearing requirement are slaughtered for meat production. Goats are reared for meat and milk as main produce while manure, skin and fiber are other products of economic importance. Medium sized goat breeds of Rajasthan maintained under extensive range management weigh 2.9, 11.0, 15.1, 18.5 and 20.0 kg at birth, weaning, 6-month, 9-month and 12-month of age, respectively. The spent does and breeding bucks on an average weigh 27 and 35 kg, respectively. Dressing yield of finisher and spent goats is reported to range from 46- 48 and 40-42%, respectively. The average lactation yield of the goats range from 150-200 kg whereas in dominant goat rearing western part of the state, goats are not routinely milked. The goat farmers in the region usually harvest manageable quantity milk for domestic consumption while majority of the milk is allowed to be suckled by the kids in a strategy to harvest better finishing weight and market price of the kids. Under farmers management, the kids surplus to the breeding requirement, are marketed for meat purpose around 9-12 months of age weighing around 19-21 kg with average carcass yield of 10 kg.

Goats maintained in institutional herd on protected natural rangeland under protocol of free grazing and top feed supplementation during lean summer season and cultivated pasture are able to achieve 20-21 kg finishing weight at 12 months of age. However, the growth response of kids on silvipasture is sizably higher: the kids maintained on two and three tier silvipasture are able to achieve 21 kg finishing weight at six month of age. The weaner kids maintained under the grazing with concentrate supplementation @ 1.5 % of BW are able to achieve 22 kg finishing weight at six months of age. As regards commercial kid production, it is recommended that the protocol of grazing with cheaper concentrate supplementation should be followed. Unlike intensive lamb production program, the protocol of intensive feeding in confinement is not effective for kids and hence is not recommended.

Rajasthan with sizable sheep and goat population has potential for commercial mutton and chevon production. Although the venture has commercial viability and profitability still entrepreneurs avoid investing in the trade due to social and religious taboos. Although the state is rich sheep and goat population still due to poor crop production the state is deficient in energy and protein supplements for intensive feeding programs whereas the regions of the state rich in crop production do not have viable small ruminant population. Hence it is suggested that entrepreneurs or progressive farmers may procure weaner lambs from dominant sheep producing zones of the state and take up intensive feeding program near some metropolitan city or exit port and market the finisher lambs in domestic or international market.

Buffalo

Rajasthan had 7.75 m buffalo population in 1992 which increased to 9.76 m by 1997 registering a very high 26 % growth rate during the

---

period: the projected figure for 2003 is 10.75 m. The extraction rate of buffalo is 9 and 41% in Indonesia and Egypt respectively whereas it is relatively low in India (11%) with clear possibility of increasing it to 30% without affecting the milk production or draught potential of the species. India has already attained sizable gain in green, white and blue revolution and it is envisaged that in decades to come, pink revolution will not be far behind and buffalo meat will sizably contribute in this direction. The buffalo meat can effectively meet the requirement of red meat as it is consumed by all religious segment of the society. In the existing extraction rate about one million buffalo are slaughtered annually and at conservative estimate of average 38 kg dressed carcass weight, 38 m kg buffalo meat is contributed to Ag GDP of the state. Moreover out of the total dressed meat export from India the share of buffalo meat is nearly 82%. At the present, 10-15% population consumes buffalo meat whereas it is envisaged more consumers will be attracted to the meat due its cheaper price and quality attributes. The districts of Rajasthan having sizable buffalo population are Jaipur, Alwar, Bharatpur, Sikar and Nagaur in that order (Table-2) where intensive buffalo meat production programmes can be taken up with commercial viability.

Male buffalo calves surplus to the breeding requirement are eliminated in their infancy in cruel process of virtual starvation. The elimination of the male calves is a compulsion for the farmers as they compete for meager feed resources and suckle some milk, which can be marketed by the farmers. Moreover virtually there is no market for the male calves and they are disposed off at an early age that too in throw away price. Entrepreneurs can enter buffalo calf fattening programme in a big way and harvest profit. The weaner male calve can be purchased from the farmers at yearling age and quarantined, dewormed and fattened under intensive feeding on high energy and protein ration to gain about 100 kg in three months period. The finisher calves can be slaughtered hygienically in abattoirs meeting domestic and international specification for the respective markets.

Organized feeding of buffalo calves on a 60:40 roughage concentrate and in the form of complete feed can sustain a growth rate of 700- 800 ADG with 5- 6 feed conversion ratio and feed efficiency ranging from 15-20%. The cost of feed intake/kg gain in live weight ranged between Rs.16-18 for different feed combinations. Dressing yield of the calves ranged between 47-52% on pre slaughter weight. Carcass fat content in all these feeding regimes was 11% providing desired carcass fat content even for calorie conscious consumers (Sharma *et al.*, 1995).

Pig

The pig population of Rajasthan was 0.25 m in 1992 and increased to 0.30 m by 1997 and would have touched 0.36 m in recent estimates. Contribution of pork was hardly 3% to total meat production of the country in the year 1975 and increased to 10% by 1995 and the quantum of increase was highest among the livestock species reared for meat production. One of the major factors contributing to the phenomenal increase in pork production in the country was the higher litter size and prolificacy of the species. Pig rearing in Rajasthan is in unorganized sector and the animals are raised mainly as scavengers by poor and backward community. Moreover majority of population avoid eating pork based on religious ground whereas it serves as source of nutritional and food security for the poorest of the poor in the state. The districts with sizable pig population are Jaipur (0.34 m), Bharatpur (0.27 m), Ajmer (0.21 m), Alwar (0.20 m), in that order.

The pigs are fast and prolific breeder with two litters in a year and 10-12 litter size at birth. The litter size at weaning depends on care in management and feeding of pre weaner piglets and lactation yield of the sows. The average carcass yield of the finisher pigs average to 75%. The information on growth profile and carcass traits of the pigs raised as scavenger are however scanty.

Table-1: Livestock census of Rajasthan (in millions).

Livestock category	Population 1992	Population 1997	Growth rate (%)	Population by 2003**
Cattle	11.59	12.16	4.91	12.89
Buffalo	7.75	9.76	25.9	10.75
Sheep	12.17	14.31	17.62	17.33
Goats	15.06	16.94	12.44	19.47
Camels	0.73	0.66	-8.55	-
Pigs	0.25	0.30	22.21	-
Poultry	3.00	4.38	22.21	-

Basic statistics Rajasthan, 2000 **Projected figure

Table-2: District wise livestock population of Rajasthan (in 00000)

District	Cattle	Buffalo	Sheep	Goat	Camel	Pig	Poultry
Ajmer	4.81	3.11	7.14	5.63	0.04	0.21	14.99
Alwar	2.04	7.64	1.09	4.95	0.20	0.20	2.42
Banswara	5.80	2.26	0.24	3.40	0.02	0.00	4.22
Baran	4.19	1.79	0.18	1.85	0.02	0.16	0.49
Barmer	5.41	0.93	15.11	18.68	1.14	0.05	0.09
Bharatpur	1.04	5.28	0.09	1.50	0.05	0.27	0.08
Bhilwara	7.52	3.79	8.45	7.05	0.09	0.12	1.36
Bikaner	5.59	1.07	11.47	6.42	0.06	0.01	0.23
Bundi	3.21	2.29	0.87	3.24	0.05	0.09	0.56
Chittorgarh	6.27	3.95	1.34	4.89	0.06	0.06	1.33
Churu	3.28	2.07	6.42	8.36	0.73	0.03	0.21
Dausa	1.43	3.48	0.48	2.28	0.09	0.11	0.21
Dholpur	0.75	2.57	0.12	0.97	0.00	0.06	0.10
Dungerpur	4.01	1.76	1.43	2.96	0.04	0.00	2.35
Ganganagar	4.11	3.25	3.50	3.01	0.28	0.05	1.16
Jaipur	4.43	7.74	3.47	6.93	0.21	0.34	1.25
Jaiselmer	3.10	0.01	12.07	8.93	0.42	0.00	0.12
Jalore	2.87	3.15	7.10	4.66	0.13	0.10	0.16

District	Cattle	Buffalo	Sheep	Goat	Camel	Pig	Poultry
Jhalawar	4.30	2.19	0.16	2.40	0.01	0.09	0.85
Jhunjhunu	1.22	3.52	2.23	5.11	0.30	0.05	1.30
Jodhpur	6.69	2.09	15.60	12.96	0.43	0.00	0.37
Kota	2.68	1.81	0.29	1.56	0.03	0.15	0.53
Nagur	4.86	4.37	11.71	10.89	0.28	0.08	0.36
Pali	4.16	3.13	13.69	6.06	0.13	0.13	0.43
Rajsamand	3.09	2.22	2.19	4.78	0.05	0.04	0.30
Sawaimadhopur	1.87	2.33	0.78	2.71	0.07	0.13	0.26
Sikar	2.14	4.59	3.06	7.77	0.27	0.09	0.82
Sirohi	2.14	1.28	3.03	2.97	0.06	0.01	0.48
Tonk	3.36	2.51	3.10	3.28	0.02	0.13	0.51
Udaipur	9.70	4.57	2.44	8.50	0.09	0.00	4.32
Hanumangarh	3.12	3.33	2.92	2.43	0.58	0.04	0.59
Karauli	1.59	3.35	0.44	2.14	0.07	0.12	0.22

Basic statistics Rajasthan, 2000

Table-3: Important sheep and goat breeds of Rajasthan for meat production.

Sheep	Goat
Malpura	Jamunapari
Sonadi	Kutchi
Marwari	Sirohi
Jaiselmeri	Marwari
Pugal	Jahkrana
Kheri	

References

CSWRI (2003). Annual Report, Jaivigyan Project on Household Food and Nutritional Security Improvement in Migratory Sheep Production Programme for Tribal Farmers of North-West, Central Sheep and Wool Research Institute, Avikanagar.

Ranjhan, S. K. 2003. Buffalo meat production in India – opportunities and challenges. Interaction Seminar on Meat Production and Export Potential: Impacts of Globalization and Quality Concepts, 17th September, CIRG, Makhdoom, pp. 33-41.

Singh, V. K., Karim, S. A. and Sureshkumar, S. (2003). Prospects of mutton and chevon production and export from India. Symposium on "Impact of globalization on Indian meat industry". 11-12 December, pp. 27-31, Rajiv Gandhi College of Veterinary and Animal Sciences, Pondicherry.

Karim, S. A. (2004). Nutrition and feeding management of sheep for mutton production. Lead paper *In:* National Seminar "Opportunities and Challenges in Nutrition and Feeding Management of Sheep, Goat and Rabbits for Sustainable Production" February 10 -12, Central Sheep and Wool Research Institute, Avikanagar, pp. 88-99.

Sharma, D. D., Sehgal, J. P., Singhal, K. K. and Ghosh, M. K. (1995). Fattening of growing male buffalo calves for quality meat production. Project report, NDRI, Karnal.

Singh, V. K. and Karim, S. A. (1994). Improvement in sheep production through technological interventions. Lead paper *In:* National Seminar "Opportunities and Challenges in Nutrition and Feeding Management of Sheep, Goat and Rabbits for Sustainable Production" February 10-12, Central Sheep and Wool Research Institute, Avikanagar, pp. 216-223.

CHAPTER 14

Blue Print for Indian Buffalo Meat Industry in Different Agro-Climatic Regions

T.R.K.Murthy and M. Muthukumar

National Research Centre on Meat, CRIDA Campus,
Hyderabad -500 059 (AP)

India is one of the most important livestock raising countries with world's bovine population of 219.6 million cattle and 94.1 million buffaloes (FAO, 2001). Livestock sector is an important component of the Indian agriculture, ranking after crop production from the viewpoint of its contribution to the gross national product (GNP) as well as employment potential in rural areas.

Meat finds a prime place in the palates of about 70% people (80 crore) in India. The annual total meat production is estimated at 4.9 million tonnes, standing eighth in world's meat production (FAO, 2001). Meat and byproducts have played an important role in man's development from the earliest time. Meat sector has large inbuilt potential for generating direct and indirect employment in wide ranging ancillary activities such as commercial breeding, manufacture of slaughter and processing equipment, animal byproducts based industries relating to hides, bones and others.

Buffalo Production in Different Agro-Climatic Regions

India being the home tract of buffaloes houses about 94 million heads. There are about 7 recognised breeds distributed in different agro-climatic regions. The buffaloes found in the northern region of the country are massive built and good in milk production. The average body weight of adult ones ranges from 450 to 650 kg. The important breeds of this region are Murrah and Nili. Medium to massive frame bodied buffaloes are found in the western region of the India.

Jaffarabadi and Surti are the two major breeds of this region. The body weight of buffaloes of this region ranges between 300 and 600 kg. Central parts of nation hold breeds like Nagpuri, and Mand. When compare to northern region, these animals are of lighter type and moderate milk yielders. Toda is the main breed of buffalo exists in the southern region of the country. It is of less economic importance (Banerjee, 1991).

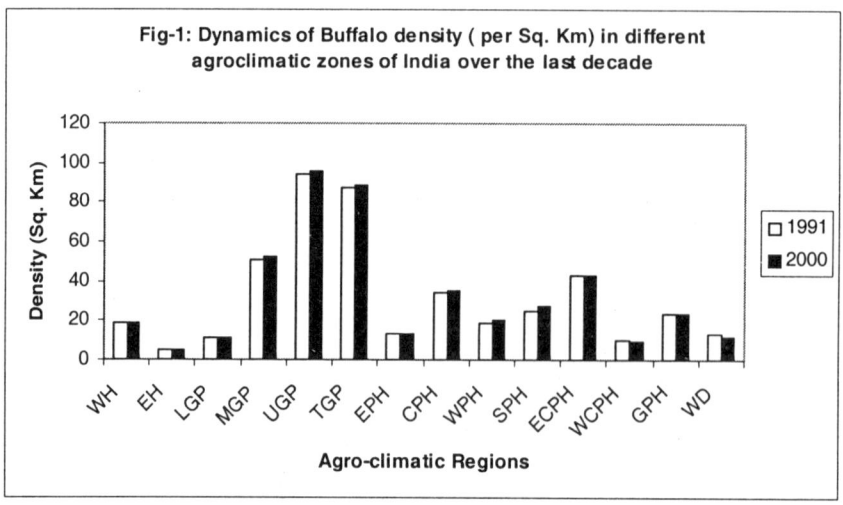

Fig-1: Dynamics of Buffalo density (per Sq. Km) in different agroclimatic zones of India over the last decade

The entirety of the milieu of a region (climate, terrain, soil, ecology, crop production) determine which species or species mix of animals will be reared by people there. Moreover, keeping one or the other species of livestock or mix of them is determined mostly on the locally desired type of livelihood security and less by the productivity or products for the market. The concentrations of buffaloes exist in the Gangetic plain regions and to some extent the East Coast plain and hill region (Table 1). The density of buffaloes in these regions ranges between 51 and 96/Sq.Km. (Fig.1). The states coming under this region are Punjab, Haryana, Western and Central U.P. In these regions, the area under fodder crops production (13%) and productivity of the crops (14 tonnes/ha) are more. The dry fodder availability in this region is more than the requirement (118.4%). The availability of the concentrates also higher in these regions. All the above said factors contribute for the higher density of buffaloes in these regions (Sastry, 2000).

Table-1: Highest and lowest priority regions of different livestock species

Agro-climatic Regions	Species Priority				Region description	Regions products
	Very high	High	Low	Very low		
WH		S	CBG	P	Sheep-goat-cattle	Wool-meat
EH		CP	G	BS	Cattle-goat-pig	Meat-Draft
LGP	CGP	S	B		Cattle-goat-pig	Draft-meat
MGP	P	CBG	S		Pig-Bovine-goat	Draft-meat
UGP	BP	G	CS		Buffalo-pig-goat	Milk-meat
TGP	B	SP	CC		Bufalo-sheep-goat	Milk-meat
EPH		C	BSGP		Cattle-All others	Draft-meat
CPH	SG	B	CP		Sheep-goat-buffalo	Wool-meat
EEPH			CBSG	A	All Livestocks	Suppl.
SPH	S		CBGP		Sheep-All others	Meat-Suppl.
ECPH		BS	CGP		Buffalo-Sheep	Milk-Suppl.
WCPG	C	G	P	BS	Cattle-Goat-Pig	Draft-meat
GPH			CBSG	P	All Livestocks	Suppl.
WD	S		BG	CP	Sheep-others	Wool-meat
I			GP	CBS	Low livestocks	Suppl.

(Source : Sastry, 2000)

B-Buffalo   C-Cattle   G-Goat   S-Sheep   P- Pig

WH - Western Himalayan   CPH - Central Plateau & Hills
EH - Eastern Himalayan   WPH - Western Plateau & Hills
LGP - Lower Gangetic plains   SPH - Southern Plateau & Hills
MGP - Middle Gangetic plains   ECPH - East coast Plateau & Hills
UGP - Upper Gangetic plains   WCPH - West coast Plateau & Hills
TGP - Trans Gangetic plains   GPH - Gujarat plains and Hills
EPH - Eastern Plateau & Hills   WD - Western dry

The agro-climatic regions viz. Western Himalayas, Eastern Himalayas, Western dry, Gujarat plain Hills, Eastern Plateau and Hill, Western Plateau and Hill, West coast Plain and Hills have low density of buffaloes (Fig. 1). The common features of these regions are less agricultural development, extreme climate (too arid, hot or cold), difficult terrain (undulating/upland) condition. Hence buffaloes population is less (Sastry, 2000).

Possible Development Priorities

The commodities priorities in buffalo research in different in different states has been presented in table 2. The success of any livestock enterprise greatly depends on the continuous supply of good quality nutritious feeds. Since inadequate feed of poor quality tend to be a limiting factor in exploiting the genetic potential of livestock. Green fodder is the crux of the problem of livestock production in India. Green fodder shortage is acute in all regions and is going to continue to be so if the present conditions prevail. Grazing is probably the only feasible source of livelihood for people living in rural areas of India. But most of the grasslands are over grazed, have poor soil and are dominated by coarse grass. It is great responsibility of livestock managers to ensure that this resource base is preserved for future generations. Mixed cropping system contributes total quantity of buffalo meat production (Sere and Steinfeld, 1996) (Table 3).

Table-2: Commodity priorities in buffalo research in different states (Percent)

State	Milk	Meat	Hide & Skin	Draught power	Total
North					
Haryana	98.18	0.00	0.35	1.47	100.00
Himachal Pradesh	98.72	0.00	1.27	0.02	100.00
Jammu & Kashmir	98.88	0.00	0.60	0.52	100.00
Panjab	99.38	0.00	0.49	0.13	100.00
Uttar Pradesh	96.29	1.48	0.52	1.70	100.00
South					
Andhra Pradesh	93.77	2.62	0.78	2.83	100.00
Karnataka	97.21	0.57	0.68	1.54	100.00

Kerala	52.93	14.17	2.88	30.01	100.00
Tamil Nadu	97.78	1.41	0.58	0.24	100.00
West					
Goa	89.13	6.15	0.58	4.13	100.00
Gujarat	98.12	0.20	1.36	0.32	100.00
Madhya Pradesh	96.74	0.20	0.48	2.58	100.00
Maharashtra	93.89	4.74	1.32	0.06	100.00
Rajasthan	98.02	0.45	0.95	0.58	100.00
East					
Bihar	84.63	3.37	2.45	9.55	100.00
Orissa	54.81	0.00	2.06	43.12	100.00
West Bengal	74.19	0.00	2.17	23.64	100.00
North East					
Arunachal Pradesh	0.00	0.00	7.42	92.58	100.00
Assam	79.87	0.79	1.23	18.12	100.00
Manipur	48.75	36.89	2.65	11.71	100.00
Meghalaya	54.50	12.01	1.07	32.42	100.00
Mizoram	0.00	15.82	0.90	83.28	100.00
Nagaland	0.00	57.05	3.50	39.45	100.00
Sikkim	76.16	23.18	0.66	0.00	100.00
Tripura	0.00	45.33	5.19	49.48	100.00
India	93.21	2.90	1.02	2.88	100.00

(Source: Birthal, *et al.*, 2002)

The waste lands which account for about 328 million ha should be converted into fodder bank by suitable production system like silvi pasture (Ragnekar, 1990). This not only provides good amount of nutritious fodder but also avoids wind erosion. The national commission on agriculture way back in 1976 emphasized that fodder is of crucial importance and recommended identification of plant species suitable for waste lands and development of technologies for improving their productivity.

Table-3: Quantity (1000 t) and percent of global livestock products produced by the three major systems.

Product	Grazing		Mixed crop-livestock		Industrial	
	1000 t	Percent	1000 t	Percent	1000 t	Percent
Beef and veal	12, 289	23.4	34,249	65.1	6055	11.5
Buffalo meat	0	0.0	2,652	100.00	0	0.0
Sheep and goat meat	2981	30.0	6860	69.0	100	1.0
Pig meat	685	1.0	42,821	59.8	28,163	39.3
Poultry meat	796	1.8	10,469	24.2	31,967	73.9
Eggs	524	1.3	12,289	30.8	27,071	67.9
Dairy milk	38,775	8.2	4,34,332	91.8	0	0.0

(Source : Sere and Steinfeld, 1996)

By considering densities and holding pattern of individual species, Sastry (2000) suggested development of cattle followed by buffalo in the vast interior areas of the peninsular India and buffalo followed by cattle all along the East coast of India should be received more priority.

Economic Importance of Buffaloes

About 11 million buffaloes are being slaughtered annually contributing 1.42 million tonnes of meat accounting for 44.5% of total world buffalo meat production (FAO, 2001) (Table 4). Except few states like Kerala, West Bengal, Assam and some north eastern states, the cow slaughter is legally banned even for domestic consumption. With imposition of gradual ban on cow slaughter, buffaloes are emerging as prosperous meat animals. Because of their ability to thrive on harsh conditions and low quality rations, the preference for buffaloes is ever increasing. There is a good demand for Indian buffalo meat because of its low fat content and less cholesterol. Moreover, higher content of myoglobin imparts desired meat colour required for processed product.

Table-4: Percentage contribution of livestock and meat by India to the world

	% of total animal	% of total meat
Cattle	14.91	2.44
Buffalo	51.74	44.50
Sheep	4.34	2.31
Goat	17.15	12.60
Pig	1.65	0.54
Chicken	4.24	1.05
Total	11.01	2.13

(Source : FAO, 2001)

Potential for Higher Meat Production

It has long been felt that meat production potential of the country has not been adequately exploited. If done, this would help in improving the nutritional level of the people, increasing the employment and cash flow in rural areas; and, using the export potential for increasing the foreign exchange earnings. Based on the targeted minimum requirement of 20g per capita per day for animal protein, the estimated demand for the present population would be 7.7 million tonnes meat as against the present production of 4.9 million tonnes (Table 5). Though India has the world's largest number of livestock population, the percentage of cattle and buffaloes slaughtered in relation to their population is less than 3 percent and 11 per cent respectively (Table 6). Hence many more number of meat animals are likely to be available for slaughter. Moreover, if the male buffalo calves which are presently neglected and allowed to die soon after birth, reared for meat, they could become a valuable meat resource. The fact that livestock in India are reared naturally on green pastures and are not fed with any growth promoters, hormones, antibiotics and other chemicals, creates good demand in international market. Low production cost also makes Indian meat more competitive in overseas market.

Table-5: Projected meat production (million tonnes) in 2020

	India		World		Developing Countries		Developed countries	
	1997	2020	1997	2020	1997	2020	1997	2020
Buffalo & Cattle	2.9	5.1	64	82.4	22.4	46.6	32.2	35.8
Pig	0.5	0.9	92.6	121.9	386	81.1	37.5	40.7
Sheep & goat	0.7	1.1	13	16.2	6.4	11.8	3.6	4.4
Poultry	0.54	1.2	51.5	82.9	21.5	41.8	25.9	34.3
All meats	4.7	8.3	221	303.3	88.9	188.2	99.3	115.2

Table-6: Extraction rates of buffaloes in India and some other countries

Country	Population	Slaughtered	Extraction rate (%)
Asia	164.9	20.2	12.3
India	93.8	10.3	11.0
Pakistan	22.7	4.0	20.7
China	22.6	3.6	11.5
Egypt	3.2	1.7	41.1
Indonesia	2.8	0.2	9.0

Indian Meat Industry

The very word Meat is taboo in this country and this social prejudice is the root of many of the problems that this age-old industry faces in it's healthy growth, scientific improvement and contribution to the speedier growth of livestock industry. The production of meat in India both for domestic and export is around 2.2% of world's total production of 210 million tonnes (FAO, 2001).

*Domestic market :*The consumption of meat is mostly in the form of fresh, without any processing immediately after slaughter. Frozen and further processed meat consumption is restricted to metro cities. Of the meat consumed in this country, mutton and chevon though expensive have enjoyed a greater appeal than the cheaper beef or buffalo meat due to certain religious prejudices and variation in food habits. The domestic meat requirement is mostly sourced from municipal slaughterhouses and retail outlets. The slaughterhouses, except for a few private ones, are operated as service abattoirs, where butchers slaughter the animals with their personnel for a fee. The marketing of meat is quite complex and includes a number of middle men, who in most cases take the major share of total price charged to the consumer. Current marketing system involves mobilization of buffaloes from large number of small holders and assembling in local/ regional markets.

There are about 3,500 recognized traditional slaughterhouses. The existing condition in the majority of these traditional slaughterhouses is far from satisfactory. Most of them are lacking basic facilities like water, electricity, drainage, over head rail and waste disposal. The slaughter and dressing is carried out on the open ground. Few have facility to dress on the rails. Carcasses are exposed to heavy contamination from soil and faceal materials. There is inadequate or no ante and postmortem examination. As a result clean meat production has remained a distant dream for the country and the public health risks are grave. Moreover the recovery of byproducts is very poor and enormous quantities of by products are wasted. Hence, these abattoirs urgently need up-gradation and modernization with a foresight on future demand of meat for both domestic and export

Export market: Meat for export is being sourced from 12 modern integrated meat complexes and 50-60 small meat processing units. Indian meat is normally exported either fresh chilled or frozen (deboned and deglanded) form. Egypt, Malaysia, UAE, Philippines, Iran, Jordan, Kuwait, Mauritius and other Arab countries as well as some countries of east and South Asia are the main importing countries for meat from India. About 70% is contributed by buffalo meat. Considering the ever growing demand for sheep/goat meat in India, the export of these product is allowed to a limited extent. However, the export of buffalo meat is at present under open general license (OGL). There has been a rapid growth in export of meat during the

Table-7:   Export of meat and meat products

Meat and meat products	Annual export value (Crore Rs.)							
	1995-96	1996-97	1997-98	1998-99	1999-00	2000-01	2001-02	2002-03
Buffalo meat	555.00	619.04	729.30	691.29	706.42	1375.04	1144.42	1305.45
Sheep meat	56.42	71.04	62.66	78.48	90.89	78.16	33.07	39.95
Poultry products	26.21	81.44	88.84	51.73	54.25	86.18	130.07	156.47
Animal Casings	0.08	12.27	11.95	13.59	11.69	12.29	9.63	14.27
Processed meat	4.07	4.44	2.22	2.99	4.58	1.58	1.29	4.8

(Source : APEDA, 2004)

last 15 years (Table7). However, it is less than 5% of the total meat production.

The major meat production for export is largely concentrated in Uttar Pradesh, Maharastra and recently in Andhra Pradesh and West Bengal. The major exporters are M/s Allana Sons Ltd., Mumbai, Al-Kabeer Exports Ltd., Hyderabad, Hind - Agro Industries Ltd., Aligarh. The exporting establishments are approved by concerned regulatory authority of state/central government and also accredited by the importing nations official agencies. These modern meat complexes are having facility for production of safe and wholesome meat and provision for efficient recovery and processing of byproducts. The major components of these establishments includes lairage (for resting animals prior to slaughter), stunning pen (making animals to unconscious state), mechanized slaughter hall, edible and inedible (rendering) by products processing section, chiller (to bring down the carcass temperature), meat processing section, freezers, cold storage and effluent treatment plant. Under the regime of WTO, it is essential that meat industries should operate in accordance with the standards laid by Codex Alimentarius Commission. Few of the export houses have implementation and accreditation of HACCP and ISO. Most of them have in house laboratory to monitor raw and finished product quality.

Value addition: As most of the buffaloes are slaughtered at the end of their productive economic life, the meat from such animals is generally tough and less palatable. Cheap and effective methods, which can be easily adaptable at household or commercial scale, have been developed at various laboratories to improve functional properties and palatability of the spent animal meats. The technology includes ageing, use of calcium chloride, papain, ginger extract, cucumis, polyphosphates, electrical stimulation etc. Moreover, development of value added products would facilitate better utilization of low value cuts and by products in addition to beneficial disposal of tough spent animal meat. At present very small quantities of processed meat products are being produced which is meager in relation to enormous quantity of available buffalo meat at cheaper prices. Further the demand for convenience meat based fast food is ever increasing due to rapid industrialization, higher standards of living and increasing number of working women. There is a tremendous potential for rapid growth in this sector, provided dissemination of knowledge generated out of extensive work down at various research laboratories.

By-products utilization: By products form an important component of slaughter of animals. The yield of by products ranges between 50-55% of live weight of buffaloes. Efficient recovery and utilization of by-products and wastes is the lifeline of meat industry. Revenue generation from by products goes up tremendously when they are processed into high value products. A few units in India are producing value added by products like di-calcium phosphate, ossein, gelatin, bone chips, bone meat, bone girt that have export value. There is a vast potential for utilizing blood, liver and certain glands for pharmaceutical purpose. Intestines could be processed to casings, catgut etc. Organs like liver, heart, lungs, kidneys, rumen, reticulam are more suitable for pet foods manufacturing. Processed and finished leather and leather products fetch good price. As a matter of fact profits from a meat factory comes from by-products when utilized properly.

Some Action Areas for Indian Buffalo Meat Industry

1. Appropriate wasteland development programme is crucial to meet acute fodder shortage.

2. Buffalo development programme should give more thrust in the interior peninsular India and East Coast regions of India.

3. There is an urgent need to upgrade and modernise the existing slaughterhouses for production of hygienic and wholesome meat. Rural and semi modern abattoirs would be appropriate for Indian situation while integrated meat processing complex are necessary for export.

4. The productivity (carcass yield) of buffaloes in India is lower due to indiscriminate breeding, inadequate feeding and animal husbandry practices. Necessary efforts should be put in to harmonize the production with germplasm conservation.

5. Salvage of male buffalo calves from early death and allowing them to reach live weight of 300 or 400 kg through adequate feeding will undoubtedly increases meat production.

6. Establishment of National meat development board for promoting development of organized meat sector to sustain animal production.

7. Issues relating to cruelty meted out to the animals need to be addressed diligently. Provision for humane method of slaughtering, transporting animals in stress less condition should be made an essential part of meat production.

8. There is a need to organize the collection centre for recovering and processing of byproducts and wastes generated in small and scattered slaughtering units.

9. Disease free zones must be strengthened and marketing of animals to be regulated appropriately.

10. Sound R & D programmes preferably need based should be evolved by various   R& D institutions in collaboration with Industry. Cost reduction, convenience, better quality, functionality, health concern, variety, long shelf life, byproducts utilization, industrialization of the process should remain the major objectives of these research endeavors.

Conclusion

The challenges before the Indian meat industry are adequate feed and fodder for livestock, improvement in hygienic status of meat, disease control in livestock, efficient recovery and utilization of the byproducts besides value addition to meat. Strong political will, defined government policies, integrated system approach coupled with development and fair execution of social mission would result in sustainable meat sector.

References

APEDA, 2004.Agricultural processed food Export Development Authority.

Banerjee, G.C., 1991. A Text book of animal Husbandry. Oxford & IBH publishing Co., Pvt. Ltd., New Delhi.

Birthal, P.S., Joshi ,P.K and Anjani Kumar, 2002. Assessment of research Priorities for livestock sector in India. Policy Paper 15.National Centre for Agricultural Economics and Applied Research, New Delhi.

FAO, 2001. Food and agriculture organization of United Nations. FAOSTAT database.

Rangnekar, D.V., 1990. Hand book of animal husbandry. ICAR publication, New Delhi.

Sastry, N.S.R., 2000. Regional considerations for appropriate livestock Development Strategies in India. Journal of Indian veterinary Association, Kerala.   Vol.5(3):16-29.

Sere, C. and Steinfeld, H., 1996. World Livestock production Systems, current status. Issues and trends, FAO Animal Production Health Paper, Rome, pp 127.

CHAPTER 15

Rearing and Feeding of Buffaloes under Middle Gangetic Plain Regions of India

D.N. Verma and Udeybir Singh

Narendra Deva University of Agriculture & Technology,
Kumarganj, Faizabad - 224 229 (UP)

The buffalo is a multi-purpose animal as it provides meat, milk and draught it besides organic waste like dung and urine for fuel and manures, dead carcass and bones for feeding of simple stomached animal like poultry and pigs and hide, horns, hooves and tallow for industrial uses. These are the some importance of buffalo for the society. Buffalo is the main pillar of India dairy industry and contributes about 50% of the total milk produced. Asia is the main habitat of buffalo which have about 161.4 millions (97.2 per cent) of the total world buffalo population (166.4 millions), whereas India has 94.13 millions buffaloes (56.6 per cent).

Buffaloes are preferred over cattle because of their superior quality of milk (high fat and protein content), better efficiency in utilization of nutrients from poor quality fibrous tropical feeds, relatively better disease resistance capacity and adaptability to tropical climates.

General Information about Buffalo

Scientific name	*Bubalus bubalis*
Type	Swamp & Reverine
Important breeds	Murrah, Nili-Ravi, Surti, Mehsana, Jaffarabadi and Bhadawari

Average daily wt gain	400 g.
Age at puberty	15-36 (Avg. 24) months
Male age of sexual maturity	30 months
Oestrus cycle length	21 days
Duration of estrus	20 hours
Gestation period	305 days
Calving season	July-January
Body temperature	99-102^0 f
Respiration rate	16-18 per minute
Pulse rate	40-45 per minute
Jaw movement	40-70 per minute

Animal Husbandry Regions in India

Animal husbandry regions in India are more or less super imposed with the agricultural regions of the country. The regions are made on the basis of the type and quality of the livestock as adopted to rainfall, temperature, soil and the feed of the regions. There are five regions on the basis of traditional animal husbandry practices of old days, which were greatly influenced by agro-climatic conditions. These are :

1. Temperate Himalayan Region
2. Dry North Region
3. Wet Eastern Region
4. Southern Region
5. Coastal Region

However as per Planning Commission of India there are 15 distinct agroclimatic Zone in India, out of which middle Gangetic plain region includes eastern U.P.

Table-1:   Floor Space Requirement for Buffaloes

	Covered area (m²)	Open paddock (m²)
Buffalo	4	8
Young calves	1	2
Older calves	2	4
Pregnant animal	12	12
Height of the shed	175- 220 cm.	

Table-2:   Requirement of energy (TDN, g) for maintenance and gain of growing buffaloes.

Live wt.	g/w 0.⁷⁵kg	g/g gain	Reference
100-200	34.5	2.76	Kearl (1982)
300	34.5	3.04	Kearl (1982)
400	34.5	3.58	Kearl (1982)
100	38.8	1.05	Sen *et al.* (1978)
200	38.8	1.41	Sen *et al.* (1978)
300	38.8	2.02	Sen *et al.* (1978)
100	29.5-34.2	1.25	Pathak and Verma (1993)
200	29.5-34.5	1.45	Pathak and Verma (1993)
300	29.5-34.2	1.65	Pathak and Verma (1993)
100	47.0	0.78	Udeybir and Mandal (2001)
200	39.9	1.56	Udeybir and Mandal (2001)
300	36.9	1.95	Udeybir and Mandal (2001)

Table-3: Protein (CP*/ DCP**, g) requirements for maintenance and gain of growing buffaloes.

Live wt.	g/w 0.75kg	g/g gain	Reference
100	5.15*/2.53**	0.42*/0.35**	Kearl (1982)
200	5.41*/2.54**	0.51*/0.41**	Kearl (1982)
300	5.23*/2.54**	0.57*/0.43**	Kearl (1982)
400	5.24*/2.54**	0.54*/0.40**	Kearl (1982)
100-400	2.95**	0.24**	Pathak and Verma (1993)
100	3.88**	0.28**	Sen *et al.* (1978)
200	3.88**	0.34**	Sen *et al.* (1978)
100	7.64*/5.05**	0.44*/0.27**	Udeybir and Mandal (2001)
200	6.12*/4.18**	0.51*/0.30**	Udeybir and Mandal (2001)
300	5.98*/3.60**	0.48*/0.28**	Udeybir and Mandal (2001)
Adult	2.48**	-	Kurar and Mudgal (1977)
Adult	2.85**	-	Ranjhan and Pathak (1979)

Feeding of New Born Calf

The main objective of good management and balanced feeding of buffalo calves is to obtain optimum growth rate in keeping with their genetic potential so that they can attain early maturity. For the rearing of buffalo calves, two systems are followed :

1. *Suckling the dam :* In this system calf directly suckle the mother and intake the colostrum just after an hour of birth.

2. *Weaning system :* This system is followed at larger organized farms in which calf is not allowed to suckle the mother but fed with colostrum by hand feeding method.

In both the systems, calf should be provided colostrum as early as possible (with in half hour of birth) because colostrum provides antibodies, which are absorbed intact as such but later antibody titre is decreased. Colostrum has high nutritive value and laxative action to remove the meconium (first faeces of calf). Colostrum per day should be provided about 1/10th of the calf body weight. Generally buffaloes calf has about 25-30 kg birth weight. So it required 2.5 to 3 kg colostrum per day upto 3 days of age. But care should be taken that it should be divided into 3-5 doses (minimum 2) other wise single dose excessive intake of colostrum may cause indigestion and diarrhoea. If colostrum is not available from the mother because of any reason like non-letdown of milk, premature parturition or death of mother, than colostrum from any other dam's may be given accordingly. If this is also not possible than boil two kg milk and mixed with 2 eggs, and 30 ml castor oil, cool it and then fed to the calf.

At 4-7th day 2.5-3.0 kg milk per day in divided doses is sufficient to meet out the requirements with a body wt. gain of 400-500 g/d.

Feeding Milk Replacer

Milk replacer is a diet fed to calves as early as at 10 days of age to replace milk from economy point of view. Milk replacer should resemble milk more or less in broad chemical composition especially in terms of quantity and quality of protein fatty acids, minerals and vitamins. It should have a biological value equivalent to that of milk, and the ingredients used for it should be low in crude fibre and free from any anti-metabolites. But is should replace milk gradually to facilitate its acceptance and to avoid drop in growth rate.

Table-4: Composition of an ideal milk replacer

Constituents	Percentage
Linseed meal	40
Fish meal	12
Wheat flour	10
Milk	13
Coconut oil	10

Butyric acid	0.30
Citric acid	1.5
Molasses	10.0
Mineral Mixture	2.0
Common salt	1.0
Rovi mix. (gl 100 kg.)	15

The dry powder milk replacer is mixed in five times more water, boil it and cool it than fed to calves. The ingredients of an ideal milk replacer may be changed as per the availability of these ingredients from area to area.

Table-5: Rate of feeding milk replacer.

Age	Milk (kg)	Milk replacer (kg)
1-3 day	2.5-3	
4-10	3-3.5	
11-20	3.5	50-200
21-40	3.5-2.5	200-400
41-60	2.5-1.5	400-700
61-90	1.5-0.5	700-1000

Feeding of Calf Starter

It is a dry concentrate mixture which is formulated to meet the requirements of calves and fed from the first week of age. After feeding milk, calves are offered a dry calf starter and a good quality of hay simultaneously in separate containers. Usually calf start nibbling from the second week of age. An ideal calf starter contains 20-23% DCP and 70-75% TDN.

Table-6: Composition of ideal calf starter

Constituents	Percentage
Maize	42%
Groundnut cake	28%
Fish meal	7%
Wheat bran	20%
Mineral mixture	2 %
Common salt	1%

The ingredients of calf starter can be changed according to the availability of the feed in the different regions of the country.

Feeding of Heifers

After 3 months of age feeding and management of calves are more convenient and involve less expenditure. Here male calves are not properly managed but ignored and are kept only for the letting down of milk and when become adult are used for draught and meat. They are fed on the residues left by productive animals and let loose on grazing which alone do not provide sufficient nutrition to support growth in early stages. Due to such neglected feeding male calves die prematurely before six months of age. Heifer calves are the future dairy buffaloes, hence they should be reared in such manner that they gain at a moderate rate in their body weight. An average 450-500 g/ day gain in live weight may be considered optimum for Indian buffalos heifers. Body wt, average daily gain, availability of feeds and fodders and composition of feed and fodders are essential factor which are helpful to formulate the ration. A 67.8g/ $w^{0.75}$kg dry matter for maintenance and 2.13 g/ g gain DM is required for 100 kg growing buffaloes (Udeybir and Mandal. 2001) wheares TDN requirements for maintenance and growth are 47.0 g/w $^{0.75}$kg and 0.78 g/g gain, respectively. Similarly CP and DCP requirement for maintenance and gain is 7.64 g/w $^{0.75}$kg and 5.05 g/w $^{0.75}$kg and 0.44 g/g gain and 0.27 g/g gain, respectively. Similarly NRC (1988) recommended 3.2 kg DM,

1.88 kg TDN and 422 g protein for 100 kg heifers growing at the rate of 500g/day.

Feeding of Lactating Buffalo

Milch animal are sophisticated machines which convert inferior type of feed into superior food for human consumption. The nutrient requirement for lactation is maintenance requirement (which will depend upon the body wt. of the buffalo) + production requirement (which will depend upon quantity and quality of milk produced). For example a buffalo of 500 kg body weight producing 10 litres of milk having 7% fat will required 10.50 kg DM, 8.3 kg TDN, and 0.93 kg DCP. This requirement can be supplied through different combinations of feed and fodders like wheat straw + concentrate mixture, green fodder + wheat straw + concentrate mixture etc. for example 20.00 kg green oat + 5 kg wheat straw + 4 kg concentrate mixture (15% DCP, 73% TDN) will meet out above requirements.

If lactating animal is in advanced stage of pregnancy also than 140 g DCP 700 g TDN is required as pregnancy allowance (Ranjhan, 1991).

Feeding of Pregnant Buffaloes

The intercalving period should be 370 days in buffaloes. So buffaloes are to be dried two months before the next calving. During dry period, buffalo should build up the body reserve lost in lactation and will require additional nutrients to support the fast growth of foetus which take place in the last quarter of the pregnancy.

In "Steaming up" of dry buffaloes in which animals are offered extra quantities of concentrates which increase gradually during the last 6-8 weeks of pregnancy. Steaming up is claimed to increase milk production, in part by preparing the buffalo for large ration of concentrates which is received in early lactation. The protein requirement in pregnancy is more than that of energy requirements. Normally 50% DCP and 25% of TDN of the maintenance are fed above the maintenance ration. These requirements are fulfilled by feeding an additional quantity of 1 to 1.5 kg concentrate mixture.

Feeding of Dry Buffaloes

Sometimes buffaloes are non-milking and non-pregnant because of one or another reason. These dry animals are needed to be fed maintenance rations. The maintenance requirements of an average 500 kg buffalo are 340 g DCP, 3.7 kg TDN, 20 g calcium and 15 g phosphorus. Thus 30 kg green fodder (maize, sorghum etc.) or 7-8 kg straws or stovers with 1 kg concentrate mixture are sufficient to maintain the dry buffaloes.

Constraint in Buffalo Feeding in Middle Gangetic Plain Region

1. Large number of nondescript breeds
2. Poor managemental practices
3. Poor feeding practices
4. Less Implementation of scientific recommendation

1. ***Large number of nondescript breeds:*** In middle gangetic plane region, large number of buffaloes are of non descript breeds. These have poor out look, less production and low milk yield per lactation than a descript breed. But it can be overcome by introducing the high producing breeds in this area either by transfer of quality animals or by introducing artificial insemination system using the semen of quality breeds like Murrah.

2. ***Poor managemental practices :*** Except some recognized dairy farm and research institutes managemental practices are poor in all classes of farmers (land less to large land holder). They mostly keep the animals in backyard of their house where facilities like proper drainage system, electric supply, water supply, optimum space per animal and exercise facilities are lacking. For successful buffalo rearing, extension worker and veterinarian can play important role by disseminating valuable information about these facilities to the farmers.

3. ***Feeding Practices:*** Due to lack of knowledge and unawareness of farmers about the feeding value of various feeds and fodders and nutrient requirements of different categories of farm animals, the supply of nutrients is unbalanced (Singh, 1998). On exclusive feeding of leguminous forages, a large amount of protein goes waste and on the feeding of the

cereal crop straw, animal suffers from protein, mineral and vitamin deficiency. Non-producing animals are over fed during the availability of green pasture whereas producing animals are underfed because of high requirements for production, which is not met out by green fodder alone until sufficient amount of concentrates are added in the ration of lactating buffaloes.

4. Less implementation of scientific recommendation : No doubt there is large gap exist between scientist and farmers interaction. Though sincere efforts are made to minimize it, even then a small portion of scientific recommendation is utilized by farmers or animal producers. They mostly follow traditional system of feeding and rearing of buffaloes which is required to be modified by implementation of scientific feeding to make the buffalo production cheap and economical.

References

Kearl, L.C. (1982) Nutrient requirement of ruminants in developing countries. Intern. Feed stuffs institute, Utah Agriculture Experiment al Station, Utah State University Logon, Utah – 84322, USA.

Kurar, C.K. and Mudgal, V.D. (1977) Feeding of Buffaloes. Publ. No. 180, NDRI Karnal.

NRC (1988) Nutrient requirement of domestic animals. Nutrient requirement of dairy cattle. 6th Edu. National Academy of Sciences, National Research Council, Washington, D.C.

Pathak N.N. and Verma D.N. (1993) Nutrient requirement of buffaloes. Int. Book Distr. Agency, Charbagh, Lucknow.

Ranjan, S.K. (1991). Chemical composition and nutritive value of Indian feeds and feeding of farm animals. ICAR, New Delhi.

Ranjhan, S.K. and Pathak, N.N. (1979). Management and feeding of buffaloes. I[st] edu. Vikas Pub. House Pvt. Ltd., New Delhi.

Sen, K.C., Ray, S.N. and Ranjhan, S.K. (1978). Nutritive value of Indian feeds and feeding of farm animals. Bull. 25 ICAR New Delhi.

Singh S.P., 1998 Ph.D. Thesis. Submitted to NDUAT, Faizabad.

Udeybir and Mandal A.B. (2001). Energy and protein requirements for growing buffaloes. Buffalo Journal 2:163-178.

CHAPTER 16

Nutritional Studies on Bhadawari Buffalo

Sultan Singh, S. S. Kundu, A. K. Misra and C. B. Sachan*

Plant Animal Relationship Division
Indian Grassland and Fodder Research Institute, Jhansi-284 003 (UP)
*Krishi Vigyan Kendra, College of Agriculture, Gwalior (MP)

Buffalo has a prominent place in livestock wealth particularly in Asian region. Domesticated buffaloes of word have been classified in two main categories namely riverine and swamp depending up variation in their habitat and genome. River buffaloes are of massive body size, have 50 chromosomes and are primarily used for milk production in addition to secondary meat and draught purpose. River buffaloes have about 18 milch breeds of India, Pakistan, Egypt and Europe. Swamp buffaloes are stocky animals with marshy land habitats and have 48 chromosomes. Swamp buffaloes are mainly confined to Southeast Asian region. About 95% of the world total buffalo population is found in 28 Asian countries. The number of buffalo's distributed in different countries across the world is given in Table (1).

Table-1: Major buffalo producing countries of the world (Unit: Million heads)

Country	Year			
	1990	1998	2000	2003
World	147.9	162.36	164.9	170.40
Azerbaijan	-	-	0.293	0.311
Bangladesh	0.772	0.850	0.828	0.830
Brazil	1.397	1.1	1.15	-

Bulgaria	0.023	0.011	0.003	-
China	21.422	22.6	22.6	22.75
Cambodia	0.736	0.71	0.71	0.62
Egypt	2.897	3.15	3.21	-
Georgia	-	0.01	-	0.033
India	78.320	91.78	93.8	96.9
Indonesia	3.335	2.86	2.85	2.35
Iran	0.44	0.47	0.50	0.55
Iraq	0.140	0.06	0.06	0.068
Italy	0.112	0.17	0.17	-
Kazakhstan	-	0.10	-	0.09
Laos	1.072	1.92	1.00	1.08
Malaysia	0.205	0.15	0.15	0.14
Myanmar	2.061	2.39	2.44	2.60
Nepal	3.012	3.40	3.50	3.70
Pakistan	17.373	22.00	22.70	24.8
Philippines	2.765	3.00	3.02	3.14
Sri Lanka	0.958	0.72	0.72	0.63
Thailand	5.094	2.20	2.10	1.84
Turkey	0.429	0.19	0.17	0.16
Vietnam	2.84	2.91	2.89	2.81

Buffalo supply 53% of the total milk production in India although their population is below 33% of the total bovine population. Buffalo milk contributes more than 6 % in world's total milk production. With 56% of word's buffalo population India contribute about 65 5 to the word's milk pool. The contribution of buffalo towards milk, meat and hides is given Table (2). Buffalo population has registered a growth of4.95% in world over the last five years. (1998-2003). Invariably some countries have registered a positive growth rate, while others have registered a negative growth rate (Table 3). The main Asian countries that have registered negative growth rate are Thailand Indonesia and Sri Lanka. The largest buffalo populated countries are India and Pakistan, of which India and Pakistan contribute about 70% of the total world population.

Table-2: **Products of buffalo in the world in 2003 (Unit: Metric tons)**

Country	Meat	Milk	Hides
World	3,179,887	72615909	856825
Asia	2,871,527	70394264	821385
Bangladesh	3500	22,400	2,250
Bhutan	32	320	7
Cambodia	13,440	-	2772
China	3.96,250	2,650,000	118770
Myanmar	22,440	116,018	5,000
Sri Lanka	5,311	68,000	1645
India	1,471,080	4,7850,000	525,000
Indonesia	45,128	-	6897
Iran	11,775	230,000	1727
Iraq	2,250	27,500	255
Laos	18,300	-	2662
Malaysia	3646	7360	502
Nepal	130,000	806694	33306
Pakistan	509,000	18520,000	85750
Philippines	81000	-	9350
Thailand	53130	-	6300
Turkey	5100	63327	560
Vietnam	98900	31,000	18,400

Source: FAO production yearbook 2003

Table 3: Growth in Asian buffalo population in some selected countries (Million)

Countries	Year		Annual Growth %	Growth %
	1998	2003		
World	162.36	170.4	0.97	4.95
Asia	-	-	-	-
Europe	-	-	-	-
South America	-	-	-	-
Bangladesh	0.85	0.83	-0.48	-2.35
Cambodia	0.71	0.62	-2.67	-12.68
China	22.6	22.75	0.13	0.66
India	91.78	96.9	1.09	5.58
Indonesia	2.86	2.35	-3.85	-17.83
Iran	0.47	0.55	3.19	17.02
Laos	1.92	1.08	-10.87	-43.75
Malaysia	0.15	0.14	-1.37	-6.67
Myanmar	2.39	2.6	1.70	8.79
Nepal	3.4	3.7	1.71	8.82
Pakistan	22	24.8	2.42	12.73
Philippines	3	3.14	0.92	4.67
Sri Lanka	0.72	0.63	-2.64	-12.50
Thailand	2.2	1.8	-3.93	-18.18
Vietnam	2.91	2.81	-0.70	-3.44

Source: FAO production yearbook 2003

Buffalo constitutes the main source of milk, meat and draft in different parts of the country. There are different recognized breeds (Murrah, Nili-ravi, Jafarabadi, Surti, Mehsana, Nagpuri, Pandharpuri, Kakahandi, Toda, Tarai rtc.) of buffalo located in the regions of their origin. Bhadawari is one of the recognized breeds of buffalo. The breed is widely known for its higher (13%) butterfat content (Kaura, 1950). The origin of this breed is around Agra and Etawah districts of

U.P. and Bhind and Morena districts of Madhya Pradesh. The region was a part of the erstwhile Bhadawar Estate from where the name of these animals originated.

The breed is typically characterized by copper color with scanty hairs, which are black at roots and reddish brown at the tips, sometimes the hairs are completely black (Ranjhan and Pathak 1979). Bhadawari animals are medium in size, wedged shape body and all four legs from horns to knees and hocks are purely brownish white (Kaura, 1950). This breed has lowest calving internal (Bhat, 1981) but the age at first calving seems longer (58.3 months) compared to other buffalo breeds (Singh and Nivasarkar, 1997).

Generally buffaloes had better efficiency of nutrients utilization from poor quality fibrous tropical feeds than cattle. Within the buffalo breeds, Bhadawari is more hardy resistance to diseases, efficient converter of poor quality feeds into production and better adapted to high temperature and harsh situations. Nutrient needs of buffaloes differ from these of dairy cattle of temperate regions because of feeds, climate and digestive physiology manifestations. Nutritional studies on farm animals were started about a century ago. However, the studies on nutrient utilization and requirements of buffaloes were carried out during the last three decades (Sharma and Talpatra, 1963, Gupta et al 1966, Negi et al 1968). In the last three decades systematic nutritional studies were carried out to determine the nutrient requirements mainly the protein and energy for maintenance and different physiological functions (growth and milk) (Agrawal, 1974, Pathak and Ranjhan 1979, Kurar and Mudgal 1980, 81, Baruah et al 1983, Sivaiah and Mudgal 1978). There is wide variation in nutrient requirements prescribed by different existing feeding standards for buffaloes. There might be variability within breeds for their nutrient needs for different physiological functions, efficiency to utilize the feeds of diverse chemical entity as well as production and reproduction efficiencies.

Though nutrient requirements of buffaloes have been determined across the country during the previous 30 years. However the work is mainly confined to few breeds i.e. Murrah, Nili-Ravi, Mehsana, Surti and others that are high milk yielder and had more population in the country compared to Bhadawari. For Bhadawari buffaloes no systematic study seems to be carried out where in nutrient requirement and efficiency of nutrients utilization form diverse feeds and forages for this breed had been documented. Sporadic nutritional studies are available on the Bhadawari buffalo.

Under the aegis of NPB (Bhadawari), at IGFRI a herd of 42 adult female and 5 breeding bulls is maintained. Total herd strength including calves is 72. The average body weight of male and female calves at birth, 3 and 6 months of age was 26.30, 48.88 and 68 .0; and 24.06, 43.20 and 66.30 kg, respectively. However the average body weight of adult females was 371.0 kg. As per one estimate the total number of heads of this breed in the whole of its breeding tract is about 25-30 thousands. Some studies on nutrition of this breed have been carried out at IGFRI, Jhansi. The results of these studies are summarized and discussed here.

Nutritional Studies

Chemical composition

The method to determine the chemical composition of feeds and fodders was proposed by Hennenberg and Stohmann as early as in 1864. This system partitions the feeds in to six components namely moisture, crude protein, crude fiber, ether extract, ash and nitrogen free extract. Subsequently ash component was fractioned into hydrochloric acid soluble and insoluble ash, which is a important attribute feed quality. Later in 1967 Van Soest proposed another method that partitioned the feeds carbohydrate in to NDF, ADF, cellulose, hemi cellulose, lignin and silica.

Proximate constituents (CP, OM and EE) and cell wall composition (NDF, ADF, cellulose, hemi cellulose and lignin) of different feeds and fodders used in feeding studies with Bhadawari buffaloes are presented in Table 4.

Table 4: Chemical composition of feeds and fodders

Feeds/fodders	CP	OM	NDF	ADF	Cellulose	Hemi cellulose	Lignin	EE
Barley straw	4.16	90.94	69.40	40.45				1.34
Wheat straw	3.97	92.09	74.45	49.30	38.60	25.15	7.05	
Gram straw	6.92	92.48	63.36	44.40	33.50	18.96	9.70	
Masoor straw	4.90	91.88	64.25	47.40	36.20	16.85	10.10	
Sorghum hay	4.74	85.43	63.71	33.21	25.10	30.50	5.00	2.46
Grass hay	4.45	92.10	70.77	45.65	36.59	25.12	7.31	1.94
Stylosanthes hay	10.95	94.47	56.4	37.59	28.51	18.84	.55	6.45
Sorghum kutti	4.90	93.25	73.08	44.64	35.46	28.44	6.96	1.33
Mixed grass	6.12	90.54	77.27	46.86	34.89	30.41	7.27	1.36
Wheat bran	12.81	89.53	44.20	10.67	-	33.53	-	2.34
Concentrate mixture-1	21.53	90.86	31.10	11.10	-	20.00	-	4.15
Concentrate mixture-2	16.49	93.18	37.90	20.11	14.86	17.84	4.61	3.70
Concentrate mixture-3	20.87	87.82	39.68	15.50	10.30	24.18	3.20	
Concentrate mixture-4	20.10	92.87	49.40	17.54	11.24	32.16	4.43	3.28

Carbohydrate and Energy Fractions of Feeds and Fodders

Recently proposed method (CNCP system) from Cornell University is the most accurate method characterizes the carbohydrates and proteins of feeds and forage into different fractions as per their availability to the animals. This system characterizes the feeds into different protein (A, B_1, B_2, B_3 and C) and carbohydrate (Sugar, fructant; Fast degradation – starch pectin, oligosaccharides; Slow degradable starch; Fast degradation fiber; and lignin) fractions. Thus methods helps to rank the feeds based on carbohydrates and proteins degradability and thus guides in dietary formulation accordingly. This will assist in optimization and in efficient utilization of feeds/forages in nutrient feeding.

Based on chemical constituents as well as potential degradation rates Pichard and Van Soest (1977) and Krishnamorthy et al (1982, 1983) fractionated feeds into various categories.

CNCP system partitions the feeds and fodder in to different protein and carbohydrate fraction. The diagrammatic representation of protein and carbohydrate fractions is given in Fig (1). These fractions

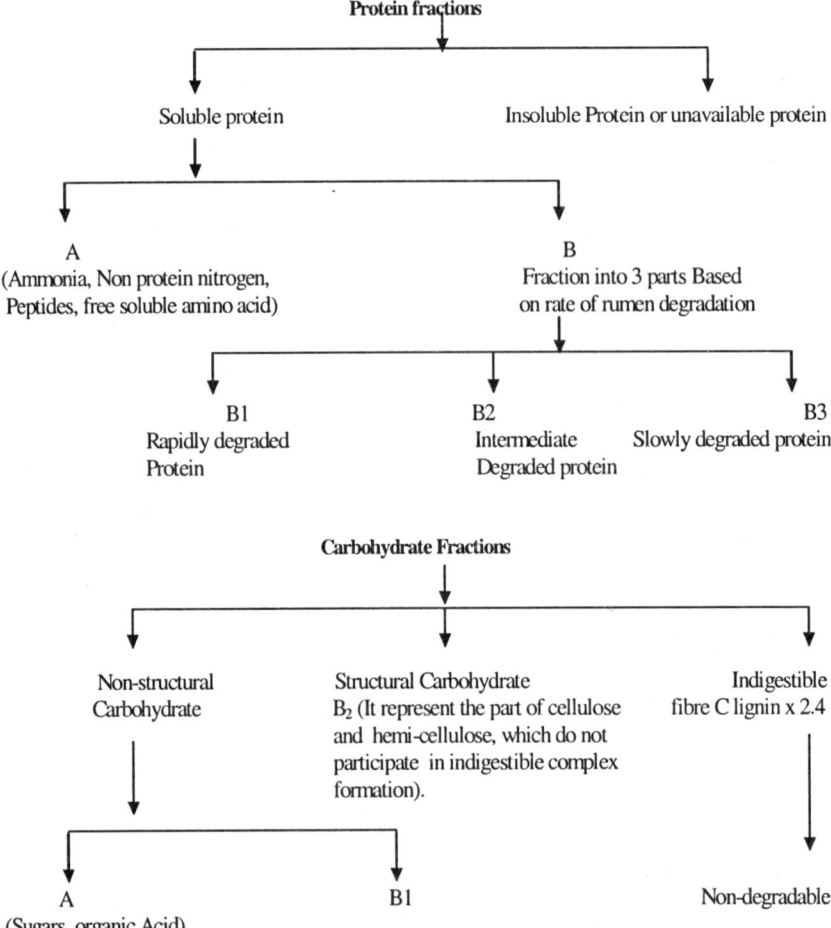

Fig 1: Diagrammatic representation of protein and carbohydrate fractions

are based on the chemical constituents as well as on potential degradability rates of carbohydrate and protein of feeds and forages.

These fractions of protein (NPN, soluble peptide, soluble protein) and carbohydrate (sugar, starch, NSP and other carbohydrate) of feeds and forages are utilized by different microbes and animals differently. Thus the estimation of these carbohydrate and proteins fractions in different feeds and fodders is of utmost importance not only for their nutritive evaluation but also as basis for dietary formulation to meet the nutritional requirement of animals. The degradation rates of feeds/forages coupled with their passage rates are also one of the determinants of net energy and protein available from feeds. At IGFRI CNCP system has been initiated to evaluate the feeds and forages for their nutritive value. The results on evaluation of different feeds (Energy and Protein Sources) fodders and tree leaves for carbohydrate and protein fractions as per CNCP system (Table 5 and 6) revealed that soybean has higher concentration of sugar as % of carbohydrate with minimum lignin % NDF and C fraction of carbohydrate, while maximum concentration of lignin % NDF and fraction C of carbohydrate was recorded in MSC and CSC (Singh et al, 2003). In respect of protein quality soybean had the minimum of NPN % SP (11.83), SP % CP (9.01) and ADFIP % CP (1.69) amongst the evaluated protein sources while the maximum concentration of these fractions was recorded in CSC. ADFIP % CP and fraction A of N was minimum (2.64) and maximum (12.39) in barley grain. Similarly lignin % NDF and undegradable carbohydrate fraction –C was minimum in barley (1.31 &and0.83%) exhibiting barley grains as better energy source for quality estimates of protein and carbohydrate fractions. Among the evaluated fodders maximum contents of sugar % CHO were observed in berseem also had maximum contents of SP % CP (48.61) and lower contents of ADFIP % CP (7.67) of protein and nitrogen fractions (Kundu et al, 2003). Proteins and CHO fractions of various feeds and fodders has been reported from IGFRI (Bhadauria *et al.,* 2002, Singh *et al.,* 2002) and other laboratories (Jeya Prakash *et al.,* 2002).

Table 5: Carbohydrate fraction of protein, energy and fodder sources

	NDF % DM	Lignin % NDF	CHO %	Starch %	CHO % DM	CC % CHO	CB2 % CHO	NSC % CHO	Starch % NSC	CB1 % CHO	CA % CHO
PROTEIN SOURCES											
Linseed C.	44.85	13.89	62.14	30.21	62.14	24.06	37.29	38.64	57.25	30.21	69.79
M.S.C.	27.46	21.62	49.57	25.72	49.57	28.74	20.6	50.65	42.13	25.72	74.28
C.S.C	40.3	20.07	65.73	47.9	69.73	29.53	28.36	42.11	92.42	47.9	52.1
Soyabean	44.06	0.86	39.18	16	39.18	2.32	84.75	12.92	24.81	16	84
G.N.C	22.85	15.74	45.97	13.58	45.67	18.9	3.42	77.67	17.07	13.58	86.42
ENERGY SOURCE											
Rice Polish	67.71	21.22	68.32	17.3	67.9	50.78	44.42	4.78	90.82	17.3	82.7
maize grain	12.52	4.84	82.86	88	82.86	1.75	11.52	86.71	99.21	86	14
Wheat bran	36.23	7.42	75.79	53.3	75.79	8.51	34.81	56.66	81.87	53.3	46.7
Barley	22.75	1.31	85.94	59.8	85.94	0.83	24.27	74.89	78.36	59.8	40.2
FODDERS											
Berseem	36.28	15.32	62.43	8.56	62.43	21.36	33.21	45.42	14.78	8.56	91.44
Cowpea	63.31	15.17	68.14	12.3	68.14	33.82	50.9	15.27	39.06	12.3	87.7
S. Hamata	54.16	11.7	77.95	38.4	77.95	19.51	45.16	35.33	84.79	38.4	61.6
A.Catechue	48.12	18.93	76.43	13.7	76.43	28.6	27.31	44.08	27.11	13.7	86.3
Sorghum	71.84	6.75	85	16.27	85	13.69	66.85	19.45	55.18	16.27	83.73
W Straw	63.22	8.35	74.43	28.9	74.59	16.98	63.45	19.56	80.98	28.9	71.1
P Maximum	76.18	6.66	80.28	5.17	80.28	15.16	75.29	9.54	70.49	17.19	82.81
Maize	62.96	4.98	83.9	38.3	83.9	8.96	60.81	30.22	95.72	38.3	61.7

Table-6: Protein and Nitrogen Fractions of Protein, Energy Feeds and Fodders.

	CP %	NPN % SP	SP % CP	NDFIP % CP	ADFIP % CP	A	B1	B2	B3	C
PROTEIN SOURCES										
Linseed C.	25.89	26.3	23.53	25.96	6.45	6.2	17.32	50.51	19.51	6.45
M.S.C	35.03	35.28	38.98	8.81	5.08	13.75	25.23	52.2	3.73	5.08
C.S.C.	30.36	76.12	48.93	12.67	5.87	36.24	11.69	38.39	6.8	5.87
Soybean	47.63	11.83	9.01	20.92	1.69	1.07	7.95	70.06	19.22	1.69
G.N.C.	28.12	76.54	29.69	29.91	4.87	22.65	7.03	40.39	25.04	4.87
ENERGY SOURCE										
Rice polish	6.75	34.35	24.14	47.28	19.23	8.68	15.85	28.57	28.05	19.23
W. Bran (NH3)	15.38	23.63	18.42	20.03	2.97	4.33	14.09	59.55	19.06	2.97
Barley	9	34.17	36.32	13.01	2.64	12.39	23.93	53.28	10.37	2.64
Maize grain	11.91	35.84	14.3	14.08	4.44	5.12	9.18	80.79	9.6	4.44
FODDERS										
Berseem	16.28	67.2	48.61	13.56	7.67	32.65	15.96	37.82	5.89	7.67
Cowpea	10.14	41.75	5.6	54.99	20.08	2.32	3.27	37.53	34.19	20.08
S hamata	13.05	37.85	18.35	47.02	12.36	5.22	15.97	31.72	35.84	11.23
A catechu	13.73	78.14	11.47	39.19	19.83	0.45	2.52	49.33	19.36	19.83
W straw	15.03	25.68	18.42	30.37	2.97	20	19.14	20.48	23.15	7.22
P maximum	7.04	90.01	24.8	50.51	24.86	23.05	1.74	24.69	25.64	24.87
Sorghum	6.47	80.78	22.18	51.39	16.21	17.88	4.26	26.42	35.18	16.21
Maize	9.39	24.62	21.2	47.07	11.23	5.22	15.97	31.72	35.84	11.23

Feeding Experiments

Barley straw with UP Agro milch ration was evaluated in 8 each of lactating Bhadawari and Murrah buffaloes having average body weight of 348±2.35 and 492.06±9.15 kg, respectively. The animals were offered barley straw ad-lib and 1 kg wheat bran. The animals were given concentrate mixture as per their requirement. Dry matter intake (kg) in group I and II was 10.448±.48 and 8.169±.41 kg respectively. However, dry matte intake (% body weight) was 2.12±

0.90 and 2.35±0.10 kg, respectively. Digestibility coefficients of dry matter, NDF, ADF, CF, CP, EE and NFE were 52.66±0.9, 52.74+1.25; 46.88±1.02, 44.69±1.22; 36.9±1.93, 32.58±1.68; 49.03±1.69 respectively in group I. and the corresponding figures for group II were 49.30±1.32; 46.78±2.61, 44.62±2.55; 71.46±1.93, 72.75±2.32, 61.69±1.30, 60.99±1.20 respectively Table 7). DCP intake (g) and TDN intake (kg) by Murrah and Bhadawari buffaloes were 538±0.05, 381±.04 and 5.734±0.32, 4.44±.0.22. DCP (%) was found to be 5.15±.39 in group 5.15 and I ±. 62 in group II. TDN% of the corresponding groups was 54.72±0.32 and 54.25±0.87, respectively. Nitrogen balance (g) was 26.28+7.47 and 23.5+4.64 in Murrah and Bhadawari animals, respectively.

Table-4: **Comparative nutrient utilization in lactating Murrah and Bhadawari buffaloes.**

Parameters	Group I (Murrah)	Group II (Bhadawari)
Body weight (kg)	492.06 ± 9.15	348.31 ± 9.35
DM intake (kg)	10.448 ± 0.48	8.169 ± 0.41
DM intake (%body weight)	2.12 ± 0.09	2.35 ± 0.10
Nutrients digestibility (%)		
DM	52.66 ± 0.90	52.74 ± 1.25
OM	57.29 ± 0.74	56.84 ± 1.09
NDF	46.88 ± 1.02	44.69 ± 1.22
ADF	36.9 ± 1.93	32.58 ± 1.68
CP	46.78 ± 2.61	44.62 ± 2.55
CF	49.02 ± 1.83	49.30 ± 1.36
EE	71.46 ± 1.93	72.75 ± 2.32
NFE	61.69 ± 0.89	60.98 ± 1.23
Nitrogen balance		
N intake (g)	182.37 ± 8.44	135.33 ± 6.5

N outgo (g)

Feaces	96.29±4.79	74.3 ± 03.45
Urine	31.32 ± 2.59	18.44 ± 3.16
Milk	28.47 ± 3.68	19.09 ± 2.93
Balance (g)	26.28 ± 7.47	23.5 ± 4.63

Nutrients density (%)

DCP	5.14 ± 0.39	5.15 ± 0.62
TDN	54.72 ± 0.32	54.25 ± 0.87

Nutrient intake

DCP (g)	538 ± 0.05	381 ± 0.04
TDN (kg)	5.734 ± 0.32	4.441 ± 0.22

In an experiment of 105 days three different sources of straw namely wheat straw, Gram straw and Masoor (Lentil) straw were evaluated in 18 lactating Bhadawari buffaloes distributed into 3 dietary groups (G1, G2 and G3) of 6 animals in each. The animals were offered 0.5 kg concentrate at feeding (9:30- 10. am), while 1 kg concentrate was offered at morning and evening milking.

Dry matter intake of animals was significantly ($P<0.05$) higher on Masoor straw group (2.83) compared to lowest on wheat straw (2.42%) diet. DMI (g/ Kg $W^{1.75}$) was 105.29, 113.33 and 122.71 in G_1, G_2 and G_3 group, respectively (Table 8). Dry matter digestibility was comparable amongst the dietary groups and ranged from 52.24 to 53.58% across the treatment groups. Crude protein digestibility was significantly ($P<0.05$) lower in wheat straw based group (36.13) than gram straw (47.83) and Masoor straw (49.42%) fed group. Fiber fractions (NDF, ADF and cellulose) digestibility was comparable amongst the dietary groups. The digestibility of NDF, ADF and cellulose varied from 44.20-47.24, 36.33-41.57 and 50.31-53.45 % across the dietary groups.

Nitrogen balance (g/d) was significantly ($P<0.05$) higher in G_2 (47.74) and G_3 (44.94) than G_1 (13.52) group, respectively.

Table-5: Dry matter intake and nutrients digestibility in Bhadawari buffaloes fed different sources of straw.

Parameter Intake	Wheat straw (G_1)	Gram straw (G_2)	Masoor straw (G_3)
Animal weight	356.2	364.8	360.8
Intake kg/d	8.57	9.46	10.11
% Body weight	2.42	2.59	2.83
G/Kg W $^{0.75}$	105.29	113.33	122.71
Nutrients digestibility			
DM	52.51	52.24	53.58
OM	54.83	59.72	57.48
CP	36.13	47.83	49.42
NDF	47.24	44.20	45.61
ADF	36.33	37.58	41.57
Cellulose	53.45	51.09	50.31
H. Cellulose	64.85	60.57	68.24
Nitrogen balance			
Intake – N	121.95	149.6	167.84
Fecal – N	80.84	78.18	84.98
Urinary – N	27.6	23.78	37.92
N – balance	13.5	47.64	44.94

The effect of roughage source on dry matter intake and eating pattern of 18 lactating Bhadawari buffaloes fed three different roughages namely Jowar hay (G1), grass hay (G2) and Stylosanthes hay (G3) along with 3 kg wheat straw and 2 kg concentrate mixture was studied by Sachan et al (2003). They observed that average total dry matter intake was 9.54, 8.90 and 9.69 kg/day in G1, G2 and G3, respectively. Dry matter intake of animals was more in stylosanthes (G3- 113.79 g/kgW$^{0.75}$) than jowar (G1- 111.74) and grass hay (G2- 103.24 g/kg W$^{0.75}$) roughage based groups (Table 9.).

Results revealed that in the first hour of feeding intake (kg) was significantly higher in stylosanthes-based group (G3- 2.80) compared to grass hay group (G2- 1.98). However, the intake in jowar hay based group (G1) was 2.36. In the forth hour of feeding intake of jowar hay was maximum (0.71 kg) compared to minimum in G2 (0.167 kg). However the pattern of dry matter intake from wheat straw was reverse in these groups (G1- 0.210 Vs G2- 0.880). Total dry matter intake in 4th hour of eating was 0.910, 1.047 and 0.617 kg in G1, G2 and G3 respectively. Intake of hay with wheat straw in 5th hour was 1.097, 1.040 and 0.719 kg in G1, G2 and G3 (Table 10) respectively. In the last sixteen hour of feeding, total intake of feed was 4.325, 3.064 and 3.875 kg in jowar, grass and stylosanthes hay based groups respectively. Total intake of feed in the first 5 hours was comparable between the G1 (4.760), G2 (4.620) and G3 (4.800) groups (Table 11). Results suggest that animals consumed stylosanthes hay based feed more in the first hour of feeding compared to other roughage sources. Animals again consumed more dry matter in fourth and fifth hour compared to second and third hour of feeding. It is evident from the study that stylosanthes hay is more palatable followed by jowar and grass hay as roughage source for feeding of lactating buffaloes. The unpublished results of same experiment further revealed that animals had higher dry matter digestibility on sorghum hay (62.47) than grass hay (58.23) and stylosanthes hay (58.68%) based diets (Table 12). Crude protein digestibility was lowest in grass hay (37.38) and highest on stylosanthes hay (55.92%) based group. However the animals exhibited higher digestibility of cell wall fractions (NDF, ADF, cellulose and hemi cellulose) on sorghum and grass hay than stylosanthes hay group. The digestibility of NDF, ADF and cellulose varied from 47.64-55.57, 40.67-50.54 and 55.25-63.19 % among the different roughage based groups. Nitrogen balance (g/d) was highest in G3 (73.59) followed by G1 (38.75) and G2 (22.68) group, respectively. DCP and TDN contents (%) were higher in G3 (5.89 and 63.75) than G1 (3.12 and 58.080 and G2 (2.43 and 52.97) groups, respectively.

Table-9: Dry matter intake of Bhadawari buffaloes in different dietary groups fed various roughage sources

Intake	G1	G2	G3
Kg/day	9.54	8.90	9.69
% Body wt.	2.53	2.33	2.58
G/Kg W $^{0.75}$	111.74	103.24	113.79

Each value is a mean of 18 observations

Table-10: Eating pattern (Kg) of buffaloes in different hours of feeding

Groups	1h	2h	3h	4h	5h	16h
Sorghum hay G1	2.36	0.430	0.016	0.910	1.097	4.325
Grass hay G2	1.98	0.467	0.050	1.047	1.040	3.064
Stylosanthes hay G3	2.8	0.417	0.200	0.617	0.719	3.815

Each value is a mean of 18 observations

Table-11: Total meal consumption (kg) of buffaloes in different groups

Groups	1h	3h	5h
Sorghum hay G1	2.36	2.80	4.76
Grass hay G2	1.98	2.50	4.62
Stylosanthes hay G3	2.80	3.40	4.80

Each value is a mean of 18 observations

Table-12: Intake, nutrients utilization, nitrogen balance and nutritive value of different roughage source based diets

Particulars	G1	G2	G3
Intake			
Kg/day	10.49	10.18	10.31
% Body weight	2.78	2.66	2.77
G/kg $W^{0.75}$	122.72	118.03	121.84
Digestibility			
DM	62.47	58.23	58.68
OM	63.23	61.45	61.93
CP	47.12	37.38	55.92
NDF	55.57	52.38	47.64
ADF	46.89	50.54	40.67
Cellulose	58.52	63.19	55.25
Hemi-Cellulose	65.97	56.07	59.13
Nitrogen balance			
Intake	111.20	105.73	174.05
Fecal output	58.81	66.28	76.65
Urinary	13.64	16.77	23.81
Balance	38.75	22.68	73.59
Nutritive value (%)			
DCP	3.12	2.43	5.89
TDN	58.08	57.97	63.75

Grazing studies

From results of one feeding study on 93 Bhadawari buffaloes maintained at Livestock farm of JNKVV, College of Agriculture, Gwalior conducted by Yadav et al (2003) reported that if the leguminous green fodder like berseem is fed *ad lib* (50 kg/day) the concentrate feeding may be avoided without any adverse effect on milk production and its composition up to 7 liters of milk/day. They further reported that soaking of concentrate mixture had no significant effect on milk production and composition.

In one study at IGFRI 12 pregnant animals each of Bhadawari and Murrah buffalo were allocated to stall-fed and grazing+ supplemental groups of 6 animals in each. Initially the intake of animals was determined by feeding both grazing+ supplement and stall fed animals for a period of 7 days in the month of September 2003. Stall fed animals of both breeds were offered available roughage along with 2 kg of concentrate. The grazing animals were supplemented with same roughage fed to stall fed along with 2 kg concentrate at evening after returning from grazing. The intake of stall-fed and grazing + supplementation animals of both breeds was determined twice in a week during the study. Results of the study revealed that the animals meet 0.4 to 1.0 % of their dry matte requirement from grazing biomass in the month of September, October and November. In September the intake of stall fed animals was 2.56 and 2.52 % in Bhadawari and Murrah buffaloes, while the intake of grazing+ supplementation animals from supplementation was 1.53 and 1.85 % in Murrah and Bhadawari buffaloes. In the month of December animals met only 0.1% of their DM requirement from grazing fields. In month of January and February grazing+ supplementation animals consumed equal or more than stall-fed animals from the supplemented roughages and concentrate. It indicates that Murrah and Bhadawari animals consumed 1 and 0.7 % DM of their body weight from grazing. In month of October Murrah and Bhadawari animals consumed 0.4 and 0.6 % DM of their body weight from grazing.

In the month of October animal start calving and in November a digestion cum metabolism trial was conducted on 16 lactating animals.

In both breeds intake of stall fed animals was less compared to grazing+-supplemented animals (Table 13). Dry matter intake (% body weight) was 1.70 and 2.49 in Bhadawari vis-à-vis 2.0 and 2.55 in Murrah buffaloes in stalled and grazing+ supplemented situations. Within breeds Murrah revealed higher dry matter intake than Bhadawari in both the feeding conditions. Dry matter digestibility of grazing+-supplemented animals in Bhadawari was more (60.91) than stall fed animals (55.82%). However Murrah animals had similar dry matter digestibility both in grazing and stall-fed. Grazing animals of Bhadawari buffalo had higher nutrients digestibility than stall fed animals. Murrah buffalo exhibited comparable nutrients digestibility except that of CP between grazing+ supplemented and stall fed animals.

Table-13: Dry matter intake and nutrients digestibility in Bhadawari and Murrah buffaloes in stall fed vis-à- vis stall fed animals

Attributes	Bhadawari		Murrah	
	Stall fed	Grazing+ Supplementation	Stall fed	Grazing + Supplementation
Animals weight	402.5	401.2	528.33	518.2
Intake Kg/d	7.62	10.33	9.68	12.88
G/Kg W0.75	78.08	112.56	90.23	120.90
% Body weight	1.70	2.49	2.0	2.55
Digestibility				
DM	55.82	60.91	58.44	57.39
OM	58.38	62.84	60.57	59.88
CP	49.86	53.47	51.20	46.13
NDF	52.23	58.70	55.72	57.53
ADF	38.57	50.88	42.98	43.74
Cellulose	52.30	63.03	59.64	59.25
Hemi cellulose	65.32	59.54	71.48	70.94

Screening of the work on Bhadawari nutrition revealed that the work is not only scanty rather unsystematic. There is need to study this breed for its nutritional requirements and its efficiency to utilize nutrients from diverse feed resources that synchronizes with genetic and physiological make up of this breed. To achieve it systematic studies on nutrition of this breed for different physiological functions (maintenance, growth milk production and reproduction) is of paramount importance. Nutritional studies needs to be conducted in the light of social and economic relevance of rearing this breed.

References

Agrawal, V. P. 1974. Ph D. Thesis, Agra University.

Baruah, K. K., Ranjhan, S. K. and Pathak, N. N. 1983. Nutrient requirement for growth of desi male

Buffalo calves. Proceedings 5th World Conference on Animal Production, National Institute Animal

Indusry, Ibarki, 305 Japan.

Bhadauria, K.K.S., Pailan, G.H., Das, M.M., Kundu, S.S., Singh, J.P. and Lodhi, G.N. (2002). Evaluation of the shrubs and tree leaves for carbohydrates and nitrogen fractions. Indian J. Ani. Sci. 72 (1): 87-90

Bhat P.N. (1981). Cattle and Buffaloes. Animal Genetic Resources in India. NDRI Publication No. 192, Karnal

Gupta, B. S., Bhargava, V. N., Raina, N. N. and Singh, S. N. 1966. Indian J. Vet. Sci. Animal

Husbandry., 36-90.

FAO. 1995-2000. Production Year Book, FAO Rome Italy.

Jeya Prakash, R., Kundu, S.S. and Das, M.M. (2002). Studies on carbohydrates and nitrogen fractionation of certain feeds and their utilization in growing buffaloes fed on total mixed ration. Indian J. Anim. Sciences (Accepted).

Kaura, R. L. 1950. Cattle development in Uttar Pradesh, pp. 21-22, Department of Animal Husbandry, Lucknow, U.P.

Kundu S.S., Singh, Sultan, Kumar Anil, Singh N.P (2003). Forage evaluation for ruminant nutrition. Lead paper presented in national symposium of RMSI, Held at IGFRI, Jhansi.

Kurar, C. K. and Mudgal , V. D. 1980. Indian J Dairy Sc., 50: 797

Kurar, C. K. and Mudgal , V. D. 1981. Indian J Animal Sci., 51:817

Negi, S. S., Joshi, B. C. and Kehar, N. D. 1968. J Res., P. A. U. Ludhiana., 5: 92

Pathak, N. N. and Ranjan, S. K. 1979. Proceedings All India Symposium on Protein and NPN utilization in ruminants., N. D. R. I. , Karnal,

India.

Ranjhan, S. K. and Pathak, N. N. 1979. Management and Feeding of Buffaloes. Ist edn. Vikas Publication House Pvt. Ltd. New Delhi.

Ranjhan , S.K. 2003. Changing role of buffalo production in the third millennium. Souvenir 4th Asian buffalo congress on "Buffalo for food security and rural employment", 25-28 Feb. 2003, New Delhi

Sachan, C. B. Kundu, S. S. Singh, Sultan, Singh, N. P. and Singh, V. P. 2003. Dry matter intake and eating pattern of Bhadawari buffaloes on different sources of roughage. National seminar on " Sustainable live stock production- a national thrust." , held at JNKVV, Jabalpur from 6-8 March, 2003, pp. 46.

Sharama, K. M. L. and Talapatra, S. K. 1963. Indian J Dairy Sci., 17: 236

Singh, K.K., Dass, M.M., Samanta, A.K., Kundu, S.S. and Sharma, S.D (2002). Evaluation of certain feed resources for carbohydrate and protein fractions and *in situ* digestion characteristics. *Indian J. Anim. Sci.* 72 (9): 794 – 797.

Singh, Sultan , Gupta, Arti and Kundu, S. S. 2003.Protein and nitrogen fractions vis-à-vis cell wall polysaccharides in commonly used protein and energy sources. 4th Asian Buffalo Congress on " Buffalo for food security and rural employment" held at New Delhi from 25-28 february, 2003, vol Ii, pp: 180

Sivaiah, K. and Mudgal, V. D. 1978. Effect of feeding different levels of protein and energy on feed utilization for growth and milk production in buffaloes. Annual report, pp. 145-146, NDRI, Karnal, India.

Yadav, S. S., Sachan, C. B. and Gupta, R. D. 1999. Studies on milk production performance of Bhadawari buffaloes as influenced by feeding management. In: Proceedings of International Conference on Sustainable animal Production and Health and Environment, Future Challenges, November 24-27, CCS Hariana Agricultural University, Hisar.

CHAPTER 17

Bypass Nutrient Technology for Growing and Lactating Buffaloes

S.S. Thakur and R.K. Raikwar

National Dairy Research Institute, Karnal – 132 001 (Haryana)

Buffalo is the main dairy animal of India. At the start of new millennium, nearly 94 million buffaloes produced more than 45 million MT of milk (56 per cent of the total milk production) as compared to 34.16 million MT of the nearly 218 million cattle population of the country. Besides milk production, contribution of buffalo to total meat production is about 30 per cent, which is growing at a rate of 3 per cent per year. These facts make the buffalo an important livestock species of our country, which is making substantial contributions to improve food and nutrition security and providing the livelihood to the rural masses engaged in buffalo husbandry however, but there is enough scope to enhance the productivity of buffalo in terms of growth, reproduction and milk production, as buffaloes exhibit lower growth rate, higher calf mortality, higher age at first calving, longer calving interval, shorter lactation period and poor reproductive performance. Improvement can be brought about in various aspects of buffalo production i.e. breeding, nutrition, health and general management, out of which nutrition is of the paramount importance because without proper nutrition neither optimum production from a superior animal can be achieved nor the breeding with superior germplasm will work. Moreover, inadequate nutrition hampers the immunity, which makes the animal less resistant to diseases. In brief that nutrition determines the overall productivity of an animal. Since buffaloes have the ability to utilize the course forages such as straws and stovers (kadbis), more efficiently than cattle, but such diets based on such forages cannot support the higher rate of growth as well as milk production.

The lower growth rate of buffalo calves is the main cause of delayed puberty observed in our country. The age at puberty is correlated with body weight gains and delay in sexual maturity increases the age at first calving (in case of she buffaloes), which decreases the average productive life of an animal. However, better growth rate can help to overcome such type of problems. Attainment of higher growth rate also opens the possibility to exploit the potential of buffalo calves as broilers for meat production. In both cases, better growth rate can be achieved through strategic feeding of certain nutrients, out of which bypass protein and protected/bypass fat hold great potential. Their use in ration of growing animals could result in better growth performance.

Like growth, milk production in buffaloes also poses characteristic challenges to the animal nutritionists. Average lactation yield of improved dairy buffaloes is 1500–2000 kg, and many individual animals even have record up to 4500 kg in 305 days of lactation. Meeting the nutritional requirements of such animals is a real challenge for Animal Nutritionists in conditions especially when good quality feeds, fodders are not available in sufficient quantity. Buffalo milk contains more fats and solids than cow's milk. The average fat content ranges between 7-8 per cent in Murrah buffalo 13 per cent in Bhadawari buffaloes. Buffalo milk also has more average protein content than cow. In addition to above factors, the stage of lactation is also equally important from nutritional point of view. During early quarter of lactation, energy and protein intake is not sufficient to meet the colossus drain of nutrients through milk. Hence, the high milk-producing buffaloes need challenge feeding.

In the present article, the importance of bypass protein and bypass/protected fat feeding for achieving desirable production levels in terms of growth rate and milk production in buffaloes is discussed.

Bypass Nutrient Technology

There are limited energy and proteins supplements available in the form of grains and cakes, which cannot economically fulfill the nutrient requirement of high yielding and high potential growing animals. However, the efficiency of utilization for milk production and growth can be substantially increased by controlled degradation in the rumen after suitable treatment of these ingredients. This technology is called bypass nutrient technology.

Bypass Nutrients

Bypass nutrient called as 'rumen protected' has been defined by the Association of American Feed Control Officials (Noel, 2000) as a nutrient (s) fed in such a form that provides an increase in the flow of that nutrient(s), unchanged, to the abomasum, yet is available to the animals in the intestine.

Bypass Protein

Rumen bypass proteins are proteins containing feedstuffs that have been treated or processed by various methods to decrease their ruminal degradation and increase the content of digestible ruminally undegraded protein (UDP). Bypass protein, as well as protein supplements that have an inherent high rate of ruminal escape, may be useful in dairy animals feeding because of lower content of digestible RUP in most feedstuffs. Conventionally, most of the basal ration contains sufficient amount of ruminally degradable protein (UDP) but are deficient in UDP.

Need of Bypass Protein

Ruminants have two types of nitrogen requirement –

I   Nitrogen requirement of the microorganisms present in rumen to ensure optimal ruminal fermentation, and

II   Needs of the host animal tissues for maintenance and production.

Satisfying the N - need of the microbes helps to satisfy the needs of the host animal, because the microbial protein produced in the rumen makes a large contribution to the amino acid pool which meets the requirement of various tissues and organs. From the quality aspect, the two requirements differ. The rumen microbes can satisfy their

N–requirements either from dietary true protein or NPN sources while organs and tissues need only amino acids. So, these differences lead to the development of bypass protein feeding concept.

How Bypass Protein is Prepared?

Various methods through which the rate and extent of ruminal degradation of feed protein can be decreased and undegradable fraction could be increased are:

i)   Heat processing,

ii)   Chemical treatment, and

iii)   Combination of heat and chemical treatment.

Heat Processing

It decreases ruminal protein degradation by denaturation of proteins and by the formation of protein–carbohydrate and protein–protein linkages. Commercial methods of heat processing include:

⊙   Cooker expeller processing of oil seeds

⊙   Additional heat treatment of solvent extracted oil meals

⊙   Roasting

⊙   Extrusion

⊙   Pressure toasting

⊙   Micronization of oil seeds

Chemical Treatments

On the basis of chemicals used, it can be divided into three groups:

i)   Chemical that combines with and introduces cross-links in proteins e.g. aldehydes. Formaldehyde is most widely used and, the most effective treatment for protein protection.

ii)   Chemicals that cause denaturation e.g. acids, alkalis and ethanol.

iii)   Chemicals that bind with protein causing no or little alteration in protein structure e.g. tannins, Xylose etc..

Combination of Heat and Chemical Agents

This may be an effective method in which chemical and heat treatments, both are combined. In this method, chemical treatments to oilseed meals are given first and then heat processing is done at predetermined temperature. Such conditions promote the occurrence of the chemical reaction. The combined treatments enhance non-enzymatic browning (Maillard reactions).

Bypass protein in feedstuffs (g/100 g CP)

Feed Stuffs	Undegraded dietary protein
Solvent extracted coconut cake	76 (70-81)
Corn grain, cracked	81 (71-87)
Rice bran	62
Sorghum grain	75
Brewers grain	53 (48-61)
Canola meal	31 (26-37)
Cottonseed cake	49 (35-70)
Groundnut cake	32 (6-38)
Aniseed meal	35 (11-45)
Safflower cake	39
Soybean meal	34 (10-50)

Figures in bracket indicate the range of values reported in literature (Sampath, 1990)

Bypass Protein and Growth

To achieve better growth rate in buffalo calves, bypass protein feeding can be used successfully. Several studies have been conducted in this regard which indicated that buffalo calves attained better nutrient utilization and average daily weight gain on feeding of diets with higher proportion of bypass protein.

Chatterjee (1998) studied the effect of formaldehyde (FA) treated mustard cake on growth and nutrient utilization in growing buffalo calves at NDRI, Karnal. Twelve male calves were divided into two groups of six each. The calves were fed concentrates mixture (having 32 parts of untreated mustard cake) with RDP/UDP ratio of 72: 28, wheat straw and green oat to fulfill the nutrient requirement as per Kearl, 1982 as control (T_0) and concentrate mixture (containing 32 parts of FA treated mustard cake) with RDP/UDP ratio of 40: 60, wheat straw and green oat as test group (T_1) for 119 days. An increase of 55.44 per cent in average daily gain was observed in calves fed FA treated mustard cake over those fed untreated mustard cake.

In another study Tiwari and Yadava (1994) investigated the effect of formaldehyde treated mustard cake on growth rate in buffalo calves. Five groups of four animals each were fed conventional concentrate mixture containing untreated mustard cake (A, control), 50% of untreated mustard cake replaced with FA treated cake (B), 75% replacement of mustard cake with FA treated mustard cake (C), 100% replacement of untreated mustard cake with FA treated mustard cake and 100% replacement of mustard cake with groundnut cake (E) for 120 days. Growth rate in control group calves was 338 g/d whereas increase of 29.59, 30.76 and 64.79 per cent was observed in calves fed 50, 75 and 100 per cent formaldehyde treated mustard cake instead of untreated mustard cake. Feed conversion efficiency was 10.23kg DMI/ kg body weight gain in control and improvement of 22.60, 22.38 and 34.70 per cent was recorded in formaldehyde treated mustard cake fed calves, respectively. Formaldehyde treatment of mustard cake may be more economical in those areas where groundnut cake is not available in abundance and is costlier, as similar results could be achieved by feeding of formaldehyde treated mustard cakes.

Bypass Protein and Milk Production in Buffaloes

Nutritionally balanced diets play an indispensable role in milk production so as to exploit the genetic potential of buffaloes as well as their reproductive efficiency (Arora, 1993). Ruminants are normally able to meet the essential amino acid requirements with microbial protein synthesized during rumen fermentation. During early lactation in high yielding buffaloes producing >10 lit of milk per day, labile tissue proteins provide amino acids for milk production as the feed intake does not commensurate with the amino acid requirement of mammary tissues. This mobilization of amino acids from tissues adversely affects the body condition score and subsequent milk yield. Feeding of bypass protein in such situation could provide extra amino acids to the mammary gland that supports high level of milk production. If protein with high biological value (BV) is supplied as such through diet to such animals, BV of high quality protein is depleted in the process of microbial degradation.

The use of bypass protein alters the RDP/UDP ratio in the diet increasing the efficiency of protein utilization, supply and absorption of amino acids like lysine and methionine in the small intestine that can help meeting the requirements of high yielding buffaloes. It also increases the quantity of metabolizable protein, which results in higher milk production.

Chatterjee (1998) studied the effect of bypass protein feeding in lactating buffaloes yielding 7.6 to 7.8 kg/d at NDRI, Karnal. Ten buffaloes were divided into two groups of five each. Buffaloes of control group were fed on green maize along with required concentrate mixture (CM-I) containing untreated mustard cake and groundnut cake whereas the second group was fed the same roughage along with concentrate mixture – II (CM-II), in which formaldehyde treated mustard cake replaced the untreated mustard cake. RDP/UDP ratio for both concentrates CM-I, CM-II was 70:30 and 43:57, respectively. Average milk yield in 105 days was 5.98kg/ day in control. An increase of 11.2 per cent in milk yield and 16 per cent in 4% FCM yield was recorded in buffaloes of second group fed with formaldehyde treated mustard cake. Milk fat and total solid contents were also higher in treated group. Feeding of formaldehyde treated mustard cake as bypass protein meal, also reduced the feed cost per kg FCM produced by 12.2 per cent as compared to control.

In another lactation study conducted by Bharadwaj and Sengupta (1999), an increase of 11.4 and 17.6 per cent was observed in milk yield and 4% FCM yield, respectively of buffaloes fed with commercially available bypass protein meal over the control group.

Bypass/Protected Fat

Bypass/ protected fats are fat sources which do not undergo hydrolysis and bio-hydrogenation in rumen showing very little negative effects on feed digestibility when fed to dairy animals at higher levels of supplementation and act as dense energy source and modify fatty acid profile of body tissue and milk. Being in a dry state rumen protected fats have the added advantage and hence are easily transportable and can be mixed into the diets without any special equipment.

Protection of fat can be done by following methods:

i) Surrounding unsaturated fatty acids by a protective capsule, such as formaldehyde treated proteins that act to shield internal fatty acids from bio-hydrogenation.

ii) Chemical modification of unsaturated fatty acids to chemical forms that resist bio-hydrogenation, such as calcium salts of fatty acids or fatty amides.

Types of protected/ bypass fat are as follows:

Crystalline Fat

Generally, it is made of saturated and mono-saturated fatty acids and are often considered as protected fat. Due to high melting point, they do not melt at ruminal temperature thereby resist rumen hydrolysis.

Direct Formaldehyde Treatment of Oilseeds

Oilseeds can be directly treated with formaldehyde. Treated proteins make the coating of lipids, which causes the protection of lipid from hydrolysis and bio-hydrogenation in rumen.

Calcium Salts of Fatty Acids

Ca salts of fatty acids do not disturb organic matter digestibility in rumen (Elmeddah *et al.*, 1991) due to specific effect of ionized calcium (Ferlay and Doreau, 1993).

Prilled Fatty Acids

Liquefying a mixture of fatty acids high in saturated fatty acid content and spraying the mixture of fatty acids under pressure into a cooled atmosphere results in a dried prilled fatty acid supplement that is inert in rumen (Grummer, 1988).

Need of Bypass Fat

In fattening calves with high growth potential and during early lactation in high yielding buffaloes, the energy intake is not sufficient to meet the requirement. Addition of high proportion of grains in the ration decreases ruminal pH, leading to ruminal acidosis and depression in fibre digestion. These conditions may lead to lower feed intake causing depressed growth rate in growing buffalo calves and lower milk production and milk fat depression in lactating buffaloes. Fat, being a dense energy source could be an alternative. But fat supplementation at more than 4-5 per cent level in the ration has been reported to adversely affect DMI and fibre digestion, as the free fatty acids liberated in the rumen inhibit the activity of rumen microbes. This has led the development of the concept of feeding of rumen inert or bypass fat that does not get hydrolyzed in the rumen.

Bypass Fat and Growth

There have been few studies conducted to study the effect of bypass fat on growth in ruminants as growing ruminants can be well fed to fulfill their energy demand but from the fatty acid profile of body tissue/ meat point of view, bypass fat feeding may be a kind of strategic nutrient. It is an established fact that dietary unsaturated fatty acids get bio-hydrogenated in rumen and ultimately deposited in body tissues in saturated form. These saturated fatty acids are not desirable for human health because of their supportive role in arteriosclerosis and other cardio-vascular diseases.

In an experiment, Kushibiki *et al.* (1997) evaluated the effect of calcium salts of palm oil on the growth response in Japanese Black calves and found that addition of 5 per cent Ca salts of palm oil of dietary dry matter resulted an increase of 30.9 kg body weight gain at 38 week of age in comparison to control.

Raikwar 2004 (unpublished) conducted a growth study of 120 days in cross bred cattle calves in which sixteen calves (6-8 month of age) were divided in to two groups and fed to attain the average daily gain of 500 g per day as per NRC, 2001. First group (C), as control fed with ration containing no bypass fat while treatment group, T, was fed ration containing Calcium salts of acid oil fatty acids as bypass fat @ 4% of dietary dry matter. Average growth rate was 490.89 ±0.01 in group C and 541.62±0.02 in group T exhibiting an increase (p>0.05) of 10.33 per cent growth treated over control group.

Bypass/Protected Fat and Milk Production in Buffalo

Because of certain peculiar characteristics of buffaloes regarding energy metabolism and milk fat content, there is a need for specific energy nutrition, especially for high yielders. Buffalo milk has an average fat content of 7-8 per cent. In addition buffaloes require 45 per cent more metabolizable energy (ME) than cattle during early stage of lactation (Lall and Tripathi, 1995). Sebastian *et al.* (1970) also, reported maintenance requirement of lactating buffaloes is 38% higher than that of lactating cattle of comparable size. So, buffaloes pose a peculiar challenge for animal nutritionists to fulfill energy requirement of high yielders.

Beyond a particular level of milk yield i.e. 10 kg/day in buffaloes; it is difficult to meet their energy requirements through ordinary ration. Usually all such buffaloes will remain under negative energy balance during first quarter of lactation. This energy deficit condition may lead to lowering of body condition score, persistency of milk production and reproduction performance. These problems can be overcome by incorporating bypass fat in the ration of such animals.

Western countries, use of bypass fat as an energy source in high yielding cows has been well established. However, in India it is yet to find acceptability due to various reasons. As a potent source of dietary energy, bypass fat can play a very important role in nutrition of high yielding buffaloes. Availability and awareness about such type of supplements are the main causes of unacceptability.

Only few studies have been conducted on the use of bypass fat in lactating buffaloes in our country. In an experiment Chauhan et al. (1999) investigated the effect of Bypass fat in buffaloes in their early lactation. Buffaloes were yielding 9-10 lit. milk/day. Thirty buffaloes were divided into five groups of six each. Buffaloes of control were offered ammoniated wheat straw and sorghum silage ad lib + 10 kg green berseem along with concentrate mixture containing formaldehyde treated GNC as CM-I. In another groups in addition to feeds offered to control, animals were supplemented with 200 g protected fat (T1), 400 g protected fat (T2), 1 kg boiled cotton seed cake (T3), and 1 kg cotton seed cake (T4), respectively. The total milk produced during 305 days per animal in control, T1, T2, T3 and T4 groups were 2270, 2402, 2307, 2444 and 2316 litres, rᵣ ꜱectively showing an increase of 5.8 per cent in buffaloes fed 200g protected fat over that of control.

In a recently conducted study at NDRI, Karnal, it was observed that the positive effect of feeding calcium salts of acid oil fatty acids (bypass fat) on milk production in Murrah buffaloes in early lactation. Twelve buffaloes were divided into two groups of 6 animals in each. First group (C) was fed control ration containing no bypass fat and the second group (T) was fed ration containing 4% (of dietary DM) Ca salts of acid oil. The average daily milk yield over 6 fortnights was 9.49 ±0.28kg/d in C group and 10.68 ±0.17kg/d in T group. The value of 4% FCM yield (kg/d) was also significantly higher for T group (13.45 ±0.020) than for C (11.87±0.21). Milk fat per cent was also higher in treated group but the difference was not significant.

Thus the literature indicates the usefulness of supplementation of bypass fat in the ration of growing and lactating (high yielding) buffaloes. However, sufficient research work has not been taken up on the use of bypass nutrient technology in buffaloes. So long term studies are needed to establish the utility of this technology for achieving higher productivity in buffaloes of India.

Major Thrust Areas of Research

◉ Feeding of bypass nutrients to buffalo calves to exploit their potential as meat animal and its effect on meat quality.

◉ Bypass nutrient feeding in high yielding buffaloes for augmenting milk production and its quality.

◉ Reproductive performance of buffaloes by feeding bypass nutrients.

◉ Scaling up of bypass nutrient production technique specially Calcium salts of non traditional oil sources such as acid oil, rice bran oil soap stock etc.

References

Arora, S.P.1997. Thrust areas of nutritional research for enhancing milk production. Perspective – 2020. VIII Anim. Nutr. Workers Conf., Chennai – 600007. pp-109.

Bharadwaj, A. and Sengupta, B.P.1999. Effect of sources of dietary unprotected and protected protein in lactating Murrah buffaloes. Buffalo J. **15(3):** 269-278.

Chatterjee, A. 1998. Ruminal and post ruminal digestibility of formaldehyde treated mustard cake protein and its effect on growth and milk production in Murrah buffaloes. Ph. D. Thesis submitted to NDRI (Deemed University), Karnal.

Chauhan, T.R.; Gupta, R; Arora, U,; Bharadwaj, A. and Lall, D. 1999. Effect of supplementing calcium salts of long chain fatty acids on milk yield and composition in lactating buffaloes. Bubalis Bubalus, **2/99:** 51-58.

Elmeddah, Y.; Doreau, M. and Michalet- Doreau, B. 1991. Interaction of lipid supply and carbohydrate in the diet of sheep with digestion and ruminal digestion. J. Agric. Sci., **116:** 437-445.

Ferley, A. and Doreau, M. 1993. Effect of lipid supply in the diet of cows on calcium and magnesium pools in the rumen. Proc. Nutr. Soc., **52:** 151A.

Grummer, R.R. 1988. Influence of prilled fat and calcium soaps of palm oil fatty acids on rumen fermentation and nutrient digestibility. J. Dairy Sci., **71:** 117.

Kearl, L.C. 1982. Nutrient Requirements of Ruminants in Developing Countries. International Feedstuff Institute. Utah Agricultural Expt. Station, Utah State University, Logan, USA.

Kushibiki, S.; Kozuhiro, U.; Tevada, T. and Hayashi, T.1997. The effect of calcium soaps supplementation on growth response in beef calves during grazing and housing period. Anim. Feed Sci. Technol., **68:** 983-986.

Lall, D. and Tripathi, V.N. 1995. Meeting the energy requirement of high yielding buffaloes. Indian Dairyman, XLVII(**9**):37-41.

Noel, R.J. 2000. Official feed terms. Pages 187–200. In Association of American Feed Control Officials, Official Publication 2000.

Sampath, K.T. 1990. Rumen degradable protein and undegradable protein content of feeds and fodders – a Review. Indian J. Dairy Sci., **43:** 1-10.

Sebestian, L.; Mudgal, V.D. and Nair,P.G. 1970. Comparative efficiency of milk production by Sahiwal cows and Murrah Buffaloes J. Anim. Sci., **30(2):** 255-56.

Tiwari, D.P. and Yadava, I.S.1994. Effect on growth, nutrient utilization and blood metabolites in buffalo calves fed rations containing formaldehyde treated mustard cake. Indian J. Anim. Sci. **64(6):** 625-630.

CHAPTER 18

Unconventional Feed Resources for Buffaloes in Various Agro-Climatic Regions

S.K.Tomar

National Dairy Research Institute, Karnal-132 001 (Haryana)

The Planning Commission aimed at the regionalisation of the Indian agricultural economy and attempted to bring integration of plans of agroclimatic regions with the state and National Plans. The final goal is to organize agricultural planning systems for 15 agroclimatic regions (Khanna, 1989) so identified and to develop policies for faster agricultural development on regional basis. The zones were delineated on the basis of the characteristics of soil topography, climate and water resources. Buffaloes, primarily reared for milk production are distributed right from Kanyakumari in south to the bugyals above Shri Kedar Nath at the height over 4000 m in north and Jammu and Kashmir in west to hilly state of north-eastern region. Local small size buffaloes are also found on Andaman and Nicobar islands.

Buffaloes in the Indian sub-continent make greater contribution to milk production i.e. 54% though they constitute only 31% of the bovine population (Aneja *et.al.*, 2002). Feeding of buffaloes for long was carried on, in lines of cattle feeding in want of separate feeding standards. Nutritional requirements of Indian buffaloes have been brought up in a series of publications (Ranjhan and Pathak 1979, Kearl, 1982, ICAR, 1985, Pathak and Verma, 1993, Ranjhan, 1998, Mandal *et al.*, 2002). During the last two decades buffalo meat production has taken an industrial shape to a great extent and has opened an additional avenue for the profitable rearing of till recently cosidered undesired male calves.

The annual demands of feeds and fodder for the year 2000 are estimated to 88.05 million metric tonnes of concentrate, 632.61 million metric tonnes of dry fodder and 830.12 million metric tonnes of green fodder. However, only 46.18 million metric tonnes of concentrates, 523.61 million metric tonnes of dry fodder and 573.50 million metric tonnes of green fodder are available (Aneja *et al.*, 2002). The gap between availability and demands of concentrate is very wide, there is a deficit of about 48% for concentrate, 31% for green fodder and 17% for dry fodder.

Feeds and fodders consumed by cattle are also eaten by buffaloes, however shortage of good quality roughage and concentrates is a chronic problem, which will likely to continue for many years if suitable steps were not taken to limit the population of non-productive cattle and buffaloes in India (Pathak, 2003). Dry crop residues will continue to constitute significant part of bovine diets, which will as usual have energy-protein depression effect due to very low digestibility besides being deficient in many essential minerals and vitamins and may lead to productive and reproductive disorders (Tomar and Arora, 1982)

To bridge the gap of demand and supply of feed resources, unconventional feeds and fodder are being widely used to meet the requirement of large ruminants. To meet the future challenges, intensive system of feeding management have been developed through concerted efforts. The objective of this paper is to explore use of unconventional feed resources in buffalo production in different agroclimatic regions of the country as well as to identify thrust areas of future research in the changing scenario.

Agroclimatic Regions In India

The pattern of rainfall, soil and cropping system all over India reflects the climatic variations in the different parts of the country. It varies from per-humid in north-east India to arid in Rajasthan. A belt of arid or semi-arid climates extends from the north to the south, dividing the humid climates of the west coast and the central eastern parts of the country, where the annual rainfall is generally less than 1000mm.

Of the 15 agro-climatic regions, few of them could be clubbed together for purposes of feeding and common available unconventional feed resources for buffaloes.

Western Temperate Himalayan Region

Small herds or groups of 1 to 10 animals are reared in the hilly region of Jammu and Kashmir, Himachal Pradesh, Uttranchal and so far no knowledge is available about their evolution and linkages with the other Indian buffaloes. These are also either domesticated or tamed buffaloes but almost all are lighter in size and narrower in shape with strong foot hold. The state of Jammu and Kashmir has three distinct climatic zones viz. Jammu, Kashmir and Ladakh. The latter two zones have cold to extreme cold climate whereas the former one has sub-temperate, therefore buffalo rearing practices is being followed only in Jammu region. Except for summer months, buffaloes are either stallfed or let loose for grazing in the barren paddy fields/ foot hill pastures. More number of animals were seen on alpine pastures during May to August than other months of the year. No migration to the high alpine pastures is observed during winter months (Tomar and Lall, 1992). The fodder production potential of J&K is higher than the fodder requirements. The estimated annual fodder production in J&K from forests, pastures and lakes was about 52.4 MT, as against the estimated annual fodder requirement of 49.2 MT (Kotwal, 1982). This clearly indicates that there is a considerable loss of feeds and fodders because of various reasons. The exact total quantity of aquatic vegetation is although not worked out but the same has not been exploited to the desired level. Lastly, the pasture herbage from forest and paddy fields also either remains unharvested in the forest due to difficult terrain or parish due to severe rainfall during autumn. Thus, there is a huge wastage of the overall available feeds and fodder resources in the state resulting in its acute shortage during lean period. The farmers have to depend upon the purchase of fodder from outside states.

Tree leaves : The potential roles of tree foliage in ruminant nutrition are such as : a) high digestibility biomass resource, b) supplement to provide nutrients deficient in the diet, c) an enhancement of microbial growth and digestion of cellulose in the rumen; sources of undergradable protein, d) source of vitamins and minerals to complement deficiencies in the basal feed resource, e) availability in and around the farm, provisions of variety in the diet, laxative effect, and f) reductions in the requirements for purchased concentrates. The use of tree forages as components of diets is a widespread practice in this region. Tree leaves have high protein contents (9-25%) and generally have low rates of degradability in the rumen.

The more important tree forages of this region are *Robinia, willow, Ailanthus* and *Celtis*. The average composition of these tree forages ranged, dry matter 37.6 to 40.4, crude protein 9.5 to 18.5, ether extract 2.4 to 7.2, crude fibre 7.5 to 27.3, nitrogen free extract 42.6 to 52.9, total ash 10.7 to 23.2 and silica 1.4 to 7.6 (Tomar and Sharma, 2000). Traditional crop residues (paddy straw) with tree leaves were more nutritive than oats/or berseem.

Banj (*Quercus incana*): It is mostly found in the Himalayan forest at the height of 1,500 to 3,000 m above sea level. Leaves contains 5.7 per cent DCP and 43.7 per cent TDN. Buffaloes can be fed with these leaves at the rate of 2 to 2.5 per cent of their body weight.

Biul (*Grewia optiva* or *G. Oppositifolia*): Mostly found at an altitude of 500 to 2,500 m. It is also found in plains (tarai), 15 to 20 kg of green leaves per year can be taken from one tree. The leaves are highly palatable and nutritious (Negi *et al*. 1979). These are lopped for animal feed during winter months. The leaves contain 17 to 23% CP, 2.5 to 5.0% EE, 17 to 24% CF, 11 to 13% ash and 35 to 45% NFE. The digestibility of crude protein is very high (75%). The voluntary intake is 3.5% of body weight.

Bhoj (*Betula utilis*): It also grows to 3,000 to 4,500 m altitude above mean sea level in the Himalayan ranges. The leaves are fed to all classes of ruminants including buffaloes.

Mulberry (*Morus indica*): The leaf fodder of mulberry can be profitably utilised as a supplement to poor quality roughage. On DM basis, the leaves contained 15.0-27.6%CP, 2.3-8.0%EE, 9.1-15.3%CF, 48.0-49.7%NFE, 63.3% total carbohydrates, 14.3-22.9% ash, 2.42-4.71% Ca, 0.23-0.97%P, (Makkar *et. al.*, 1989). The feeding value of mulberry leaves is rated high by the livestock owners. 6 kg of leaves per day can be fed to milch buffalo/cow without adversely affecting the health of animal or the yield and butter content of milk.

Other important fodder trees in Himalayan ranges are bamboo (*Dendrocalamus strictus*), bahera (*Terminalia bellerica*), jharkeri (*Ziziphus nummularia*), sal (*Shorea robusta*), having 1 to 9% DCP and 42 to 55% TDN.

Aquatic vegetations: Feeding of water weeds at flowering stage specially to large ruminants during summer to autumn season is a unique and popular animal feeding practice in rural J&K. These vegetations belonged to genus *Nymphaea, Carex* and *Phragmites* and obtained from fresh water bodies. Proximate analysis of these vegetations revealed that dry matter contents were higher, (20.5-40.0%) during August-September than May-June harvest (11.1-41.8%). *Nymphaea* spp. seems highly nutritious as revealed by proximate analysis (CP 19.4, NFE 55.4 and total ash 11.7%). Although the nutritive values of *Carex* and *pharagmites* spp were not comparable with *Nymphaea* spp, yet these were far better than conventional feeds and fodder offered to livestock (Tomar and Sharma, 2000).

Apple Waste *(Malus pumila)* : Apple is a premier table fruit and cultivated in J&K, Himachal Pradesh and Uttranchal. Apple pomace is the main byproduct obtained after juice extraction. Apple pomace and damaged apples are high in energy (TDN 62-65%) and low in proteins (DCP 1.6-2.6%) and can be preserved as silage which can be safely fed to dairy buffaloes and cattle from 15 to 30% of their total DM. Urea treated apple pomace replaced 40% maize grain without affecting DMI, feed efficiency, digestibility of nutrients and milk production (Bakalkar, 1987).

Salseed meal *(Shorea robusta):* About 5000 km^2 of area is under sal in the Sub-Himalayan forest. About 55 million tonnes of sal kernel and 6 million tonnes of fat are available every year with the extraction of oil, a large quantity of deoiled sal seed-meal is available for livestock feeding. Sal seed meal contains about 8-10 per cent crude protein but the digestibility of protein is negligible due to tannin contents (8-13%). The TDN value is about 45% per cent, hence it is only a source of energy.

Salseed meal in the diet significantly decreased crude protein digestibility due to presence of tannins, however ammoniated salseed meal has improved protein digestibility. It, as such, is not palatable but when added to a level of 20-30%, the feed intake was not significantly affected. The general recommendations are to use deoiled salseed meal (DSSM) as energy source to maximum level of 10% in the rations of growing, dry and lactating cattle and buffaloes. About 30% of the tannins of DSSM were made inactive by treating salseed with NH_3 treatment seems to be most practicable (Maheshwari *et al.*, 1984).

Eastern Himalayan and Lower Gangatic Plain Regions

Probably no attempt has been made to differentiate the buffaloes of Sikkim, Arunachal Pradesh, Mizorum, Manipur, Nagaland, Tripura, Assam, Meghalaya and West Bengal, although all states are having one or more types of buffaloes. The native buffaloes of these states are being classified as swamp buffaloes but large number of dairy buffaloes are also found. Assamese and Sikamese are small sized hill buffaloes of the state of Assam and Sikkim.

The prime constraints for buffaloes production in this region is scarcity of feed and fodders. Hardly any farmer directly produces any feed or fodder for animal feeding. Mostly crop residues and crop byproducts, which are nutritionally very poor, are available for the animals to thrive. Per capita availability of such poor quality feeds is also limited in this region. Accordingly, farmers feed their productive cattle and buffaloes with kitchen /vegetable /fruit wastes, unconventional feeds collected from forests, field grass and tree leaves collected from all possible sources and thus try to satisfy the nutritional requirement of the animals.

Fodder Tree Leaves: A very large number of tree leaves are used as ruminant feed in both plains and hills. The comprehensive survey work of Gupta (1981) and many others (Balaraman, 1981) indicated that several hundreds of varieties of tree fodders are commonly and traditionally fed to livestock in north east region. Dry matter intake of fodder tree leaves varies from 1.0 to 2.5 kg/ 100 kg body weight. Low intake of dry matter appears to be a limiting factor for energy supply from tree fodders to dairy cows/buffaloes of high producing ability. TDN values of fodder tree leaves vary from as low as 26.90% in *Cordia dichotoma* to very high value of 70.22% in *Leucaena leucocephala*. Approximately, 60% of the fodder tree leaves evaluated so far have been found to contain less than 50% TDN on dry matter basis (Saha *et.al.* 1997).

Byproducts of Rice and Pulses: Eastern region is essentially a rice zone. Rice *Kura* (a mixture of rice bran and mixed husk) and rice grit (small particles of rice grain mixed with dirt and generally unfit for human consumption) are two common byproducts of rice crop which are locally available every where in this region. Saha *et al.* (1997) showed that rice kura has 5.75-12.69% CP, 0.78-14.73% EE, 20.25-32.53%CF and 33.96-59.37%NFE, rice grit has 3.70-10.72%CP, 0.81-3.52% EE, 1.08-34.79% CF and 48.54-85.64%NFE.

Various pulse *chunis* are also available in some places where pulses are cultivated but the total quantity is very low and restricted to a few pockets only. Arahar *chuni*, Kalai *chuni*, and Moth *chunies* are mainly available and has 10-20% CP, 0.5-2.0%EE, 10-27%CF, 33-60%NFE and good amount of Ca and P. Pulse husk is also available having 17.27% CP, 2% EE, 17.91%CF, 52.28% NFE and 10.54% total ash.

Horticultural Waste

Pineapple wastes: From pineapple (*Ananas camosus*) cultivation and canning of its fruits, large volume of waste materials of nutritional value are available for livestock feeding in this region. Pineapple wastes are of two types, viz cannery wastes and plant residues. Pineapple cannery waste consists of outer peels (shell), the crown and bud ends of the fruit, fruit trimmings, the inner core and the pomace of the fruit from which the juice has been extracted. The pineapple waste contains 80-95% moisture and 5% CP. Buffalo can consume about 2 per cent of dry matter through this waste.

Jackfruit wastes: The waste from ripen fruits is more palatable than the waste from raw fruits. It contains 13.3% CP, 5.09%EE, 12.4%CF, 0.21% Ca and 0.35%P. It is a rich source of energy having 60.5%NFE. It can be incorporated in buffalo ration upto 10-15%. Banana peels, orange peels, mango peels, lemon peels have 6-13% CP and 55-74% NFE and can be included in the ration of buffaloes. Cauliflower leaves, papaya leaves, cabbage leaves and pea pods have 12-27% CP, 38-55% NFE, 1.0-4.50% Ca, 0.05-0.60% P and can be included in the ration of ruminants.

Aquatic plants: Water hyacinth (*Eichhornia crassipes*) is a growing weed of terrestrial as well as running water. The plants provides 2.8, 24.9 and 44.7 per cent digestible crude protein, starch equivalent and total digestible nutrients, respectively, on DM basis to cattle and buffaloes when fed as silage, and 4.4, 39.9 and 51.9 per cent DCP, SE and TDN, respectively when fed as hay. An increase of 10-15% in milk yield is reported when it was incorporated in the ration of buffaloes.

Tea industry wastes: India produces nearly 400 million kg of tea every year and main tea growing states are West Bengal and Assam. Tea wastes in the form of fluffs, stalks and sweepings become available during production, storage, handling of tea garden factories, warehouses, packing factories and shipment sheds at ports. About 2-3 per cent tea is wasted, this amounts to 10,000 tonnes in India. Caffein can be isolated from waste tea leaves. It contains about 18 per cent of crude protein and can be used as feed ingredient in the ration of cattle and buffaloes. Spent tea leaves (STL) contain 25 per cent CP and can be used upto 20% in the concentrate mixture (Ranjhan, 1990).

Sunnhemp: Sunnhemp (*Crotolaria juncea*) is grown throughout India mainly is West Bengal and adjoining states. In most cases it is used as manure, however in some parts it is fed as fodder. After crushing this can be fed to cattle and buffaloes but feeding, as such is not palatable. This can be mixed with other palatable feed stuffs in a concentrate mixture and fed to buffaloes. It contain 13.9%CP. 2.6%EE, 34.1%CF, 40.6% NFE, 8.8% total ash. Sunnhemp contains 4.7% lysine and 1.7% methionine (on protein basis).

Middle, Upper and Trans Gangatic Plain Regions

The hot tropical region comprising Punjab, Haryana, Delhi and Uttar Pradesh is the home tract of most of the established high yielding milch breeds of buffaloes i.e. Murrah, Nili-Ravi. In most of the parts, wheat straw, paddy straw and stovers are the staple roughage. Cultivation of green fodder is more popular is Punjab, Haryana, Delhi and western U.P. as compared to other regions. Mostly farmers (59%) burnt the paddy straw in the field and cattle and buffaloes were fed with fresh chaffed sorghum, berseem and oats in Karnal district of Haryana (Tomar and Thakur, 2002).

Generally the buffaloes are fed on crop residues like wheat straw, sugar cane tops, cultivated green fodder and concentrates. The unconventional feeds and fodders of this region are as below:

Rice Hulls: Rice hulls or husk is the largest byproducts of rice milling industry and most of it is either burnt or used as bedding for poultry and animals. It contains 3.2%CP, 1.4%EE, 32.7% CF, 38.7% NFE, 23.6% Ash, 16.0% silica, 69.4% NDF and 38.0%ADF (Ranjhan, 1990). Rice husk as such is not palatable. Ground rice husk is often mixed to adulterate the rice polishing/bran (Singh and Kundu, 2003). Husk when substituted at 0, 33, 66 and 100 per cent of rice straw in growing crossbred calves ration. The growth rate was depressed with the increase of rice hulls. However at 33% level, the depression was not significant as compared to that in the control. The DCP and TDN of rice hulls were 0.16 and 11.7 per cent respectively.

Maize husk: It is superior to even gram husk in its nutritive value. It has 8.1% CP, 1.5% EE, 15.7%CF, 72.5% NFE and 2.3% ash. Fifty per cent of the dry roughage portion in the ration of adult non-producing cattle and buffaloes can be replaced by this product.

Sugarcane Bagasse: Bagasse is one of the most important byproduct of sugar industry. It is a source of cellulose but poor in protein and high in lignin. As such it is not very much relished by the animals. Cattle and swamp buffaloes can consume about 80 per cent of their expected dry matter intake through bagasse impregnated with 1 and 2 per cent urea. Buffaloes appeared to have significantly higher apparent digestibility of crude protein of ration consisting 60% bagasse and 1% urea, than cattle. In India, bagasse has been used as famine roughage for ruminants. About 60 kg ration consisting of 20 kg bagasse, 6 kg molasses, 0.34 kg common salt, 0.34 kg mineral mixture and 65 liters of water forms a maintenance ration along with 800 g of groundnut cake for adult animals weighing 400 kg. The optimum steam pressure for treatment of bagasse was 7 kg per cm^3 for 30 minutes and fortification with 1 per cent urea and molasses. The steam treated bagasse + urea + molasses costs about Rs. 300 per tonne, this compares favourably with conventional roughage. Sengar *et al.* (1993) reported that sodium chlorite treatment of bagasse had decreased the lignin content and increased *in situ* dry matter disappearance.

Sugarcane molasses: Molasses is about 3% of the sugarcane or about $1/3^{rd}$ of the quantity of sugar produced. It contains about 61% sugar and 3% protein and has been used in animal feeding at 5-10% levels. Molasses is mainly used as (a) carrier for urea impregnation of poor quality roughages (b) as binder for commercial pelleted feed and (c) as sweetner for increasing the voluntary intake of unpalatable feeds.

Fruit and Vegetable Wastes

Neem seed cake (*Azadirachta indica*): Neem is a large evergreen tree, which is native of India. Neem seed cake is produced from seeds after extraction of oil. Neem seed kernel cake has 35-40% CP, but found unsuitable for animal feeding due to presence of bitter and toxic triterpenoids (azadirachtin, salanin, nimbin, nimbidiol etc). Neem seed cake besides being unpalatable is also adversely affecting the growth, male reproductive system and even sometimes leading to haematurea. The cake in alkaline medium without water washing was tried either by soaking in water (1:5 w/v) containing either NaOH for

24 h or by ensiling with 2.5% urea (w/w) for 5-6 d. The sun dried and ground alkali treated and urea ammoniated neem seed kernel cake was found suitable in the feeding of cattle and buffalo calves without affecting their growth, nutrient utilization, blood profile, rumen/ caecal fermentation pattern indicating that this cake after treatment/ ammoniation can be substituted as wholesome protein to spare costly vegetable protein sources in the buffalo feeds (Sastry and Agrawal, 1998).

Mango seed kernels: The annual availability of mango seed kernels is about 5,00,000 tonnes (SEA, 2000). It has a high content of starch i.e. 92% of its total carbohydrates content. Dhingra and Kapoor (1985) reported that it has about 68-70% neutral lipids of the total lipids. Stearic and oleic acids are the main components and saturated and unsaturated fatty acids were 40 and 60%, respectively. Histidine and cystine were in greater amount as compared to other amino acids in the mango seed kernel (Barman and Rai, 2003). *In vitro* proteins digestibility varied from 26to 29% but when soaked in water *in vitro* digestibility increased by 50%. Growth rate was similar (440 g/d/ animal) when used 20% mango seed kernel in the concentrate mixture of calves. Digestibility of CP, NFE and EE were 17.66, 17.72 and 58.54 per cent, respectively in mango seed kernel based diet.

Potato waste: Potato waste is the by-products remaining after extraction of starch with cold water. 16-20% of starch is produced and 3-3.5% dried potato waste is obtained for feeding to cattle and buffaloes (Boucque and Fiems, 1988).

Tomato pomace: Tomato pomace accounts for about 20% of the total production. It consists of skin, pulp and seeds. Dried and ground pomace contains 22.36% CP and 28.11% CF with a nutritive value of 13.7-15.8%DCP and 60.8-73.3% TDN. Tomato waste can be incorporated upto 35-40% of the concentrate mixture in the diet of cattle and buffaloes. Other vegetable wastes of regional importance are cabbage, carrots, cauliflower, lady's finger, empty pea pod wastes which can be included in the diet of cattle and buffaloes (Rai and Barman, 2004).

Central Belt of India

Extending from Gujarat to Orissa and also some part of Andhra Pradesh, most of the buffalo breeds are medium sized, good milk yielder, widely adaptable to harsh agro climatic conditions, are present. Well recognized breeds in this group are Mehsana, Zaffarabadi, Surti, Nagpuri, Bhadawari and relatively less known breeds of Orissa like Sambalpuri, Bheda, Kujang, Parikad and Manda. In this region hardly any farmer grow fodder for livestock except in part of Rajasthan and Gujarat. Generally animals thrive on fodder tree leaves, leguminous pulse straws and chunies and unconventional oil seed straws and meals.

Fodder Tree Leaves

Ardu leaves: Ardu (*Ailanthus grandis* and *A. grandulosa*) are commonly found in Rajasthan. These species grow into enormous trees. A fully grown tree gives 6 to 7 quintals of edible leaves twice a year. The leaves are palatable and voluntary intake is 1.5–2.0 per cent of the live weight. It contains 13% DCP and 63% TDN. Cattle and buffaloes can be maintained exclusively on ardu leaves.

Beri: Beri or Chinese date (*Ziziphus jujuba*) is a very common shrub commonly found in the desert region of India. In Rajasthan and Gujarat leaves are dried and stored for use as protein supplement with the normal grazing. It contains about 18-20% CP but has poor digestibility due to presence of tannins.

Acacia: They are mostly found in the arid region of tropical countries. Acacia is small to medium sized with thorns and pods. Acacia leaves are lopped for feeding ruminants.

Subabul: Subabul (*Leucaena leucocephala*) is a perennial shrub. The leaves contain 21-25% protein. When grown for fodder, the first cut can be taken within 6-9 months of sowing and subsequent cuts at intervals of about 4 months. Subabul can make up the protein requirement for maintenance of cattle and buffaloes weighing 400 kg, if given at 25-30% in the ration. Garacia *et al.* (1996) reported that optimum harvesting interval or age at harvest should be about 8 weeks or just before the flowering. Rai *et al.* (1994) concluded that subabul leaf meal can be used as sole protein replacement for cakes in the ration for economical milk production.

Leguminous Pulse Straws: Leguminous crops having fibrous residues are: black gram (Phaseolus radiatus), green gram, (P. mungo), moth (P. aconitifolius), cow pea (Vigna sinensis), masoor (Lens esculenta), pigeon pea (Cajanus indicus) and ground nut (Arachis hypogea). The residues are composed of husk of the pod with leaves and tender stems which are more nutritious than the cereal straws and stovers. They contain 8.0-12.5% CP and 40-50% NFE. The voluntary intake varies from 1.8-2.5 per cent of the body weight in cattle and buffaloes. Groundnut straw along with wheat straw (35%) and wheat bran (1 kg) can meet the nutritional requirement of a buffalo weighing 450 kg and producing 5 kg milk.

Cotton straw: Cotton straw in ground form could be used as roughage source (30%) in complete diets for lactating buffaloes. The palatability and nutrients utilization of cotton straw was improved by expander method without enhancing the cost of feed on production basis, such feeds could prove to be useful during fodder scarcity (Nagalakshmi *et al.*, 2004).

Oil Seed Meals/Cakes

Babul pods: Babul (*Acacia* spp) pods are produced in India about 600,000 tonnes annually in the states like southern parts of UP, AP, Maharashtra, Rajasthan, MP and South Haryana. Babul pods contain about 13% CP and 59% TDN (Punj, 1988). The use of babul pods as animal feed is limited due to its high tannin content (Barman and Rai, 2003 a). Babul pods contain almost all the essential amino acids in good proportion i.e. histidine content of pods was higher than that of whole egg protein in terms of g/100 g protein whereas methionine content was similar with that of whole egg protein (Barman and Rai, 2003 b). Berman *et al.* (2003) reported that goats are more resistant to *Acacia nilotica* tannins followed by cattle and buffalo. The main degraded products of *Acacia nilotica* tannin in buffalos are phloroglucinol and resorcinol. The concentrations of these products decreased noticeably from 18 to 42 h in buffalo rumen and 4 and

6% tannin levels (Siddhuraju *et al.*, 1996).

Mahua seed cake: Mahua (*Bassia laifolia*) belongs to the genus *Bassia* under order sapotaceae, is widely distributed in the deciduous forests of India, producing 0.12 million tonnes of cake from 0.21 million tonnes of seeds collected annually. Mahua seed cake is a moderate source of protein (15-24% CP) and contains reasonably good amount of NFE (53.3-60.1%). It contains saponin (mowrin) varying from 4.6-20.0% and hydrolysable tannins upto 6.4% which renders the cake unpalatable and toxic. Ensiling the cake with fertilizer grade urea @ 6.7% (w/w) for 30 d converted the cake into a palatable product and improved feed intake and nutrient utilization in large ruminants (Gowda, 1991). Groundnut cake N of concentrate mixture at 0, 25, 50, 75 and 100% levels were replaced with processed mahua seed cake and fed to buffaloes, Milk yield and milk composition remained normal (Tiwari and Patle, 1997).

Castor bean meal: The castor (*Ricinus communis*) production, during the year 1990, was 5,00,000 MT (FAO, 1990). The caster bean meal (CBM) is one among inedible protein rich byproduct of oil industry which contains 30-40% CP but it has toxic principles viz. Ricin, ricinine, allergen and chlorogenic acid. Incorporation of processed CBM (after suitable detoxification) as protein source in the diet of ruminants had no ill effect. The CBM contained 20.3% DCP and 57.4% TDN (Sastry and Agarwal, 1998).

Karanj cake: More than 111,000 tonnes of karanj (*Pongamia glabra*) seeds are available in India under wide range of agro climatic conditions. The seed contains on an average 28.3% oil. The cake left after oil extraction was found to be unpalatable and toxic even at lower level (4%) of dietary incorporation. The toxic principles Karanjine, pongaglabrone and pongapin are associated with oil fraction. The deoiled (solvent extracted) karanj cake is a rich source of protein (33% CP, 25.5% DCP), low fibre (5.4%) and sufficient TDN (62-71%). This cake can be incorporated in the concentrate mixture of cattle and buffaloes.

Kosum cake: Kosum (*Schleichara oleosa*) is available widely in MP, where more than 100,000 tonnes of cake is available (Singh, 1987).

It contains cyanogenic glucosides. Autoclaving, boiling, deoiling or even storing reduces cyanide content by 65% . Its DCP, TDN values are 16.25 and 78%, respectively.

Tobacco seed cake: After USA and China, India (Orissa and Gujarat) ranks third in world tobacco (*Nicotiana tobaccum*) production and produces about 9,000 tonnes of tobacco seed cake. It contains about 26% CP and 20% CF. It can partly replace conventional concentrate in the ration of cattle and buffaloes, if replaced gradually.

Citrus waste: Large ruminants can be fed citrus waste upto 40 kg/ head/day without any harmful effects (Umoh, 1982). After pressing oranges, lemons or grape fruit for juice, the left over which comprises the pod seeds and rag etc amounts to 40-60% by weight of the fruit could be very palatable to buffaloes. Citrus pulp contains a relatively high amount of pectin and soluble carbohydrates, and for this reason dried citrus pulp has bee n used to replace cereals in ruminants diet (Bhattacharya and Hard, 1973)

Southern Region

It includes Andhra Pradesh, Tamil Nadu, Karnataka and lower parts of Maharashtra. Small scattered groups of similar buffaloes were found in all the states down the Nagpur but except the description of Toda and South Kanara, probably attempt has not been made to describe the native buffaloes of this region. In Andhra Pradesh, buffaloes of Godawari breed were found. Sorghum, pearl millet, coarse rice and other millets are the cereal crops. Unconventional feeds available in this region areas below-

Cocoa pod husk: The dried cocoa (*Theobroma cacao*) pods are brittle and the husk contain 6% CP, 31% CF, 3% EE, 12% ash. It can be used upto 10-15% in the diet of ruminants. Cocoa pod husk contains alkaloid thiobromine in traces, hence the alkaloid does not produce any deleterious effect on animals, but in quick fermentation condition cocoa pod husk thiobromine may create problem. Sun dried pods are ground for cattle and buffalo feed.

Coconut meal: Coconut (*cocos mucifera*) is cultivated mainly in Karnataka, Kerala, Tamil Nadu and Andhra Pradesh. Coconut availability in Asia is about 3.75 m MT and 35-40% left as meal. Its DCP content is 21% and ME/Kg is 1.9 Mcal (Sastry and Agarwal, 1998). The storing quality is poor as rancidity develops due to its oil content (3.8%) which may lead to diarrhoea in animals.

Tamarind seed: Tamarind (*Tamarinus indicus*) seeds abundantly available in MP and Maharashtra (about 500,000 tonnes). The seed contain 30-45% red hulls and 60-65% white kernel. The hull and kernel have 9.1 and 15.4% CP and 11.3 and 1.5% fibre, respectively. The seed hulls can replace 10-15% of maize in concentrate mixture of buffalo calves. Kernels on the other hand can replace 95% of the maize component of concentrate mixture in buffalo calf ration (Singh 1987).

Coffee waste: Coffee (*Coffea arabica* and *C. robusta*) is cultivated commercially in Karnataka, Tamil Nadu, Kerala and Andhra Pradesh. In these states, coffee is available about 70,000 tonnes, which yield about 45,000 tonnes of coffee husk. It has about 9% CP, 40% CF, 1.0% EE and 5% ash. It has 3.4% DCP and 42.2% TDN, and can be incorporated in buffaloes ration.

Kapok seed meal: Kapok (*Ceiba pentandra*) or silk cotton seed meal is grown in M.P. and Tamil Nadu. It has 20% CP and 1.3 Mcal ME/kg. It has cyclopropenoid acid (18%) and tannins, even then during scarcity it is incorporated in the diet of ruminants.

Coastal and Island Regions

Narrow strips on east and west border of South India are coastal regions. Andaman & Nicobar Islands and Lakshadweep are also included in this region. Buffaloes of these regions are not described but there are small sized native buffaloes. Here livestock farming also depends on the unconventional feeds.

Rubber seed cake: In India, rubber (*Hevea brasiliensis*) trees are grown over an area of 224,428 hectares of which 202,320 hectares are in Kerala with an estimate potential of 150,000 tonnes of rubber seed cake. The digestible crude protein is about 18% and TDN is 54%. Rubber seed protein is poor in methionine but rich in lysine and cystine. Feeding of this cake upto 30% level in the concentrate ration of lactating cows showed no significant effect on milk yield, butter fat, melting point, iodine and saponification values .

Cassava (Tapioca) waste: Cassava (*Manihot esculenta*) feeds and roots are poor in CP (2%) but rich in NFE (85%) whereas tapioca starch waste has 2% DCP and 64% TDN. It has been recommended that tapioca waste can be safely included in the concentrate mixture with a considerable economy for maintaining body weight, milk yield and butter fat production in buffaloes.

Sea-weed: India has the coastline of more than 7000 km and about 70 species with more than 0.2 MT/year wet harvestable biomass of seaweeds from different parts of the Indian coasts is available (Sahoo *et al.,* 2001). Historical records show that the use of seaweeds in agriculture is very old and some species are used as feed and fodder for pig, horse, chicken, cattle and buffaloes. Vitamin content is relatively higher as compared to wheat and milk. About 624 marine algal species of various groups are recorded from Indian ocean, with a maximum number of 302 species in Tamil Nadu, 202 in Gujarat, 152 in Maharashtra, 89 in Lakshadweep islands and 75 in Goa. These underutilized or unutilized seaweed resources can be used as fodder or feed for cattle and buffaloes. These seaweed resources grow best in the tidal or intertidal waters along peninsular coastline and Andaman-Nicobar islands. There is possibility of using these seaweeds either after drying or preserved with chemicals before that can be used as ruminant feed. Agar-agar is being extracted from *Sargussam* seaweed and its byproduct can be used for livestock feeding. This product contain about 10% CP and 33% ash.

Conclusion

Buffaloes fed low quality forages require supplementation with the critically deficient nutrients to optimize productivity. Tree leaves and unconventional oil seed meals being rich in nitrogen and minerals, their implementation could increase the efficiency of microbial protein synthesis in the rumen leading to higher microbial protein supply to the intestine. Combination of tree leaves, unconventional oil meals and poor quality roughages would obviously be a desirable development and synergistic for buffalo production. Strategic supplementation is justified because of regular feed shortages that occur and the fact that buffaloes subsist for most of their life on fibrous crop residues on small forms. Most of the work has been concentrated for feeding these unconventional feeds as such without any suitable and economic detoxification to convert them into a wholesome protein or energy substitutes. Added costs of detoxification and supplements often discourage their use. There is a need for future systematic research on the optimization of the use of unconventional feeds and develop strategies for their optimum supplementation in different agro-climatic regions.

References

Aneja, R.P., Mathur, B.N., Chandan, R.C. and Banerjee, A.K. 2002. Technology of Indian Milk Products. A Dairy India Publ., Delhi, India, pp- 34.

Bakalkar, R.K. 1987. Nutritional evaluation of chemically and biologically treated apple pomace and groundnut shells for cattle. Ph.D. Thesis, submitted to NDRI, Karnal, Haryana.

Balaraman, N. 1981. Feeds and fodder resources for livestock in Sikkim. Res. Bull. No. 13, ICAR, Research Complex for NEH Region, Sikkim Centre, Gangtok, India, pp-1-71.

Barman, K. and Rai, S.N. 2003a. Potential of mango seed kernel (Mangifera indica) as an animal feed. *Indian Dairyman*, **55(8)**: 59-62.

Barman, K. and Rai, S.N. 2003b. Utilization of tanniferous feeds in animal ration : 2. Chemical composition, amino acid profile and tannin fractionation of certain Indian agro industrial byproducts. *Anim. Feed Sci. Technol.* (Accepted).

Barman, K., Rai, S.N. and Mueller-Harvey, I. 2003. *In vitro* digestibility and tannin metabolites of total mixed rations containing *A. nilotica* pods in buffalo, cattle and goat rumen fluid. *Br. J. Nutr.* (submitted).

Bhattacharya, A.N. and Harb, M. 1973. Dried citrus pulp as a grain replacement for Awassi lambs. *J. Anim. Sci.* **36**: 1175-1180.

Boucque, C.H.V. and Fiems, L.O. 1988. Vegetable byproducts of agro industrial origin. *Livestock Product Sci.* **19**: 97-135.

Dhingra, S. and Kapoor, A.C. 1985. Nutritive value of mango seed kernel. J. Sci. Food Agric. 36 : 752-756.

FAO. 1990. FAO Production Year Book. Vol. 41, Food and Agriculture Organization of United Nations, Rome.

Garcia, G.W., Ferguson, T.U., Neckles, F.A. and Archibald, K.A.E. 1996. *Anim. Feed Sci. Technol* **60**: 29.

Gowda, S.K. 1991. Performance of crossbred calves on urea ammoniated mahua (*Bassia latifolia*) seed cake. M.V.Sc. Thesis, Deemed University, IVRI, Izatnagar.

Gupta, B.N. 1981. Dairy cattle feed resources and feeding practices in the North East States, NDRI Bull., Karnal, pp 1- 161.

ICAR. 1985. Nutrient Requirement of Livestock and Poultry, ICAR, New Delhi.

Khanna, S.S. 1989. The agrocimatic approach. *In* : Survey of Indian Agriculture, 1989. *The Hindu,* Madras, India, pp 28-35.

Kearl, L.C. 1982. Nutrient Requirements of Ruminants in Developing Countries. Utah State University, Logon, Utah, USA.

Kotwal, D.N. 1982. Report on 13th quienquennial livestock census, financial commissioners office, Statistical Department, Govt of J&K, India.

Maheshwari, P.K., Gupta, B.S., Sinha, R.R.P. and Gupta, R. 1984. Symp on feeding system for maximizing livestock production. HAU, Hisar p – 31.

Makkar, H.P.S., Singh, B. and Negi, S.S.1989. Relationship of rumen degradability with microbial colonization cell wall constituents and tannin levels in some tree leaves. *Anim. Prod.* **49**: 299-303.

Mandal, B.C., Paul, S.S. and Pathak, N.N. 2002. Nutrient Requirements and Feeding of Buffaloes and Cattle. International Book Distributing Co., Lucknow.

Nagalakshmi, D., Reddy, D.N. and Kumar, M.K. 2004. Performance of Murrah buffaloes fed expander pelleted cotton straw based diets. *Anim. Nutr. Feed Sci.* **4**: 23-32.

Negi, S.S., Pal, R.N. and Ehrich, C. 1979. Tree fodders in Himachal Pradesh. *Tzveriagsgesellschaff mbh. Bruchweiesenweg,* **19**: 6101, Robdorf, West Germany.

Pathak, N.N. 2003. Buffalo production systems in India. *In* : Proc. 4th Asian Buffalo Congress on Buffalo for Food Security and Rural Employment. Vol. I, New Delhi, India, Feb 25-28, 2003, p- 36-49.

Pathak, N.N. and Verma, D.N. 1993. Nutrient Requirements of Buffalo. International Book Distrubuting Co., Lucknow.

Punj, M.L. 1988. Availability and utilization of non-conventional feed resources and their utilization by ruminants in South Asia. *In* : Non-conventional feed resources and fibrous agricultural residues, strategies for expanded utilization, Hisar, India, March 21-29, 1988, p- 50-81.

Rai, S.N. and Barman, K. 2004. New Animal Feed Resources : Problems and Potential. *In* : Proc. Nutritional technologies for commercialization of animal production systems, XI Anim. Nutr. Conf., Jabalpur, India, Jan 5-7, 2004, p- 1-14.

Rai, S.N., Walli, T.K., Srivastava, A. and Verma G.S. 1994. Effect of replacement of groundnut cake protein by leucaena leaf meal on milk production performance in goats during early lactation. *Indian J. Anim. Nutr.* **11**: 149-154.

Ranjhan, S.K. 1990. Agroindustrial Byproducts and Nonconventional Feeds for Livestock Feeding. ICAR, New Delhi, p- 7-91.

Ranjhan, S.K. 1998. Nutrient Requirements and Feeding of Farm Animals. ICAR, New Delhi.

Ranjhan, S.K. and Pathak, N.N. 1979. Feeding and Management of Buffaloes. Vikas Publ. House, Pvt Ltd., New Delhi.

Saha, R.C., Singh, R.B., Saha, R.N. and Chaudhry, A.B. 1997. Feed resources and milk production in the Eastern region, NDRI Bull. Publ. No. 282, Karnal India.

Sahoo, D., Nivedita and Debasis. 2001. Sea weeds of Indian Coasts, APH Publ. Corp., New Delhi, p- 283.

Sastry, V.R.B. and Agarwal, D.K. 1998. Utilization of unconventional oil cakes in the feeding of livestock. *In :* Proc. Feeding Strategies for sustainable Livestock Production with Emphasis on Non-conventional Feeds and Systems for Protein Evaluation in Ruminants. (Invited Papers), June 19-20, 1998, IVRI, Regional Centre, Palampur (HP) p -72-78.

SEA (Solvent Extraction Association of India) 2000. Millennium Hand Book on Indian Vegetable Oil Industry and Trade (7th ed.) SEA Publication.

Sengar, S.S., Tomar, S.K. Lall, D. and Mehra, U.R. 1993. Effect of sodium chlorite treatment of straws on their cell wall constituents and *in situ* dry matter disappearance. *Indian J. Dairy Sci.* **46:** 130-132.

Siddhuraju, P., Vijaykumari, K. and Janardan, K. 1996. Chemical composition and nutrition evaluation of an under exploited legume, *Acacia nilotica* (L) *Food Chem.* **57:** 385-391.

Singh, R. and Kundu, S.S. 2003. Quality evaluation of some animal feedstuffs available in Haryana. *Anim. Nutr, Feed Technol.,* **3:** 143-150.

Singh, U.B. 1987. Advanced Animal Nutrition for Developing Countries. Indo Vision, Pvt Ltd., Ghaziabad (UP) p – 219-235.

Tiwari, D.P. and Patle, B.R. 1997. Protein requirement of lactating buffaloes fed ration containing processed mahua seed cake. *Indian J. Anim. Nutr.* **14:** 98-103.

Tomar, S.K. and Arora, S.P. 1982. Effect of plane of nutrition on estrous cycle length and progesterone profile during estrous cycle in crossbred cows. *J. Nuclear Agric. Biol.* **11:** 29-30.

Tomar, S.K. and Lall, D. 1992. Feeding practices and livestock productivity in Kashmir valley. *Int. J. Anim. Sci.* **7:** 61-65.

Tomar, S.K. and Sharma, R.L. 2000. Fodders and feeding practices of cattle and sheep in Kashmir (India). *Trop. Agril. Res. Extn.* **3(2):** 1-5.

Tomar, S.K. and Thakur, S.S. 2002. Feed resources, feeding practices, milk production and disposal pattern in Karnal district. *Indian J. Dairy Sci.* **55(5):** 306-309.

Umoh, J.E. 1982. The use of concentrate and byproducts in beef cattle production: Emphasis on use of concentrate on pasture. Proc. National Conf on beef Production, July 27-30, 1982, Kaduna.

CHAPTER 19

Present Status of Manufacturing Densified Feed Block and Future Scope for Commercialization and Adoption

A. K. Pathak[1] and S. S. Zombade[2]

[1]General Manager Technical, [2]Managing Director
Poshak Feeds India Pvt. Ltd., 71/3 Mile Stone,
G. T. Road, Karnal -132 001 (Haryana)

India habitats largest livestock population and achieved the highest milk production in the world. The world currently produces more food per head than ever before and the global rate of increase in food production is still greater than the increasing rate of population. However a demand for animal food (milk and meat) has been incessantly growing in order to achieve nutritional satisfaction. It has been roughly estimated that global population growth will demand a three-fold increase in food production by 2050 and by then consumption of livestock products may increase five-fold in Asian countries including India. Such heightened demand will consequently need increased and sustainable livestock productivity.

In India, ruminant animals play a significant role in conversion of low quality plant materials (Straws, Stover's, Agriculture byproducts, waste, tree leaves, grasses, aquatic vegetations and horticulture waste etc.) into high quality protein rich food besides playing greater role in conserving fertility of soil, through organic manure. But alarming increase in the livestock population has lead us to a situation where the deficit of feedstuffs is becoming enormous. The area under fodder and feed production cannot be further increased due to culminating human pressure for land use to grow more cereals. For meeting out increased nutrient requirement it is advocated to adopt the concept of complete and balanced feeding system so that the animals may effectively utilize the available resources. Due to lack of knowledge or ignorance the majority of farmers are not aware towards balanced feeding. The only viable alternative seems to feed the dairy animals with densified complete feed block (Bheli) so that the productive performance of the ruminants is enhanced.

In India there is annual availability of 300 million MT of crop residues, out of which roughly 40% is used for feeding dairy animals. The total energy (TDN) availability from 120 million MT crop residues is hardly 48 million MT when it is fed alone. However, the TDN availability will be increased to 60 million MT when it is incorporated in densified complete feed blocks. The increased TDN content by 12 million MT could support 24 million MT additional milk productions in the country. By reducing particle size of the crop residues to optimum, the surface area is increased which efficiently get exposed to chemical and biological interactions in the digestive tract. Dry matter and nitrogen utilization are reported to improve in the densified complete feed blocks.

Densified complete feed block has been defined as an intimate mixture of processed ingredients (grains, oil cakes, byproducts, minerals and vitamins supplements and roughage) in a mixed form, which precludes selection and which is designed to be the sole source of food. Bheli feeding has been introduced with the aim of simplifying the feeding of animals. This practice is not only cheaper but also increases the milk production and butter fat content of the milk.

Present Status and Availability of Fodder in India

In India, the diets of ruminant animals are composed of greater proportion of dry fibrous crop residues of low feeding value like straws of wheat, paddy and minor millets and Stover's of maize, sorghum and pearl millet. The feeding values of these coarse fodders are further degraded by deficit of nitrogen, vitamins and some essential minerals. Excessive accumulation of selenium, silica and manganese further deteriorates the nutritive value of paddy straw. On certain occasions when there is acute shortage of fodder due to extensive damage of crops by natural calamities, other coarser fibrous by-products like sugarcane trash, sugarcane bagasse, husk and hulls of rice, groundnut and cottonseed and other crops are used for the feeding of animals for survival. Similar to fibrous crop residues, several agricultural byproducts and industrial waste materials are used in the concentrate mixture of farm animals.

One of the many contradictions that play livestock production in the country is the uneven availability of fodder, even crop residues that contribute the bulk of roughage. Many parts of the country such as Rajasthan, Gujarat, Andhra Pradesh, Orissa and Karnataka suffer from its shortages even in normal rainfall years while other areas like Punjab, Haryana and some parts of M.P. and U. P. are surplus in it to the extent that crop residues such as wheat, paddy stalks and sorghum and maize Stover's are burnt in the absence of any demand with in the state. The farmers find it uneconomical to collect it due to the cost and difficulties involved in its cutting, storage and transportations etc.

This situation also holds good in reserve forests, grassland near railway lines and pastures where enormous volume of dry grasses and biomass are also burnt. During any average year the volume of dry grasses and crop residues that needs to be transported to the areas of Andhra Pradesh, Maharashtra, Gujarat, Karnataka and Rajasthan amount to some 65.4 million tons. These crop residues, straws, Stover's and sugarcane tops are large in volume but low in value and so cannot be economically transported to distant places. To move fodder to a distance of 500 kms is estimated to cost Rs. 2.5/kg when wheat bhoosa sales in the market for almost half of this price. Economic movement of dry fodder is facilitated with densification process by making densified fodder block.

Present Status of Manufacturing Densified Complete Feed Blocks

POshak Feeds (India) Pvt. Ltd. Karnal is the only company in our country to produce and market more than 30000 MT of Densified complete feed blocks i.e. TMR (Bheli) as *one day packed ration* for dairy animals. In addition to TMR (Bheli) another unit *Indian Fodder Care and Technologies Pvt. Ltd. Delhi* plans to manufacture densified fodder blocks (Pendi) for feeding of animals in fodder scarcity areas / draught areas and to facilitate transportation to the manufacturing TMR (Bheli) plants in fodder scarcity areas of the country.

Company is constantly engaged in research and product development as well as in machinery improvement for the effective and commercial level of production. Similarly the formulated feed is so effective that the farmers reported that the milk yield increased by 20-30% and fat per cent increased 0.5 to 1 % when compared to their traditional feeding. With the efforts of technical and marketing staff, the Bheli is being fed to the cattle and buffaloes in Delhi, Dehradun, Rishikesh, Hardwar, Roorkee, Palwal, Ajmer, Jaipur, Bikaner, Manali, Kullu, Ludhiana, Kerala and Surat etc.

Collection of Fodder

Wheat and paddy straw is abundantly available in north / eastern regions. It is surplus in some of the states like Punjab and Haryana to the extent that in absence of demand, it is burnt. The existing system of local mandies or an established permanent village level centres can be set to procure the fodder at the harvesting time. Reserve forests have huge quantities available and unutilized of dry forest grasses. These can be procured with the help of state government at a token price.

Processing of Fodder

Most of the fibrous crop residues and non-conventional feeds require processing for the improvement of palatability, digestibility and feeding values. The use of feed processing technology is therefore more relevant. Crop residues and grasses are chopped to get particle size between 0.5 to 4.0 cm. Hard and thick items are to be crushed and chopped. Fodder could be chopped and crushed through specially designed chaff cutter. The added fodder should not be ground to power. Hay prepared from green fodder could also be used after drying.

For preparing hay, the harvested material should be firstly chopped then dried to get moisture content below 10%. Dairy animals derive fibre from, fodder hence adequate particle size in the ration to stimulate chewing activity and saliva production both of which are necessary for maintenance of rumen pH, rumen health and avoid digestive upset and low milk fat production. The level as well as particle size of fibre greatly influences the effectiveness of a fibre source in the ration of dairy animals.

Enrichment of Fodder

The Indian livestock economy is exclusively dependent upon the efficient use of crop residues as roughage source. Straws and Stover's have low nitrogen and high ligno-cellulose complex. The low digestibility of fibre is also associated with low intake of feed. Thus the intake of energy by the animals remains too low and is not enough to provide nutrients even for the maintenance purpose. The straws along with concentrates create better environment for the micro flora to grow in the rumen and ensure to ferment straw effectively.

Certain products are not palatable and animals refuse to eat but by disguising them with other palatable ingredients the animals can be made to eat. The animals on the complete feed blocks thus consume a lot of cheaper ingredients and waste products. Stability of the ration composition, reduction in nutrient selection, increase in feeding frequency and continuous buffering effects of the forage in complete feed blocks result into increased milk production as well as butter fat content.

Dried leguminous fodder (Berseem and Lucern hay) supplementation to poor quality roughage enhances the supply of nutrients and essential minerals to the animals and also increases the digestibility of poor quality roughage. The beneficial effect of incorporation of high digestibility dried leguminous fodder in low digestibility roughage diet could be due to the reason that this exerts a large effect on digestibility by providing a highly colonized fibre sources to seed bacteria on to the low digestible fibre. Its incorporation improves the dry matter intake as well as milk production.

Forage concentrate ingredients and feed additives / supplements have been blended together for preparing densified complete feed blocks so as to get the required quantity of protein, energy, minerals and vitamins. The crop residues as well as feed ingredients and supplements are being selected with the help of experts to suit the requirements of different productive traits of the dairy animals. For animals of higher productive traits there is a need to add more protected protein, fat as well as carbohydrate in the form of bypass nutrients. It is not only a blend of fodder and concentrate but also a wholesome and nutritious complete feed.

This is a very practical approach towards enriching poor quality roughages such as wheat, and paddy straw with concentrate ingredients. These are mixed mechanically in a mixer and thereafter densified using a press into a square block of desirable weight and size. This feed block provides all the essential nutrients in a balanced proportion to meet the nutrient requirements of the rumen microbes, thereby optimizing the rumen fermentation. The advantages of these feed blocks are: cost of transportation and storage is reduced by 1/3rd and animals avoid selective eating. The proportion of wheat straw to concentrate in the feed block depends on the targeted production level of the animal. For meeting the nutrient requirement of animals producing around 15-18 liters milk per day, hay of leguminous fodders, bypass protein and bypass fat can be added in required proportions. Such feed blocks are in great demand in suburban areas of metropolitan cities.

Future Scope For Commercialization and Adoption

Awareness towards the use of densified complete feed block (Bheli) in the dairy nutrition is increasingly spreading to all over the country and being adopted by the progressive dairy farmers. Commercial dairy owners are fast relying upon this new concept of feeding dairy animals. In Indian scenario where animal clusters are not so large and the animals are kept in Mega cities in densely populated areas the concept facilitates the space utilization. The supply of voluminous fodders is quite expensive due to high transportation cost. The off-season purchases are often difficult due to insufficient space for the storage of bulky fodder.

In draught prone areas like Rajasthan, Andhra Pradesh, Karnataka, Maharashtra and Gujarat where severe shortage of animal feed, especially roughage is encountered. These states can take up joint venture programmes to create fodder bank.

Metropolitan cities like Delhi and Bombay alone have a large number of crossbred cattle and improved buffaloes (approximately 5 lacs). This livestock is kept in dairy colonies and are organized on scientific and commercial basis. Besides procurement of roughage and concentrate items, space, transportation and labour is a problem of these areas. To minimize feed cost and labour and to maximize profit is the need of time. If these livestock owner offered one densified complete feed block of 14 kg per animal per day the demand is estimated to 8000 MT/day.

The idea behind developing a joint venture network in the country is to encourage promising entrepreneurs to prepare cost effective ready to eat feed based upon locally available feed stuffs for farmers so that milk production is up to genetic potential of dairy animals. This is not only a golden opportunity to new entrepreneurs but also a boon to those who want to diversify and sustain in cutthroat competition due to globalization.

CHAPTER 20

Feed Resources and their Development for Raising Buffaloes in Arid Regions of the Country

N. P. Singh and O. H. Chaturvedi

Central Sheep and Wool Research Institute,
Avikanagar- 304 501 (Rajasthan)

India possesses about 2.27 % of the World's 13062 million hectares of total land area. About 10 % of the total land area of the country is arid and 30 % semi-arid. The arid zone is mostly confined to the States of Rajasthan and Gujarat while the semi-arid zone is spread over the States of Maharashtra, Karnataka, Andhra Pradesh, Rajasthan, Tamilnadu, Gujarat and Uttar Pradesh. Rajasthan has the maximum (73.60%) of the total arid and semi-arid area followed by Gujarat (29.50 %) and Andhra Pradesh (21.50 %). The annual rainfall in the arid region varies from 10 to 40 cm, quite often erratic, so much so that the entire rainfall of the year may fall on a single day and rest of the year may go dry. The summer temperatures may be as high as 49 °C during day and fall to less than 20 °C during night. In winters, the day temperature is higher than that of the night temperature may be near freezing point. The annual rainfall in the semi-arid region varies from 50 to 60 cm. In semi-arid regions, the summer and winter temperatures are not as extreme as in the arid region but may reach 45 °C during summers and 8 °C during winters.

The crop production in these regions is a gamble mainly due to low and erratic rainfall. The farmers of these areas have, therefore, diversified from crop production to livestock production to counter the risks of crop failures as the livestocks in general are more tolerant to harsh climatic conditions as compared to the crops. Cattle, sheep and goats are the preferred ruminant livestock and buffalo rearing is not very popular in most of the arid and semi-arid regions except a few districts in Rajasthan where rainfall is comparatively higher or irrigation facilities have been developed.

Livestock Population

World's current population of cattle, buffaloes, sheep and goats is around 1368.0, 170.5, 1028.5 and 764.5 million respectively (Table 1). The Asian region possesses about 35.49, 97.06, 40.38 % and 63.78 % and India 16.53, 56.85, 5.74 and 16.28 % of the total world population of the four respective livestock species (FAO, 2003). India ranked first in cattle, buffalo and goat population in the world. Although the population of all the four livestock species has shown increasing trend since 1951 the buffalo and goat population has increased at a much faster rate than that of cattle and sheep in India. The buffalo population which had been 22 % in 1951 is now more than 30 % of the total bovine population. Rajasthan state with 53.17 million has about 11.50 % of the country's total livestock population. This includes about 12.16 million cattle, 9.76 million buffaloes, 14.31 million sheep and 16.94 million goats. Rajasthan state has about 10 % of the total buffaloes of the country. The semi-arid region has significantly higher population of buffaloes as compared to the arid region. Jaipur district has the highest (7.75 lakhs) population of buffaloes followed by Alwar (7.64 lakhs), Ganganagar (7.50 lakhs), Bharatpur (5.25 lakhs), Sawai madhopur (5.10 lakhs), Udaipur (4.75 lakhs) and so on indicating positive correlation with the amount of rainfall received and feed resource availability.

Table-1: Livestock Population and Production

Attributes	World	Asia	India
Population (Million)			
Cattle	1,368.05	485.49	226.10
Buffalo	170.46	165.45	96.90
Sheep	1,028.59	415.35	59.00
Goat	764.51	487.59	124.50
Milk Production (MMT)			
Cow	507.38	104.78	36.50
Buffalo	72.62	70.39	47.85
Sheep	7.89	3.40	–
Goat	11.82	6.29	2.61
Meat Production (MMT)			
Beef and veal	58.74	11.84	1.49
Buffalo meat	3.18	2.87	1.47
Mutton and lamb	7.73	3.78	0.23
Goat meat	4.09	3.00	0.47
Fresh Skin Production (MMT)			
Cattle hides	7.41	2.45	0.48
Buffalo hides	0.86	0.82	0.52
Sheep skins	1.63	0.77	0.05
Goat skins	0.90	0.71	0.13
Greasy Wool Production (MMT)	2.14	0.71	0.05

Source: FAO (2003)

Livestock Production

The total annual World production of milk is 599.71 MMT (Table 1). The Asian region contributes 30.82 % and India with 14.50 % of the total milk production ranks first. Total buffalo milk production in the World, Asia and India is 72.62, 70.39 and 47.85 MMT, respectively. Buffaloes provide more than 55 % of the total milk produced in our country. The milk productivity of cattle and buffaloes in Rajasthan is low owing to their poor nutrition and feeding management. They provide less than 10 % of the total milk produced in the country. World's annual production of total meat is around 73.74 MMT. The Asian countries contribute about 29.14 % and India about 4.96 % of the World meat production. The buffaloes contribute 4.31 % to the World, 13.36 % to Asia and 40.16 % to India of the total annual meat production. Buffaloes also provide 7.96 % in the world, 17.26 % in Asia and 44.07 % in India of the total skins and hides produced (FAO, 2003). Thus buffaloes are the most important livestock species in Asian countries in general and in India in particular. Although the buffaloes are more suitable for humid, damp and high rainfall areas, however, they are becoming popular and important in arid and semi-arid areas also with the improving irrigation facilities and advancing· crop production.

Land Holdings and Utilization Pattern

The total land area of the country is around 304 million hectares. Of the total area, about 22.5 % is under forest cover, 3.6 % is under permanent pasture, 1.2 % is under tree crops and groves, 6.2 5 % is barren and uncultivable land, 4.5 % is under waste lands and 8.0 % is under fallow lands. The total area available for livestock grazing works out to be about 46 % of the total land in the country. About 40 % of the total area in Rajasthan is available for livestock grazing (Pathak and Gupta, 2003). Almost all the states of Central and Peninsular India and part of Rajasthan, Gujarat, Punjab and Haryana come under semi-arid region. Permanent pastures occupy over 5.0 %, forest cover over 14 % and wastelands over 12 % in the semi-arid areas of the country. A survey conducted in the semi-arid area of Rajasthan revealed that the average land holding of the marginal, small, medium and large size farmers was 0.37 and 0.21, 3.60 acre with percentage distribution of 5.05, 6.80 and 19.06 and 20.77 and 75.68, respectively (Table 2). The proportion of irrigated land with marginal, small, medium and large farmers was 0.0, 21.25, 47.90 and 17.74 % respectively. The overall irrigated land in the survey area was 27.19 % (Chaturvedi *et al.*, 2002). The proportion of irrigated land was slightly lower in the area as compared to the national figure (35%) because the area is typically semi-arid and mostly rain fed. The average family size in the area was 9.67. It was higher in case of large farmers followed by marginal, small and medium farmers (Table 3). The (number of persons engaged and dependent on agriculture was higher in case of large farmers followed by medium, small and marginal farmers. However, due to shortage of resources like land and agricultural inputs, marginal and small farmers were found taking other jobs including hired agricultural labourers for their livelihood. Thus, crop production and animal husbandry were the main occupation for livelihood of 89.5 % population. The overall human to land ratio in the semi-arid region of Rajasthan was 1.37. It was highest in case of marginal farmers followed by small, medium and large farmers (Chaturvedi *et al.* 2002). The overall human to land ratio in Bundelkhand region (semi-arid) of Uttar Pradesh as reported by Saran *et al.* (2000) was also similar to that of semi-arid region of Rajasthan. The arable land under fodder cultivation was only 2.4 % and most of it was used for cultivation of maize, sorghum, pearl millet, cowpea, groundnut, sesamum, pigeon pea, green gram and black gram during kharif and mustard, chickpea, wheat, barley and vegetable crops during rabi season. Major fodder crops grown were M.P. chari, sorghum, pearl millet and cowpea in kharif and lucerne, berseem, kasni, senji and oats in rabi season.

Table 2. Land holdings of different classes of farmers

Category of farmer	Number of households	Total land holding (Acres)	Average size of holding (Acres)	Proportion of holding (%)	Average irrigated land/farmer (Acre)	Average unirrigated land/farmer (Acre)	Proportion of irrigated land (%)
Marginal (<2.5 Acre)	2	0.75	0.37	0.21	0.00	0.37	0.00
Small (2.5-5 Acre)	5	18.00	3.60	5.05	0.70	2.90	21.25
Medium (5-10 Acre)	10	68.00	6.80	19.06	3.17	3.62	47.90
Large (>10 Acre)	13	270.00	20.77	75.68	3.92	16.85	17.74
Overall	30	356.75	11.89	100.00	2.87	9.02	27.19

Source: Chaturvedi *et al.* (2002)

Table-3: Family size and proportion of persons engaged and dependent on agriculture

Category of farmer	Average family size	No. of persons engaged in agriculture	No. of persons depend on agriculture	No. of persons engaged in other job	% of persons in agriculture	Human : land (No./Acre)
Marginal (<2.5 Acre)	8.50	1.00	3.50	4.00	57.86	9.33
Small (2.5-5 Acre)	8.00	2.60	4.40	1.00	88.73	2.27
Medium (5-10 Acre)	7.80	3.20	3.90	0.70	92.35	1.10
Large (>10 Acre)	11.92	6.07	4.92	0.92	92.36	0.63
Overall	9.67	4.20	4.40	1.07	89.45	1.37

Livestock Holdings

The average holding of livestock and ACU in the area were found to be 39.23 and 12.67, respectively (Table 4). The average livestock holding was higher in case of marginal farmers followed by medium, large and small farmers whereas the average number of ACU was higher in case of large farmers followed by marginal, medium and small farmers (Chaturvedi *et al.* 2002). The livestock: land (ACU/acre) ratio was highest in case of marginal farmers followed by medium, small and large farmers. It was revealed that the large farmers having larger family size maintained more livestock compared to small, marginal and medium farmers because larger families having more manpower could manage larger livestock herds. Further, an acre of land with a marginal farmer sustained more livestock and human lives compared to that of large farmers. This pattern of ACU held by the farmers and livestock: land (ACU/acre) is similar to that of Bundelkhand region of Uttar Pradesh as reported by Saran *et al.* (2000). The proportion of cattle, buffaloes, sheep and goats among the livestock was 22.47, 17.10, 31.39 and 29.04 %, respectively (Table 5). The proportion of large (cattle and buffaloes) and small ruminants (sheep and goats) was 39.6 and 60.4 % indicating that among the livestock, sheep and goats have major contribution in the economy of the farmers of the arid region. Among the livestock, sheep husbandry fits in better with economical and climatological choice of the people in Rajasthan (Sirohi and Rawat, 2000). The proportion of small ruminants was higher in case of marginal farmers followed by medium, large and small farmers (Chaturvedi et al., 2002).

Table-4: Livestock holdings of different classes of farmers

Category of farmer	Average No. of cattle	Average No. of buffaloes	Average No. of sheep	Average No. of goats	Average No. of total livestock	Average No. of ACU holding	Average .ivestock: land (ACU/ Acre)
Marginal (<2.5 Acre)	2.00	0.00	63.0	0.50	65.50	12.58	2.89
Small (2.5-5 Acre)	1.60	1.00	0.40	7.40	10.40	4.63	1.32
Medium (5-10 Acre)	2.30	2.80	31.5	7.50	44.10	12.53	1.90
Large (>10 Acre)	4.85	4.08	22.85	10.77	42.54	15.87	0.80
Overall	3.27	2.87	24.67	8.43	39.23	12.67	1.34

1 ACU = 1 cow =1 ox =0.75 bull= 0.75 buffalo= 4 calves= 6 sheep =6goats (Chaturvedi et al. 2002)

Table-5: Proportionof different species (%) reared by various classes of farmers

Category of farmer	Cattle	Buffaloes	Sheep	Goats	Sheep Male Female		Goats Male Female	
Marginal (<2.5 Acre)	34.11	0.00	49.22	16.67	0.79	99.21	0.00	100.0
Small (2.5-5 Acre)	24.50	18.67	1.43	55.40	0.00	100.0	7.44	92.56
Medium (5-10 Acre)	18.58	19.02	43.61	18.79	3.70	96.30	3.01	96.99
Large (>10 Acre)	22.89	17.65	30.77	28.69	2.59	97.41	16.90	83.10
Overall	22.47	17.10	31.39	29.04	2.76	97.24	10.69	89.31

Source: Chaturvedi et al. (2002)

Feed and Fodder Resources

The major feed resources in our country are grasses, grazing, crop residues, cultivated fodders, edible weeds, tree leaves and agro-industrial by-products. Ranjhan (1994), Kumar and Mathur (1996), Ramachandra *et al.* (2001) and Ranjhan (2003) have reported potential availability and requirements of livestock feeds and fodders in the country. In arid and semi-arid regions of our country, the buffaloes are mainly fed on crop residues in addition to grazing of the available rangelands with little or no supplementation with concentrates that include agro-industrial by-products. Crop residues and by-products constitute the main feeds accounting for 40 % of the total consumption of different livestock. Green fodders contribute 26 %, the concentrates 3 % and the rest is coming from grazing (Mathur, 2004). A large gap exists between the requirement and the availability of feeds and fodders in the country as a whole. A shortage of about 31 % dry fodder, 23 % of green fodders and 47 % of concentrates has been estimated for meeting the requirements of existing livestock population in the country. Recent surveys conducted by the National Institute of Animal Nutrition and Physiology, Banglore, have revealed 45 % shortage of dry roughages, 44 % shortage of concentrates and 38 % shortage of green fodders in the country (Singh and Ramachandra, 2004). The situation in arid and semi-arid regions is much more serious. The feed scarcity is mostly due increasing human and livestock population, deterioration of common grazing lands both in quality and quantity, lack of adoption of feed and fodder production and processing technologies and low priorities given for identifying, improving and utilizing the newer feed and fodder resources (Pradhan, 2003). The feed resources available for buffaloes in arid and semi-arid areas include seasonal and perennial grasses, stubbles, shrubs, bushes and forbs while grazing on the community range lands, roadsides, canal banks and harvested fields. Top feeds, crop residues, agro-industrial by-products and agricultural waste materials are also fed to the buffaloes depending on their availability (Singh and Karim, 1997).

Feeding Systems for Buffaloes

The buffaloes in arid and semi-arid areas are reared under all the three viz. extensive, semi-intensive and intensive systems of feeding management. Buffaloes in arid regions are mostly reared under extensive system and play only a secondary role to crop as well as other livestock production. The livestock population as ACU was 36.2 million against 5.5 million hectares of grazing land in the arid region of Rajasthan. Thus the grazing intensity works out to 6.63 ACU per hectare and the grazing area per ACU to less than 0.15 hectare (Roy, 2003). Sheep, goats, idle bullocks and non-producing cattle were mostly grazed in mixed grazing on community/public rangelands for about 8-10 hours a day. The livestock were grazed from 0600-1800, 0700-1700 and 0800-1600 hr during summer, rainy and winter season, respectively. Buffaloes are either stall-fed or supplemented with dry roughage after grazing. Soaked oilseeds and their cakes as well as concentrates mixed with wheat straw, sorghum or pearl millet kadbi were also offered to the milch animals during the periods of green fodder scarcity. The rate of concentrate feeding to milch cattle and buffaloes was 2.5-3.0 kg per animal per day. Similar amount of concentrate was offered to working bullocks. The extensive system is principally one of low resource use and a low level of productivity emerges from poor nutritional availability. There is marked fluctuation in feed availability and its nutritive value among different regions, years and seasons. The greatest limitation in our rangelands and natural pastures is on the availability of adequate energy throughout the year and adequate amounts of protein for more than half the year. During the period from January to June the problem of energy is very acute and the intake is much less than even the maintenance requirements. Although the grazing on the rangelands is considered to be the cheapest method of buffalo production, the over grazing of the available lands has caused serious problems of vegetative destruction, soil erosion and land degradation. Due to long dry summers, erratic rains and light textured soils deficient in organic matter content having tendency to salinity, the natural potential of these lands caused short of the nutritional requirements of the buffaloes. The adult body weights, reproduction rates and milk production are very lower and morbidity and mortality rates are very higher. There are great prospects of improving these natural rangelands. Reseeding with *Lasiurus sindicus* in arid areas, with *Cenchrus ciliaris* in semiarid areas and *Dicanthium annulatum* perennial grasses in higher rainfall areas increased the grass yield from 0.5 ton to 2.0 ton per hectare per year. Intercropping of Cowpea in *Cenchrus ciliaris* pasture improved the nutritive value and increased the dry matter yield of the pasture by three times. Introduction of *Dolichos lablab* legume improved the nutritive value and increased the dry matter yield from 1-1.5 ton to 2.5-3.0 ton per hectare. Plantation of 50 fodder trees of *Prosopis cineraria, Ailanthus excelsa* and *Leucaena leucocephala* in a hectare had no adverse effect on the pasture growth and provided an additional yield of one ton dry matter when fully grown and lopped twice a year. Areas in arid and semi-arid regions may be identified and utilized for buffalo production with low investment. Rural common grazing lands should be improved through reseeding with nutritious perennial and high yielding grasses and legumes. Fodder tree plantation should be taken up on pasturelands, wastelands, riverbanks and roadsides and bunds of ponds, canals and agriculture fields on large scale.

A kind of compromise between extensive and intensive systems is referred to as the semi-intensive system of buffalo feeding management. This system is mostly followed in semi-arid regions of the country. It is a combination of limited free range grazing on available pasturelands and feeding in the stalls with feed and fodder supplements. Integration of buffalo rearing with arable cropping and tree cropping is also included and cut and carry system of available fodders is employed. The buffaloes are grazed for 4-6 hours and supplemented with varying amounts of natural and cultivated green and dry grasses, fodders, crop residues, straws, grains, oil seeds, oil cakes and agro-industrial by-products etc. The buffaloes utilize all available feed resources under this system. The level of nutrition is just optimum or little low but surely far better than that of extensive system. Production performance of the buffaloes depends on the quantity and quality of the grazing pastures and supplementary feeding. The dry grasses, fodders, crop residues and straws are mostly fed to cattle and buffaloes as whole without chaffing in most of the arid region. It was interesting to note that there were in general no chaff cutters, mangers and tying chains available in the area for even large ruminants for providing supplementary feeding. Under such circumstances, there was a great scope for improving the productivity of buffaloes through increasing feeding values of the poor quality roughages available in the arid and semi-arid areas by adopting simpler, cheaper and feasible feed and fodder processing, enrichment and conservation technologies.

Enrichment of Low Quality Roughages

Some of the improved technologies developed and recommended for improving the feeding and nutritive value of low-grade roughages and available for adoption by the livestock owners are detailed below. Poor quality straws like Bajra kadbi, Jawar kadbi, Mustard straw and Wheat straw are available in sufficient quantity in arid and semi-arid regions of the country. These straws are nutritionally poor and can be enriched by different treatments and utilized in the dietary of buffaloes.

Utilization of Mustard Straw

Approximately 1.7-2.0 MMT of mustard straw is annually available in the country. It is not being used in feeding of animals and is normally burnt in the fields. It contains 3-4% crude protein and its DMD is 25-30%. This straw can serve as maintenance feed for ruminant animals during scarcity period when fed as follows-

- ⊙ *Ad libitum* straw supplemented with little quantity of concentrate mixture

- ⊙ Urea-NH_3 treated straw

- ⊙ Complete Feed Block having 60 kg of mustard straw, 35 kg of concentrate mixture and 5 kg of molasses

Treatment Methods for Low Quality Roughages

S. No.	Treatment method	Description
1.	Soaking	The straw is soaked in water or boiling water for 6 or 12 hrs and fed as such or after sun drying
2.	Urea	The 4 kg urea is dissolved in 55 litre of water and sprayed on 100 kg straw. The urea sprayed straw is mixed thoroughly and air tightly stacked with polythene sheet for 21 days.
3.	NaOH	The straw is soaked in 2 % (w/v) NaOH solution for 6 hr. The excess of NaOH solution is drained overnight, sun dried and fed to sheep.
4.	AHP	The straw is soaked in 2 % (w/v) NaOH and 1.5 % (v/v) H_2O_2 solution for 6 hr. The excess of NaOH and H_2O_2 solution is drained overnight, sun dried and fed to sheep.
5.	Fungal	The straw is treated with urea as above and inoculated with *Coprinus fimetarius* fungus for 7 days and fed to sheep.
6.	Densification	The 5 kg molasses and 1 kg urea is dissolved in 5 litre of water and thoroughly mixed with 65 kg of mustard straw and 29 kg of concentrate mixture. This premix is fed into CFB making Machine and feed blocks prepared at 4000 PSI. These blocks are fed to sheep.

Enrichment Levels of Low-grade Roughages by Different Methods

Treatment	Improvement		Advantages
	CP %	DMD (unit)	
Soaking	-	24 to 16	Soaking of straw in water or boiling water strengthens its structure and reduces digestibility.
Urea	2.6 to 9.7	21.8 to 27.6	*Adlib* feeding of UTS with 200g of conc. Mix. or all UTS and Khejri leaves (75:25) can maintain the animals during scarcity period.
NaOH	2.6 to 3.2	21.8 to 31.0	Animals consumed 33% more DM from CFB based on 2% NaOH treated straw.
AHP	2.6 to 3.2	21.8 to 26.0	Improved dry matter intake, decreased N excretion in urine, improved nitrogen balance.
Fungal	2.6 to 7.7	21.7 to 28.0	Fungal treated straw with 200g of concentrate mixture can maintain the animals during scarcity period.
Densification in to complete feed blocks (CFB)	-	46.0 to 52.0	Increased density (g/cm^3) 1.8 to 2.3 times on straw based CFB, improved DMI by 25% in sheep, also improved gain in body weights.

The intensive system of buffalo rearing includes complete feeding in the stalls totally on cultivated fresh or conserved fodders, crop residues and concentrates. This system requires high labour and capital investment and is most suitable for both milk and meat production from buffaloes. In addition to providing higher body weights, reproduction rates, survival rates, milk yield and quality, this system also removes pressure from the grazing lands. Studies conducted in semi-arid areas of Rajasthan have revealed that the stall fed lactating buffaloes yielding 6-8 kg milk per day were offered 15-20 kg green lucerne, berseem or oats per head per day in addition to *ad libitum* Wheat straw or Bajra kadbi during winters and 8-10 kg green Sorghum fodder per head per day besides *ad libitum* Wheat straw or Bajra/ Jowar kadbies during summer. The quantity of Sorghum and /or Pearl millet green fodder is increased to free choice in monsoon season.

The grazing or stall-fed lactating buffaloes are daily supplemented with 1.0 to 1.5 kg per head boiled cottonseed in the morning and 1.0 to 2.0 kg per head soaked concentrate ingredients viz. crushed cereal grains, guar *churi*, cottonseed cake and/or mustard cake in the afternoon just before milking. The lactating buffaloes grazing on community pastures are also offered dry roughage during winter and summer seasons. Like lactating buffaloes, the pregnant buffaloes are also either grazed or stall-fed in arid and semi-arid regions. The advance pregnant buffaloes are supplemented with green fodder of Lucerne, Berseem, and /or Oats during winter season, while green Sorghum or Pearl millet during summer. In rainy season, the advance pregnant buffaloes are fed with *ad libitum* green fodder. The buffaloes are supplemented with the 2.0-3.0 kg concentrate ingredients viz. crushed cereal grains, oilseeds, cakes and grain by-products during later part of pregnancy. The dry buffaloes are not generally supplemented with concentrates. The buffalo calves and heifers are supplemented with concentrates at the rate of 0.5-1.0 kg per head per day. Depending upon the availability, the straws of leguminous crops like black gram, green gram, guar, groundnut, moth, cowpea etc. are also fed to various classes of buffaloes.

Feed Resource Development for Buffaloes

Feed and fodder resources for buffaloes may be developed through renovation of degraded rangelands, establishment of perennial grass, grass legume and silvi-horti-pastures, introduction of fodder crops, intercropping of fodder legumes in cereal crops, fodder tree plantation and enrichment, improvement and utilization of crop residues, agro-industrial by-products and unconventional feed resources.

Pasture Improvement and Management

The lands unfit for crop production due to edapho-climatic conditions and these may be best utilized for developing perennial pastures and silvipastures. The improvement of pasturelands. is possible by protecting them from biotic factors, grubbing unwanted bushes and weeds and preserving good natural grasses and legumes. *Cenchrus ciliaris* pasture can be established by mixing seeds in wet soil in a 1:1 ratio and then putting in open furrows spaced at 50 cm such that the seeds do not go beyond 1-2 cm of soil depth. Increasing the level of nitrogen from 0 to 60 kg/ha linearly increases the fodder yield of *Cenchrus*. Application of sheep manure. @ 10 ton/ha once in 3 years was most economical for maximization of pasture production under rain fed condition. Sheep manure @ 15 ton/ha once in 4 years and 30 kg N/ha in 4th year maintains the production of *Cenchrus* pasture above 50 q dry fodder per hectare on light sandy loam soils. *Cenchrus* pasture in the first year of establishment should be protected from grazing. Intercropping with Cowpea may be adopted during first year to increase DM yield to five times. However, Cowpea should be harvested earlier to avoid suppressing effect of vines.

A mixed pasture of grass and legume, *Cenchrus ciliaris* and *Dolichos lablab* sown in 1:1 ratio as alternate strips of 6 to 8 rows each, provides maximum dry fodder and CP per hectare. Another grass legume mixture of *Cenchrus ciliaris* and *Clitoria ternata* was found suitable. A better method of introducing *Clitoria* in *Cenchrus* pasture is through broadcasting followed by cultivator to mix into soil prior to sowing of *Cenchrus* in lines at a spacing of 50 x 30 cm. Application of sheep manure at the rate of 60 kg N equivalent/ha in a mixed pasture of *Cenchrus* and *Dolichos* or *Clitoria* is most optimum. The grasses and legumes may be harvested at 50% bloom stage to get maximum yield and quality of the forage.

The Bur (*Cenchrus biflorus*), because of its thorny inflorescence gets embedded in the fleeces of grazing sheep. The Bur infested wool is considered of low quality as it makes wool processing complicated. It looks like *Cenchrus ciliaris* or *Cenchrus setigerus* at early stage, but can be identified by its spikelets at a later stage. Infestation of Bur in the pasture may be managed by tiller cultivation. Application of Giberallic acid @ 25 ppm on a *Cenchrus- Dolichos* mixed pasture resulted in increase of 62% in grass yield, 29% in mixture yield and 42% in legume yield.

Fodder and Fruit Tree Plantation

Fodder and Fruit trees provide green fodder and fuel wood, check soil erosion, improve soil texture and fertility and provide shade to the grazing animals. Trees can be planted under silvipasture, agro-forestry, farm forestry and horti-pasture systems.

Silvi-pastoral studies involving ten multi purpose trees and bush species have shown a strong interaction between tree species and vegetation growing under them. The dry forage yield increases with increase in the relative distance from various trees. Hence, the choice of suitable tree or bush and their appropriate number per unit area is important to maintain the forage productivity of ground cover and tree leaf fodder. Fodder trees and bushes viz. Khejri (*Prosopis cineraria*), Ardu (*Ailanthus excelsa*), Babool (*Acacia nilotica*), Neem (*Azadirachta indica*), Siris (*Albizia lebbek*), Zinja (*Bauhinia racemosa*), Mulberry (*Morus alba*) and *Dichrostachys nutans* may be planted under 2 or 3 tier silvi-pasture system for maximizing biomass and fodder production. A multi tier silvipastoral system with Ardu + Zinja + Cenchrus can provide 50 q dry grass-legumes- tree leaves/ha in semi-arid areas. Trees should be planted at a distance of 10 x 10 m and the bushes inserted in between two trees along with perennial Cenchrus grass. Arable fodder crops in association with Ardu (*Ailanthus excelsa*), Babool (*Acacia nilotica*) and Khejri (*Prosopis cineraria*) under agro-forestry and farm forestry provides as good yields as without them. Plantation of 50 to 100 trees/ha is recommended under cultivation of Bajra, Sesamum, Guar, Cowpea, Moong, Moth, Dolichos and Clitoria in semi-arid conditions.

Ber (*Zizyphus mauritiana*) was well adapted for hortipastoral or agri-horti systems. Cenchrus, Bajra and Moong under rain fed and Lucerne and Berseem under irrigated conditions can be successfully grown under Ber plantation. 40 kg N and 40 kg P_2O_5 in cereal and grass crops and 25 kg N and 40 kg P_2O_5 in legumes may be applied for higher biomass and fruit yield. Protection from biotic factors and watering at the seedling establishment stage be taken up to increase the survivability and enhance the growth. Jalshakti @ 20 g/plant reduces the moisture depletion and promoted the plant growth and can be a practical and economic solution for moisture conservation.

Trees like Ardu, Khejri, Babul and Neem provide quality top feeds and their proper lopping is recommended for higher yields. Ardu trees provide higher dry fodder yield per year when lopped at six months interval. Khejri young trees produce higher dry fodder in annual lopping whereas fully-grown trees give higher yield when lopped at six months interval.

Inter cropping of fodder legumes in cereal crops

Inter cropping of legume does not affect the grain yield of cereal crops adversely and gives additional yield of fodder. Inter cropping of Dolichos with Bajra spaced at 45 x 15 cm can be adopted as an ideal combination. Legume may be introduced in line sown cereal crops under rain fed conditions. An erect type of legume like carpet legume (Dolichos) variety IGFRI. S-2214 may be sown between two rows of Bajra spaced at 45 cm. Rhizobium inoculation of Dolichos seeds can increase its fodder production by 20 %. Bajra and Dolichos mixture in 1:2 ratio shown at 30 cm spacing provides higher dry matter in low rain fall year and Maize and Dolichos in high rain fall year. Higher dry fodder was obtained by introducing Cowpea, Dolichos or Cluster Bean with Bajra crop.

Introduction of fodder crops

Clitoria ternatea may be sown at 50 x 30 cm spacing with 20 kg N and 40 kg P_2O_5/ha. The fodder may be harvested at an interval of 50 days for higher yield and better quality. Russian Giant, EC 4216, FOS-I-C-25 and NP3 varieties of Cowpea may be sown with 15 kg N and 60 kg P_2O_5/ ha for higher fodder production. *Dolichos lablab* may be sown @ 20 kg seed/ha at 30 x 30 cm spacing with the application of 20 kg P_2O_5/ha and harvesting at early to late flowering stage for higher production. FOS-277, Durgapura Safed and HFG-128 varieties of Guar may be sown with 20 kg N and 60 kg P_2O_5/ ha for higher fodder production. Other crops like Lucerne (T-9), Oats (Kent), Berseem and Methi + Oats for Rabi and Napier (NB-21), Hybrid Sorghum (Hara Sona), Bajra (Rajko, H-74) and Maize (African-tall) for summer and Kharif seasons are recommended.

Summary and Recommendations

Current world population of buffaloes is about 170.5 million. The Asian region has the highest and over 165.5 million buffaloes of the World. India possessing more than 56.80 % of the world and more than 58.50 % of the Asian population ranks first both in the Asia as well as the world. The buffaloes greatly contribute to the livelihood and economy of Indian farmers. Buffaloes have distinct economical, managerial and biological advantages over other livestock species. Most of the buffaloes in arid and semi-arid regions in India are maintained under extensive system. But the density of livestock per unit grazing area is increasing due sharp increase in livestock number and shrinkage of grazing lands resulting in vegetative destruction, ecological degradation and desertification. It is, not possible to obtain optimum production from buffaloes managed under free range grazing management. The extensive management system should be gradually replaced by semi-intensive or intensive system and new technological innovations should be adopted to improve the productivity of indigenous buffaloes. Appropriate systems of management for different purposes may be identified. Production performance and economics of buffaloes vis-a- vis other livestock species may be studied in arid and semi-arid regions. The biomass production of the common grazing lands may be improved from 2.5-3.5 quintal to 25-30 quintal per hectare through protection, reseeding with perennial grasses and legumes and plantation of fodder trees. The areas in arid and semi-arid regions that can not support crop production may be identified, developed through reseeding with perennial grasses and legumes and plantation of fodder trees and utilized for buffalo production with low investment. Both the quantity and quality of buffalo milk and meat can be improved through improved nutrition and feeding management. There is also very good scope for commercial meat production from buffaloes through semi-intensive and intensive feeding management. Efforts to identify and utilize the locally available crop residues and agro-industrial by-products for compounding complete rations and supplementary concentrate mixtures in the form of pallets and feed blocks should be intensified. Feed compounding plants capable of incorporating crop residues, tree leaves and natural vegetation in the complete feed mixtures, pallets and blocks should be set up in the rural areas. Packages of practices developed for commercial milk and meat production from buffaloes may be disseminated under field conditions duly supported by the Government, financial Institutions and the non-governmental organizations engaged in development of livestock industry in the country.

References

Chaturvedi, O. H., Tripathi, M. K., Mishra, A. S., Verma, D. L., Rawat, P. S. and Jakhmola, R. C. 2002. Land as well as livestock holding pattern and feeding practices of livestock in Malpura taluk of semiarid eastern Rajasthan. Indian J. of Small Ruminant. 8 (2):143-146.

F.A.O. 2003. Food and Agriculture Organization. FAO Production Year Book. Rome, Italy

Kumar, P. and Mathur, V.C. 1996. Agriculture in future: demand and supply perspective for the Ninth five-year plan. Economic and Political Weekly, 31(39):131-139.

Mathur, B. K. 2004. Drought proofing through utilization of low-grade roughages and non-conventional feed resources for small ruminants. Proceedings of the National Seminar on Opportunities and Challenges in Nutrition and feeding Management of Sheep, Goat and Rabbit for Sustainable Production held at CSWRI, Avikanagar on February 10-12.pp.191-201.

Pathak, P S. and Gupta, J.N.2003.Grazing resources in India: Problems and prospects. *In: Sustainable Animal Production* Edited by R. C. Jakhmola and R. K. Jain. Pointers Publishers, S.M.S. Highway, Jaipur 302 003 (Rajathsn), India.

Pradhan, K.2003. Feed resources and technological interventions for enhanced livestock productivity. Proceedings of Workshop on Current Status of Feed Processing and Future Strategies at ANGR Agricultural University, Hyderabad on July 22-23.

Ramachandra, K. S., Anandan, S. and Raju, S. S. 2001. livestock feed resources availability in India and need for a national database. Proceedings of 10th Animal Nutrition Conference held at Karnal on December .pp. 134-144.

Ranjhan, S.K. 1994. Consultancy report on availability and requirement of feed and fodder for livestock and poultry. Department of Animal Husbandry and Dairying, Ministery of Agriculture, Government of India, New Delhi.

Ranjhan, S.K. 2003. Livestock and feed situation in India. *In: Sustainable Animal Production* Edited by R. C. Jakhmola and R. K. Jain. Pointers Publishers, S.M.S. Highway, Jaipur 302 003 (Rajathsn), India.

Roy, M. M. 2003. Grassland and pasture management systems in arid and semi-arid region. *In: Sustainable Animal Production* Edited by R. C. Jakhmola and R. K. Jain. Pointers Publishers, S.M.S. Highway, Jaipur 302 003 (Rajathsn), India.

Saran, S., Singh, R. A., Singh R, Rani, S. I. and Singh, K. K. 2000. Feed resources for rearing livestock in Bundelkhand region of Uttar Pradesh. *Indian Journal of Animal Sciences* 70: 526-29.

Singh, N. P. and Karim, S. A. 1997.Utilization of non –conventional feed resources for small ruminant production. Proceedings of the VIIIth Animal Nutrition Research Workers Conference held at TNVASU, Chennai on December 12-14.

Singh, K. and Ramchandra, K. S. 2004. Feed and fodder availability, requirement and strategy for meeting the shortage in different regions for small ruminants. Proceedings of the National Seminar on Opportunities and Challenges in Nutrition and feeding Management of Sheep, Goat and Rabbit for Sustainable Production held at CSWRI, Avikanagar on February 10-12.pp.1-9.

Sirohi, Smita and Rawat, P. S. 2000. Resource use efficiency in sheep farming: A case study of district Tonk, Rajasthan. *Indian Journal of Small Ruminants* 6: 42-47.

CHAPTER 21

Forage Crops for Different Regions of the Country, their Productivity and Feeding Value

R.N. Choubey

Indian Grassland and Fodder Research Institute,
Jhansi - 284 003 (UP)

The total geographical area of India having about 328 million hectares comprises of eight agroclimates (hot arid, hot semi-arid, cold arid, cold semi-arid, sub-humid, humid, coastal with humid to per-humid and coastal with sub-humid to semi-arid). However, for testing and selection of suitable forage species and their improved varieties under the aegis of the All India Co-ordinated Research Project on Forage Crops, the entire country has been divided into 5 Agro-climatic zones viz. North-West, North-East, Central, South and Hill comprising 5, 9, 4 and 4 states, respectively. In addition, there is a specific site for island ecosystem in Andman & Nicobar Islands (Hazra, 1995).

Keeping in view the importance of forage crops for livestock population of the country, the research work for improvement of these crops is being persued by several organizations of SAUs and Indian Council of Agricultural Research. In the last three decades especially after the launch of IGFRI and All India Coordinated Research Project on Forage Crops, significant achievements have been made in respect of varietal improvement in forage crops. A number of improved varieties for forage crops have been released and recommended for cultivation in different zones/specific farming situations of the country. Earlier, gross yield received the maximum attention in breeding programmes but the efforts are also going on since past one decade to breed varieties as per specific need of the zone/location and cropping situations.

In a country like India consisting of diverse agro-climatic zones, suitable forage crops and their improved varieties with yield level are mentioned in Table-1. Crops and varieties for their specific traits for use under different situations are given in Table 2. Enough scope exists to grow crops like sorghum, maize, pearl millet, cowpea and guar in Kharif and oats, berseem, lucerne and barley in Rabi season and guinea grass, stylosanthes, napier-bajra hybrid and anjan grass as perennial forage. Crops like berseem, oats, maize and napier-bajra hybrid are suitable for irrigation intensive production areas while sorghum, pearlmillet, Lucerne and clusterbean are suitable for moisture stress areas. With respect to dry arid climates, grasses like *Lasiurus indicus, Panicum antidotale* and *P. Turgidum* are suitable for sand-dunes and sandy plains, while *Dichanthium annulatum* for clay loam to clay soils and *Cenchrus ciliaris* and *C. setigerus* for well drained sandy alluvial soils (Yadav and Rajora, 1995).

A substantial number of forage crop varieties have been released till date. Some of the important varieties require a mention due to their merits and suitability for a particular situation. Oat variety JHO-851 is an ideal multicut cereal fodder with high protein and leafiness. A variety Bundel Berseem-2 has high dry matter and resistance to root rot and stems rot. Similarly, a variety BL-10 of berseem for summer persistence, JHB-ISB-86 for high yield under temperature range of 10° C to 20° C, Anand-3, a Lucerne variety shows maximum promise at ten thousands feet from sea level, the cowpea varieties UPC 5286, UPC 4200 and Bundel Lobia-1 have prooved their worth to show good yield potential from rainfall as low as 300-400 mm in Rajasthan deserts to 1600-1800 mm at Assam hills and coastal areas. Varieties of buffel grass, Bundel Anjan-1 and CAZRI-75 and Dhaman grass, CAZRI-76 have special recognition for rainfed and dryland situations. The forage varieties have also been classified as per 15 agro climatic regions of the country (table-3 and 4)

Considering the importance of feeding value of these forage

species and their varieties, studies have been made with respect to their intake, degree of digestibility and utilization by the animal system. But the aspect has got limited attention in relation to a group of animals i.e. buffalo, which is theme of this symposium. Feeding value of some common forage in buffalo is mentioned in Table-5.

Table-1: Important forage crop varieties with area of adaptation and yield levels

Crop and Variety	Areas of adaptation	Green forage (q/ha)
I Cultivated Fodder – Legumes		
Berseem (*Trifolium alexandrinum* L)		
Mescavi	Northern and Central India	800-900
Wardan (S-99-1)	All India	900-1500
BL-1	Punjab and H.P.	100-1200
BL-10	Punjab, Haryana and H.P.	1100-1150
BL-22	Temperate zone of India	900-1000
JB-2	Northern and Central India	900-1000
Bundel Berseem –2 (JHB 146)	North West and Central zone	580-850
UPB-110	Southern zone	500-650
BL-2	Northern India	650-900
Bundel Berseem –3 (JHTB 96-4)	North eastern Zone, Eastern UP, Bihar, Orissa and W.B.	600-700
UPB-103	Northern, Central and part of South India	1000-1150
Lucerne (*Medicago sativa* L)		
Type-9	Whole of India	900-1000
Anand-2	Gujrat, Rajasthan, Haryana, M.P. ,U.P.	850-900
LL composite-3	Punjab	900-950
LL composite-5	Punjab	900-950
Anand –3	Himachal Pradesh	600-900
RL –88	Whole of India	700-1000
CO-1	Tamil Nadu and Karnataka	600-800
SS-627	Haryana, Punjab, Delhi, U.P., Rajasthan, H.P. and M.P.	800-950

Cowpea (*Vigna Unguiculata* (L) Walp.)

Russian Giant	Northern India	350-400
NP-3 (EC-4216)	Northern, Western and Central India	300-350
CO-1	Tamil Nadu, A.P., Kerala and Karnataka	275-325
UPC-287	Whole of India	350-400
UPC-5286	Whole of India	350-450
Gujarat Lobia-3	Gujarat	250-400
UPC-4200	North east zone	270-420
Lobia-88	Punjab	250-350
Bundel Lobia-2 (IFC-8503)	North West Zone	220-350
Haryana Lobia –88	North West zone	280-350
UPC 8705	Whole of India	300-420
UPC-5287	Northern India	350-400
C-30	Whole of India	300-350
Kohinoor (IGFRI-S-450)	Whole of India (for summer)	250-300
Shweta (No.998)	Whole of India	300-350
Bundel Lobia – 1 (IFC-8401)	Whole of India	250-300

Guar (*Cyamopsis tetragonoloba* (L) Taub.

HFG –156	Guar growing area of India	200-250
FS-277	Guar growing area of India	175-250
Bundel Guar-1(IGFRI –212-1)	Guar growing area of India	220-350
Bundel Guar-2 (IGFRI –2395-2)	Guar growing area of India	280-400
Bundel Guar –3	Guar growing area of India	
HFG-119	Guar growing area of India	250-300
Guara-80	Punjab	300-320

Rice been (*Vigna umbellata*)

RBL-1	Punjab	250-400
RBL-6	Punjab	220-450
K-1	Bihar, West Bengal, Orissa, N.E. region of A.P. and Kerala	200-300

K-15	Bihar, West Bengal, Orissa, N.E. region of A.P. and Kerala	200-300
Lablab Bean (Sem)		
Bundel sem –1 (JLP – 4)	Whole of India	220-350
Shaftal (Persian clover)		
SH –48	Himachal Pradesh	800-1050
White clover		
Palampur Composite –1	Himchal Pradesh	350-500
Gobhi Sarson		
GSL-1	Punjab	250-350
Sheetal (HPN-1)	Himachal Pradesh	180-300
Fenu greek (Metha)		
ML-150	Punjab	270-350

II Cultivated Fodder – Cereals

Oats (*Avena sativa*_L.)

Kent	Whole of India	450-500
OS-6	Whole of India	400-500
UPO-212	Whole of India	370-520
OL-125	Whole of India	350-480
UPO-94	Whole of India (Multicut)	450-500
OL-9	Northern and North-Western India	450-550
JHO-810	Kashmir valley	500-600
JHO-822 (Bundel Jai-822)	Central India (Multicut)	450-550
JHO-851	Whole of India (Multicut)	500-550
Sorghum (*Sorghum bicolor* (L) Moench)		
Pusa Chari-6	Whole of India	400-450
Pusa chari-9	Whole of India	400-450
SL-44	North India	350-450
HC-136	Whole of India	400-500

M.P. Chari	North India	400-500
Meethi Sudan (SSG-59-3)	Whole of India (Multicut)	500-550
IS-4776	Maharashtra, Tamil Nadu and A.P.	350-400
Jawahar Chari-6	M.P.	350-450
Jawahar Chari-69	M.P. (Multicut)	350-450
JS-20	Punjab, Haryana, Delhi	350-400
JS-263	Punjab, Haryana, Delhi	400-450
GFS-4	Gujarat	320-500
Pro-Agro chari (SSG-988)	Whole India	600-900
855 F	Whole India	600-900
HC-308	Whole India	350-550
PCH –106	Whole India	650-900
Pantchari –3 (UPFS-23)	U.P.	350-450
Gujarat Forage Sorghum –1 (AS-16)	Gujarat	400-700
MFSH-3	Whole India	500-850
LS-250	Punjab	600-950

Bajra (*Pennisetum americanum* (L) Leek.)

Giant Bajra	Entire bajra growing tract	350-400
K-677	Entire bajra growing tract	400-500
Raj Bajra Chari-2 (UUJ-IV-M)	Entire Bjara growing tract	300-450
L-72	Entire bajra growing tract	400-550
Rajko	Gujrat and Rajasthan	400-450
Fooder cumbu-8 (TNSC-1)	Entire Bajra growing tract	270-400

Maize (*Zea mays* L.)

African tall	Whole of India	500-600
Vijay composite	Whole of India	350-450
Jawahar	Whole of India	350-450
Moti composite	Whole of India	350-425
J-1006	Punjab	350-450

Manjari Composite Teosinte	Whole of India	400-450
TL-1	Punjab	380-500

III Cultivated fodder - perennial grasses

Hybrid Napier (Napier- Bajra hybrid) (*Pennisetum.purpureum x P. americanum*)

Pusa Giant	Whole of India and tropics abroad	1000-1300
NB-21	Whole of tropical humid part	1200-1500
CO-1	Tamil Nadu and Karnataka	1100-1200
Swetika-1 (IGFRI-3)	U.P., M.P., NE hills, Punjab and hills of North India	1100-1200
IGFRI-6	U.P., H.P., NE hills, Punjab and hills of North India (intercropping)	1200-1300
PBN-83	Punjab	1250-1700
CO-2	Tamilnadu and Southern part	1200-1800
CO-3	Tamilnadu and Southern part	1300-2000
IGFRI-7	Whole of India (acid soils sub-temperate regions)	1300-1500
IGFRI-10	Whole of India (acid soils sub-temperate regions)	1300-1600
Yeshwant (RBN-9)	Whole of India	1300-1400

Guinea grass (*Panicum maximum*) Jacq.

Macuenii	Kerala	600-700
Hamil	Kerala, Tamil Nadu, A.P., West Bengal, Bihar and North- Eastern States	700-800
PGG-1	North-West	900-1100
PGG-19	Punjab	750-1300
PGG-101	Punjab	800-1450
PGG-3	Northern, North-West and Central India	800-1000
PGG-9	Norther, North-West and Central India	900-1100

12. Deenanath (*Pennisetum. pedicellatum*)

P.S.-2	Bihar, West Bengal, Orissa, North region of Maharashtra and A.P.	600-700
Bundel-1 (IGFRI-43-1)	Bihar, West Bengal, Orissa, North region of Maharashtra and A.P.	600-700

IGFRI-3808	Bihar, West Bengal, Orissa, North region of Maharashtra and A.P.	500-600
Bundel-2 (IGFRI-4-2-1)	Bihar, West Bengal, Orissa, North region of Maharashtra and A.P.	550-700
TNDN-1	Tamil Nadu	550-780
Buffel Grass *(Cenchrus ciliaris)*		
Bundel Anjan –1 (IGFRI- 3108)	Arid and semi-arid region	220-400
Neel kolu kattai (Co-1) Blou buffel (FS-391)	Tamil Nadu and other semi-arid areas	280-470
Setaria Grass (Nandi Grass)		
PSS-1	Sub –temperate hill region	750-1100
Tall Fescue Grass		
Him –1	Himachal Pradesh	400-450

Table-2: Crops and varieties with their specific characters for use under different situations

Crop	Variety	Specific exploitable characters
Berseem	BL –2	Long extended growth period (Oct.-May), suitable north western part and subtropical region
	BL –22	Long extended growth and low temperature and frost resistance suitable sub-temperate hills
Lucerne	Co.1	A good perennial variety which can remain in field up to 5-6 years suitable for irrigated arid and semi arid environment
Persian Clover	SH –48	Persian clover as a crop is suitable for late planting and rice fallow. Suitable for northeast, tarai region stretches in the entire length of Himalayas and also a better crop than berseem in sub-temperate hilly areas. Most suitable crop for alkali soil.

Cowpea	UPC –4200	Suitable for humid, temporary water logged and high humid and strong acid soil areas.
	S –450	Drought resistant and suitable for summer season
Rice bean	K –1	This is a good legume material for humid and sub humid region including hills. Rice bean is much tolerant crop to strong acid soil than cowpea, hence a preferred crop in such environment.
Oats	OL –125	A good variety for calcareous soil, rocky-gravelly soil and also under shaded conditions (Agro forestry use)
	JHO –851	A good multicut (4-5 cut) with excellent herbage quality.
Pearl millet	Giant bajra	A very good regenerating material capable of giving five cuts, wide adaptability from high pH to low pH, salinity and alkalinity conditions of soil.
Maize	A-de-Cuba	Non- lodging type, most suitable for north region. Maize as a crop most suitable for acid soils.
Teosinte	Improved	Good material for temporary water logged soils (10-15 days), saline soil, calcareous soils.
Sorghum	PC –9	Most suitable variety for low land and withstands temporary water logging.
Barley	DL –454 & Azad	Most suitable variety for dual purpose with first cut for fodder followed by grain (Fodder cum grain)
	DL –36	Most suitable variety for saline, alkali and water stress environments as fodder-cum-grain crop.
Brassica	Chinese cabbage	Suitable for shaded conditions (Agro-forestry use).
Napier bajra	IGFRI –3 & IGFRI -6	Erect type and most suitable for intercropping. IGFRI –6 is also suitable for agro forestry as it is shade tolerant.

	IGFRI –7	Specially suitable acid soils and hilly areas.
Guinea grass	Hamil, PGG-13 and KVKHPGG – 14	Guinea grass as a crop is most suitable for shaded conditions. These varieties are best suited for agro forestry use.
Dinanath grass	IGFRI-4-2-1	Most suitable variety for use in degraded lands including forests.
Setaria	Kazungula	Suitable for agroforestry use.
	PSS –1	Most suitable for sub-tropical grasslands, frost resistant and suitable for shaded conditions.
Yellow anjan	CAZRI – 76	Best suited for arid areas.
Anjan grass	CAZRI – 75	Best suited for arid areas.
Dharaf grass	GAU –D –1	Suited for ravines (Earth mounds) and hill out crops.

Table-3: Important cultivated legume fodder varieties for different agro-climate regions of India

Sl. No.	Agro-Climate Region	Important Varieties
1.	Western Himalayan Region	Berseem: - Wardan (S-99-1), Mescavi, BL-1 (Lower HR), BL-10 (Lower HR), BL-22, JB-2, JHB-146, BL-2, UPB-103
		Lucerne: - Type-9, Anand-3, RL-88, SS-627 (lower HR)
		Cowpea: - Russian giant, NP-3 (EC-4216), UPC-287, UPC-5286, UPC-8705, UPC-5287, C-30, Kohinoor, Sweta (No.998), Bundel lobia-1 (IFC-8401)
		Lablab bean:- Bundel sem-1
		Shaftal :-SH-48 (Lower HR)
		White clover:- Palampur composit-1(Lower HR)
		Gobhi sarson:-Sheetal (HPN-1) (Lower HR)

2. Eastern Himalayan Region

Berseem: - BL-22, Bundel berseem-3 (JHTB96-4), Wardan (S-99-1)

Lucerne: - Type-9, RL-88

Cowpea: - UPC-287, UPC-5286, UPC-4200, UPC-8705, C-30, Kohinoor, sweta (No.998), Bundel lobia-1 (IFC-8401)

Rice bean:- K-1, K-15

Lablab bean:- Bundel sem-1

3. Lower Gangetic Plain Region

Berseem: - Wardan (S-99-1), Bundel berseem-3 (JHTB-96-4)

Lucerne: - Type-9, RL-88

Cowpea: - UPC-287, UPC-5286, UPC-8705, C-30, Kohinoor, Sweta (No.998), Bundel lobia-1 (IFC-8401)

Rice bean:- K-1, K-15

4. Middle Gangetic Plain Region

Berseem: - Wardan (S-99-1), Bundel berseem-3 (JHTB-96-4)

Lucerne: - Type-9, RL-88

Cowpea: - UPC-287, UPC-5286, UPC-8705, C-30, Kohinoor, Sweta (No.998), Bundel lobia-1 (IFC-8401)

Rice bean:- K-1, K-15

Lablab bean:- Bundel sem-1

5. Upper Gangetic Plain Region

Berseem: - Mescavi, Wardan (S-99-1), JB-2, JHB-146, BL-2, UPB-103

Lucerne: - Type-9, Anand-2, RL-88, SS-627

Cowpea: - Russian giant, NP-3 (EC-4216), UPC-287, UPC-5286, UPC-8705, UPC-5287, C-30, Kohinoor, Sweta (No.998), Bundel lobia-1 (IFC-8401)

Lablab bean:- Bundel sem-1

6. Trans Gangetic Plain Region

Berseem: - Wardan (S-99-1), BL-1, BL-10, Mescavi. JB-2, JHB-146, BL-2, UPB-103

Lucerne: - Type-9, Anand-2, LL composite-3, LL composite-5, RL-88, SS-627

Cowpea: - Russian giant, NP-3 (EC-4216), UPC-287, UPC-5286, Lobia-88, Bundel Lobia-2 (IFC-8503), Haryana Lobia-88, UPC-8705, UPC-5287, C-30, Kohinoor, Sweta (No.998), Bundel lobia-1 (IFC-8401)

Guar:- HFG-156, FS-277,Bundel guar-1(IGFRI-212-1) Bundel guar-2(IGFRI-2395-2), Bundel guar-3, HFG-119,Guara-80

Rice bean:- RBL-1, RBL-6

Lablab bean:- Bundel sem-1

Gobhi sarson:- GSL-1

Fenu greek(Metha):-ML-150

7. Eastern Plateau Hills Region

Berseem: - Wardan (S-99-1), Bundel berseem-3 (JHTB-96-4)

Lucerne: - Type-9

Cowpea: - UPC-287, UPC-5286, UPC-8705, C-30, Kohinoor, Sweta (No.998), Bundel lobia-1 (IFC-8401)

Rice bean:- K-1, K-15

Lablab bean:- Bundel sem-1

8. Central Plateau & Hills Region

Berseem: - Mescavi, Wardan (S-99-1), JB-2, JHB-146, UPB-103

Lucerne: - Type-9, Anand-2, RL-88, SS-627

Cowpea: - NP-3 (EC-4216), UPC-287, UPC-5286, UPC-8705, C-30, Kohinoor, Sweta (No.998), Bundel lobia-1 (IFC-8401)

Lablab bean:- Bundel sem-1

9.	Western Plateau & Hills Region	Berseem: - Wardan (S-99-1), UPB-110, UPB-103
		Lucerne: - Type-9, RL-88
		Cowpea: - UPC-287, UPC-5286, UPC-8705, C-30, Kohinoor, Sweta (No.998), Bundel lobia-1 (IFC-8401)
		Lablab bean:- Bundel sem-1
10.	Southern Plateau & Hills Region	Berseem: - UPB-110, UPB-103, Wardan (S-99-1)
		Lucerne: - Type-9, RL-88, CO-1
		Cowpea: - CO-1, UPC-287, UPC-5286, UPC-8705, C-30, Kohinoor, Sweta (No.998), Bundel lobia-1 (IFC-8401)
		Lablab bean:- Bundel sem-1
11.	East Coast Plains & Hills Region	Berseem: - Wardan (S-99-1), UPB-110, UPB-103,
		Lucerne: - Type-9, RL-88
		Cowpea: - UPC-287, UPC-5286, UPC-8705, C-30, , Sweta (No.998), Bundel lobia-1 (IFC-8401)Kohinoor
		Lablab bean:- Bundel sem-1
12.	West Coast Plains & Ghats Region	Berseem: - Wardan (S-99-1), UPB-110, UPB-103
		Lucerne: - Type-9, RL-88
		Cowpea: - UPC-287, UPC-5286, UPC-8705, C-30, Kohinoor, Sweta (No.998), Bundel lobia-1 (IFC-8401)
		Rice bean:- K-1, K-15
		Lablab bean:- Bundel sem-1
13.	Gujarat Plains & Hills Region	Berseem: - Wardan (S-99-1)
		Lucerne: - Type-9, Anand-2, RL-88
		Cowpea: - NP-3 (EC-4216), UPC-287, UPC-5286, Gujarat Lobia-3, UPC-8705, C-30, Kohinoor, Sweta (No.998), Bundel lobia-1 (IFC-8401)
		Guar:- HFG-156, FS-277,Bundel guar-1(IGFRI-212-1) Bundel guar-2(IGFRI-2395-2), Bundel guar-3, HFG-119
		Lablab bean:- Bundel sem-1

14.	Western Dry Region	Berseem: - Wardan (S-99-1)
		Lucerne: - Type-9, Anand-2, RL-88, SS-627
		Cowpea: - NP-3 (EC-4216), UPC-287, UPC-5286, Bundel Lobia-2 (IFC-8503), Haryana Lobia-88, UPC-8705, C-10, Kohinoor, Sweta (No.998), Bundel lobia-1 (IFC-8401)
		Guar:- HFG-156, FS-277,Bundel guar-1(IGFRI-212-1) Bundel guar-2(IGFRI-2395-2), Bundel guar-3, HFG-119
		Lablab bean:- Bundel sem-1
15.	Islands Region	Berseem: - Wardan (S-99-1)
		Lucerne: - Type-9, RL-88
		Cowpea: - UPC-287, UPC-5286, UPC-8705, C-30, Kohinoor, Sweta (No.998), Bundel lobia-1 (IFC-8401)
		Lablab bean:- Bundel sem-1

Table-4: Important cultivated cereal fodder varieties for different agro-climate regions of India

Sl. No.	Agro-Climate Region	Important Varieties
1.	Western Himalayan Region	Oat:-Kent, OS-6,UPO-212,OL-125,UPO-94,OL-9,JHO-810,JHO-851
		Sorghum:-PC-6,PC-9,,HC-136,Mithi Sudan (SSG-59-3), Pro-agro chari(SSG-988),855F,HC-308,PCH-106, MFSH-3
		Maize:- African tall, Vijay composite, Jawahar, Moti composite, Manjari composite
2.	Eastern Himalayan Region	Oat:-Kent, OS-6,UPO-212,OL-125,UPO-94,JHO-8851
		Sorghum:-PC-6,PC-9,,HC-136,Mithi Sudan (SSG-59-3), Pro-agro chari(SSG-988),855F,HC-308,PCH-106, MFSH-3
		Maize:- African tall, Vijay composite, Jawahar, Moti composite, Manjari composite

3. Lower Gangetic Plain Region

Oat:-Kent, OS-6,UPO-212,OL-125,UPO-94,OL-9,JHO-810

Sorghum:-PC-6,PC-9,,HC-136,Mithi Sudan (SSG-59-3), Pro-agro chari(SSG-988),855F,HC-308,PCH-106, MFSH-3

Bajra:-Giant Bajra,K-677 Raj bajra chari-2(UUJIM-M),L-72, fodder cumbu-8 (TNSC-1)

Maize:- African tall, Vijay composite, Jawahar, Moti composite, Manjari composite

4. Middle Gangetic Plain Region

Oat:-Kent, OS-6,UPO-212,OL-125,UPO-94,OL-9,JHO-810

Sorghum:-PC-6,PC-9,,HC-136,Mithi Sudan (SSG-59-3), Pro-agro chari(SSG-988),855F,HC-308,PCH-106, MFSH-3

Bajra:-Giant Bajra,K-677 Raj bajra chari-2(UUJIM-M),L-72, fodder cumbu-8 (TNSC-1)

Maize:- African tall, Vijay composite, Jawahar, Moti composite, Manjari composite

5. Upper Gangetic Plain Region

Oat:-Kent, OS-6,UPO-212,OL-125,UPO-94,OL-9,JHO-810

Sorghum:-PC-6,PC-9,,HC-136,Mithi Sudan (SSG-59-3), Pro-agro chari(SSG-988),855F,HC-308,PCH-106,Pant chari-3(UPFC-23) , MFSH-3

Bajra:-Giant Bajra,K-677 Raj bajra chari-2(UUJIM-M),L-72, fodder cumbu-8 (TNSC-1)

Maize:- African tall, Vijay composite, Jawahar, Moti composite, Manjari composite

6. Trans Gangetic Plain Region

Oat:-Kent, OS-6,UPO-212,OL-125,UPO-94,OL-9,JHO-810

Sorghum:-PC-6,PC-9,,HC-136,Mithi Sudan (SSG-59-3), Pro-agro chari(SSG-988),855F,HC-308,PCH-106,JS-20,JS-263, LS-250, MFSH-3

Bajra:-Giant Bajra,K-677 Raj bajra chari-2(UUJIM-M),L-72, fodder cumbu-8 (TNSC-1)

Maize:- African tall, Vijay composite, Jawahar, Moti composite, Manjari composite, J-1006

Teosinte:- TL-1

7. Eastern Plateau Hills Region

Oat:-Kent, OS-6,UPO-212,OL-125,UPO-94,OL-9,JHO-810

Sorghum:-PC-6,PC-9,,HC-136,Mithi Sudan (SSG-59-3), Pro-agro chari(SSG-988),855F,HC-308,PCH-106, MFSH-3

Bajra:-Giant Bajra,K-677 Raj bajra chari-2(UUJIM-M),L-72, fodder cumbu-8 (TNSC-1)

Maize:- African tall, Vijay composite, Jawahar, Moti composite, Manjari composite

8. Central Plateau & Hills Region

Oat:-Kent, OS-6,UPO-212,OL-125,UPO-94,OL-9,JHO-822,JHO-810

Sorghum:-PC-6,PC-9,,HC-136,Mithi Sudan (SSG-59-3), Pro-agro chari(SSG-988),855F,HC-308,PCH-106, Jawahar Chari-6, Jawahar Chari-69, MFSH-3

Bajra:-Giant Bajra,K-677 Raj bajra chari-2(UUJIM-M),L-72, fodder cumbu-8 (TNSC-1)

Maize:- African tall, Vijay composite, Jawahar, Moti composite, Manjari composite

9. Western Plateau & Hills Region

Oat:-Kent, OS-6,UPO-212,OL-125,UPO-94,OL-9,JHO-810

Sorghum:-PC-6,PC-9,,HC-136,Mithi Sudan (SSG-59-3), Pro-agro chari(SSG-988),855F,HC-308,PCH-106, MFSH-3

Bajra:-Giant Bajra,K-677 Raj bajra chari-2(UUJIM-M),L-72, fodder cumbu-8 (TNSC-1)

Maize:- African tall, Vijay composite, Jawahar, Moti composite, Manjari composite

10. Southern Plateau & Hills Region

Oat:-Kent, OS-6,UPO-212,OL-125,UPO-94,OL-9,JHO-810

Sorghum:-PC-6,PC-9,,HC-136,Mithi Sudan (SSG-59-3), Pro-agro chari(SSG-988),855F,HC-308,PCH-106, MFSH-3

Bajra:-Giant Bajra,K-677 Raj bajra chari-2(UUJIM-M),L-72, fodder cumbu-8 (TNSC-1)

Maize:- African tall, Vijay composite, Jawahar, Moti composite, Manjari composite

11. East Coast Plains & Hills Region

Oat:-Kent, OS-6,UPO-212,OL-125,UPO-94,OL-9,JHO-810

Sorghum:-PC-6,PC-9,,HC-136,Mithi Sudan (SSG-59-3), Pro-agro chari(SSG-988),855F,HC-308,PCH-106, MFSH-3

Maize:- African tall, Vijay composite, Jawahar, Moti composite, Manjari composite

12. West Coast Plains & Ghats Region

Oat:-Kent, OS-6,UPO-212,OL-125,UPO-94,OL-9,JHO-810

Sorghum:-PC-6,PC-9,,HC-136,Mithi Sudan (SSG-59-3), Pro-agro chari(SSG-988),855F,HC-308,PCH-106, MFSH-3

Maize:- African tall, Vijay composite, Jawahar, Moti composite, Manjari composite

13. Gujarat Plains & Hills Region

Oat:-Kent, OS-6,UPO-212,OL-125,UPO-94,OL-9,JHO-810

Sorghum:-PC-6,PC-9,,HC-136,Mithi Sudan (SSG-59-3), Pro-agro chari(SSG-988),855F,HC-308,PCH-106,GFS-1,GFS-4, MFSH-3,

Bajra:-Giant Bajra,K-677 Raj bajra chari-2(UUJIM-M), RajkoL-72, fodder cumbu-8 (TNSC-1)

Maize:- African tall, Vijay composite, Jawahar, Moti composite, Manjari composite

14. Western Dry Region

Oat:-Kent, OS-6,UPO-212,OL-125,UPO-94,OL-9,JHO-810

Sorghum:-PC-6,PC-9,,HC-136,Mithi Sudan (SSG-59-3), Pro-agro chari(SSG-988),855F,HC-308,PCH-106, MFSH-3

Bajra:-Giant Bajra,K-677 Raj bajra chari-2(UUJIM-M), Rajko,L-72, fodder cumbu-8 (TNSC-1)

Maize:- African tall, Vijay composite, Jawahar, Moti composite, Manjari composite

15.	Islands Region	Oat:-Kent, OS-6,UPO-212,OL-125,UPO-94,OL-9,JHO-810

Sorghum:-PC-6,PC-9,,HC-136,Mithi Sudan (SSG-59-3), Pro-agro chari(SSG-988),855F,HC-308,PCH-106, MFSH-3

Maize:- African tall, Vijay composite, Jawahar, Moti composite, Manjari composite

Table-5: Feeding value of some common forage for buffalo

Forage	DMI (% body weight	DCP (%)	TDN (%)
Berseem (2nd cut)	2.41	17.4	70.2
Cowpea	2.29	-	-
Guar (flowering)	-	13.2	61.3
Bajra/ Pearl millet			
Pre-bloom	-	5.51	52.0
Bloom	-	3.57	52.1
Mature	-	3.03	52.5
Sorghum (dough)	-	2.97	63.3
Hybrid Napier			
NB-21	-	2.98	63.5
Gajraj	1.86	5.45	58.4
Pre-flowering	1.92	4.68	48.9
Maize (Green stalk)	2.27	4.11	53.3
Oats	2.01	6.61	62.4
Sudan grass	3.22	4.13	52.7
Teosinte (pre-flowering)	1.90	6.00	57.9
Lucerne	1.88	9.35	50.4
Oat hay	-	3.65	64.3
Oat silage	-	3.53	63.4

Suggested Readings

Dairy Year Book (2001). All India Dairy Business Directory (special millennium issue) 2nd edn. Sadana Publishers & Distribution, Ghaziabad (U.P.).

Hazra, C.R. (1995). Advances in Forage Production Technology. Technical Bulletin, ICAR, New Delhi, pp. 51.

Katiyar, R.C. and Bisth, G.S. (1988). Nutrient utilization in Hariana cattle and Murrah buffalo. I. comparative study with oat hay based rations. LProceed. 2nd World Buffalo Congress, IDRC/ICAR, Dec. 12-16, 1988, New Delhi.

Lal, M., Khan, M.Y., Jaikishan, Katiyar, R.C. and Joshi, D.C. (1987). Comparative nutrient utilization by Holstein Friesian, Crossbred cattle and buffaloes fed on wheat straw based ration. Indian J. Anim. Nutr., 4 : 177.

Poonappa, C.G., Nooruddin, M.D. and Raghavan, G.H. (1971). Role of passage of feeds and its relation to digestibility of nutrients in Murrah buffaloes and Hariana cattle. Indian J. Anim. Sci., 41 : 1026.

Singh, C.B. (1988). Contribution of buffalo in Asia and economic analysis of buffalo keeping. Proceed. 2nd World Buffalo Congress, IDRC/ICAR, Dec. 12-16, 1988, New Delhi.

Singh, B.K. and Mudgal, V.D. (1967). The comparative utilization of feed nutrients from lucerne hay in buffaloes and crossbred zebu heifers, Indian J. Dairy Sci., 70 : 142.

Yadav, M.S. and M.P. Rajora (1995). Pasture development research in Western Arid region. In : New Vistas in forage Production (ed. C.R. Hazra and Bimal Misri, ICAR, IGFRI, Jhansi, 1995), P. 97-112.

CHAPTER 22

Agroforestry in Relation to Buffalo Production

M. M. Roy

Indian Grassland & Fodder Research Institute,
Jhansi - 284 003 (UP)

Domestic ruminant – buffalo is considered an essential component of Indian farming system as they provide power, rural transport, manure, cooking fuel, milk, meat etc. Thus, they have a major role in rural economy by providing income and employment in India. There has been a shift towards buffaloes rearing in the country since independence and now about 56 per cent of world's buffalo population is in India. The present population (90.80 million) is growing at a rate of 1.90 per cent (higher than cattle, sheep and camel).

The current estimates about productivity of Indian buffaloes indicate that although there has been impressive increase in milk and meat production, the per head production is quite low Although, several factors like genetic make up, environmental conditions, health practices and nutrition play important role in realizing peak output from the livestock, chronic feed deficits represent a major constraint to livestock production in the country. The livestock owners who own most of the low productive animals do not have sufficient resources to feed them and there is a tremendous pressure on land resources, leading to a variety of environmental problems.

Agroforestry – a multiple use concept of land management is capable of meeting the present challenges of forage shortages besides environmental amelioration. It has immense potential to ensure stability and sustainability in production and thus provide the much needed economic and ecological security. The National Agricultural Policy 2000 has recommended augmentation of biomass production through agroforestry. The Planning Commission in its report for vision 2010 for Greening India has also proposed a two-pronged national action plan to re-vegetate 43 million ha of wasteland area in a period of 10 years; out of which 28 million ha are to be developed through agroforestry.

The livestock that is often considered associated with ecological degradation and deforestation, may be key components of sustainable farming systems if managed properly under agroforestry systems. Agroforestry systems that incorporate animals with tree crops can enhance important cycle *viz.*, nutrient cycling besides balancing insect populations and reducing energy intensive management techniques. Also incorporation of tree leaves of *Leucaena leucocephala* as a forage supplement in the diet of buffaloes results into better live weight gain and milk yield besides reducing the cost of feed (Akbar and Gupta 1985; Dharmaraj *et al.*, 1985).

In this paper an attempt has been made to discuss the potential of agroforestry in relation to buffalo production.

Feeding Systems

In the livestock industry about 55-75 per cent of the inputs are through feeding alone. Thus management of feeding systems is an important single factor affecting the input: output ratio of the industry. Under traditional system, buffaloes are either tethered or let loose in the field at day time and brought to the enclosures/stalls at the night time or they are confined to the stalls (mostly lactating animals) or a combination of both is followed in different parts of the country. Feeding systems are primarily based on grazing the animals on native pastures of low productivity which are steadily degrading. The grazing grounds in many areas serve just as exercise ground during the major part of the year to a fairly good pasture during rainy season. During wet season some live weight gain (milk/draught) is achieved but it is followed by variable losses during dry season depending on the pressure on grazing land and availability of the quantity and quality of the vegetation.

Buffaloes receive about 50-66 per cent of dry matter requirement from crop residues (straws/stovers) supplemented by grass available during dry season. Usually no concentrate is provided to growing, working, pregnant or dry animals. Only lactating animals are given better feeding through supplementation of byproduct concentrates (oil cakes, brans, chunnies etc.) as the farmers receive immediate return on their investment.

The requirement of feed and fodder for the buffalo population of the country has been shown in Table-1. It is based on the consideration that the animals are properly fed. The current level of deficit in feeding is estimated to be over 50 per cent (Rajora 1998).

Table-1: The required average rate of feeding and annual requirement of various types of buffaloes in India

Types	Daily rate of feeding (kg)			Av. Requirement/yr (MT)		
	1	2	3	1	2	3
Improved Milch (6 kg)	3.0	15.0	8.0	10.87	54.14	28.85
Others, Milch (3 kg)	1.0	5.0	6.0	11.61	57.41	68.70
Males & young females	1.0	5.0	6.0	21.13	104.49	125.05
Total	-	-	-	43.61	216.04	222.60

(1=Concentrate mixture; 2=Green fodder; 3=Dry fodder)
(Ranjhan 1998; Ranjhan 2003)

There are three major sources of fodder supply *viz.,* crop residues, cultivated fodder and fodder from common property resources like forests, permanent pastures and grazing lands. The quantum of fodder obtained from grazing varies from region to region and the extent of forest areas. The availability of crop residues, constituting a major source of fodder is on a decline due to extensive use of combined harvester - a large portion of straws/stovers are either left in the field or burnt, especially in Punjab and Haryana. The area under fodder crops has also decreased to about 3.3 per cent (Kelly and Rao 1994). Insufficient utilization of crop residues and lack of feeding macro and micro elements (mineral mixtures) also lead to lower productivity of livestock.

The present practice of feeding buffaloes in major five regions of India and also strategies to bridge the gap is depicted in Table 2. The marked differences in soil, climate, cropping and other topographical conditions in various parts of the country affect the type of livestock and feeding strategy. Normally, farmers raise buffaloes for draught and milk and do not practice scientific feeding. However, in some regions *viz.*, northern, western and coastal there is a shift towards buffaloes and better feeding is practiced.

Livestock based Agroforestry

Agroforestry is a dynamic, ecologically based natural resource management system that through the incorporation of trees on farms and in the agricultural landscape, diversifies and sustains production for increased social, economic and environmental benefit. The agroforestry systems may be divided into many variants depending on the relative emphasis. An emphasis on timber and energy production, specifies agroforestry as silvicultural systems; when crops (including forage crops) are integrated, the system is developed into an agrisilvicultural system; when fodder shrubs and trees, pasture grasses and legumes and livestock are integrated, silvicultural systems is developed into a silvopastoral system and when agrisilvicultural system, silvicultural system and crop residues are integrated, it is classified as an agrisilvopastoral system (Singh and Roy 1991; Nitis 1997). A number of fruit trees (mangoes, tamarind, cocoa, cashews etc.) have potential for livestock integration as substantial ground area is available under them. Similarly, rubber and oil palm plantations also offer habitats for growing shade tolerant forage species for livestock gains (Gutteridge and Shelton 1994).

Integration of livestock with proper planning and management leads to sustainable farming systems. The key is to integrate the natural needs, behaviours and products of animals with the environment provided by the agroforestry system in a way that maximizes the benefits to the animals and to the system as a whole. The products and services of livestock based agroforestry include grazing/weed maintenance, cleaning of fallen fruits/nuts and organic matter and spreading manure. Comprehensive land utilization in agroforestry systems is expected to provide a relatively constant income from livestock sale and selective sale of trees and timber products. Also, grazing can enhance tree growth by controlling grass competition for moisture and nutrients. Well managed grazing may be used as an effective tool in controlling weeds and bushes without herbicides, maintenance of fire breaks and reduction in habitat for rodents. Livestock manure recycles nutrient to trees and other forage species. In addition, the fertilizer applied to forage crops is shared by the trees. Several forage species tend to be lower in fibre and more digestible when grown in a tree protected environment. The trees that provide shade or wind protection can have a climate stabilizing effect to reduce heat stress and wind chill of livestock. Protection from trees can cut the direct effect by 50 per cent or more and reduce wind velocity by as much as 70 per cent. In such a situation, livestock requires less feed energy, so their performance is improved and mortality is reduced (Reynolds 1991; Huxley 1999).

Agroforestry has the potential to enhance species diversity and improve water quality. Most of the forage species like grasses protect the soil from water and wind erosion, while adding organic matter to improve soil properties. Such systems provide an attractive landscape compared to concentrated livestock operations.

Forage Resources

Fortunately, India is endowed with a rich genetic diversity of forage plant species due to diverse eco-climatic conditions, variety of habitats and niches superimposed with tribal and ethnic diversification, plant usage and religious rituals. The choice of forage species may be classified into four major groups – forage crops, pasture grasses, pasture legumes and multipurpose trees and shrubs (MPTS).

Forage Crops: Forage as a group of crop differ from food and commercial crops in several aspects. As forage crops are grown to feed animals to achieve better production, besides high tonnage the herbage obtained should have better physical and nutritional qualities for preferential intake and utilization in the animal body. Over 250 varieties of forage crops have been released. Important varieties of major 6 forage crops are depicted in Table-3.

Pasture Grasses: Good pasture grass species have the potential to provide more protein and starch equivalent than most of the crops. Effective grazing management offer higher levels of live weight increase in livestock and also maintenance of pasture quality. Grasses exhibit a wider range of adaptability; especially in humid tropics, arid areas and alpine peaks. They have the advantage of reproducing fresh shoots through tillers, recovering from grazing or cutting. The root system binds the soil particles together forming a sod. Thus surface layer nutrients are utilized that otherwise leach into subsoil through rainfall. Many species having the capacity to spread by rhizomes or stolons, provide rapid ground coverage. A list of some pasture grasses is provided in Table-4 that could be incorporated for developing forage resource base for buffaloes under agroforestry.

Table-2: Feeding and strategies to improve buffalo production in 5 major regions of India.

Region	Buffalo Types	Feeding Status & Strategy
Hot Tropical Region Punjab, Delhi, Haryana, Eastern Rajasthan, Western Uttar Pradesh, Madhya Pradesh, Gujarat	Home tract of most of the established high yielding milch breeds of buffalo i.e. Murrah, Nili/Ravi, Mehsana, Zaffrabadi, Surti, Nagpuri	In most part wheat straw and stovers are the staple roughages. Cultivation of green fodder is more popular in this region compared to other regions.

Region	Buffalo Types	Feeding Status & Strategy
Wet Eastern Region Eastern part of Uttar Pradesh, Bihar, West Bengal, Lower Assam, parts of Maghalaya, Tripura, Mizoram, Manipur, Nagaland, Orissa, parts of Andhra Pradesh	Mostly entire male buffalo and in some parts sterile/infertile female buffalo are used for agricultural operations. Local buffaloes are mostly non-descript.	In most part paddy straw forms the staple roughage for the livestock. The area has shortage of feeds and fodder. Traditionally, the farmers do not grow fodder and depend mostly on forest grazing and cut and carry grasses as greens. Feeding of balanced concentrate mixture along with the straws and grazing is desirable.
Coastal Region The narrow strips on east and west border of Southern India	Local animals are generally unproductive and poor in health. However, in Coastal Andhra Pradesh a large quantity of good quality buffaloes are found.	Poor quality paddy straw is the main roughage for the livestock. Local grazing management and supplementation of crop residues is desirable.
Southern Region It includes parts of Madhya Pradesh, Andhra Pradesh, Maharashtra, Tamil Nadu and Karnataka states.	The buffaloes are mostly non-descript.	Sorghum, pearl millet, coarse rice and other millets are used as straw for livestock. Supplemental feeding through balanced concentrate mixture, the ingredients may be imported from other regions, is desirable.

Region	Buffalo Types	Feeding Status & Strategy
Temperate Himalayan Region It comprises of mountainous area and *Terai* of Himalaya from Kashmir to Arunachal Pradesh.	In *Terai* region light weight and low producing buffaloes are found.	The area has shortages of feed and fodder. Conservation of surplus green grass cut during monsoon for use in winters and fodder supplementation through tree leaves are desirable.

(Source: Gopalakrishnan and Lal 2000; Ranjhan 2003)

Pasture Legumes: In order to improve the productivity and quality of natural grasslands and also developing sown pastures, incorporation of legumes having forage value is considered very important.

Table-3: Important varieties of 6 major forage crops

Forage Crop	Variety & Area of adoption	Green Forage Yield (t/ha)
Fodder Sorghum (Single cut)	Pusa Chari-6 & 9 (Northern India); Pusa Chari-3 (UP); Jawahar Chari (MP); Rajasthan Chari 1 & 2 (Rajasthan); GFS-1 (Gujarat); HC-136, 260, 308 (Entire Country)	30-60
Fodder Sorghum (Multicut)	Meethi Sudan, SSG-59-3; PCH-106 (Northern India); CSV-15 & CSH-13 (Maharashtra); LX-250 (Punjab); MP Chari, ProAgro Chari-SSG988 (Entire Country)	40-95
Fodder Maize	Ganga Safed-2, Pratap Selection (North Zone); J-1006 (Punjab); A-De Cuba (North Wast); African Tall & Manjari (Entire Country)	25-50
Fodder Bajra	Rajko (Gujarat and Rajasthan); Giant Bajra, DSRB-1 (Entire Country)	30-100
Oats	JHO-851 (North Zone); OL-9 (North, North West and Southern Hill); OL-125 (Central & North West); Bundel-Jai 822 (Central India); Kent (OS-6 & O-S7); UPO-94, UPO-212 (Entire Country)	35-52

Forage Crop	Variety & Area of adoption	Green Forage Yield (t/ha)
Bajra-Napier Hybrid	PBN-83 (Punjab), Yeshwant (RBN-9) (Maharashtra)	110-160
Cowpea	Charodi, Gujarat Cowpea-3 (Gujarat); Sweta (No. 988), DFC-1 (Maharashtra); CS-88 (Haryana); Cowpea-88 (Punjab); UPC-9202 (Central Zone); UPC-42 (North West Zone); UPC-42 (North West Zone); UPC-287, UPC-8705, Country), Bundel Lobia-1 (IFC-8401) (Entire Country)	22-45
Guar	Guara-80 (Punjab); Bundel Guar-3 (IGFRI-1019-1) (Arid and Semiarid Regions); HFG-156, Bundel Guar-1 (IGFRI-212-1), Bundel Guar-2 (IGFRI-2395-2) (Entire Country)	22-35
Berseem	BL-2 (Northern India); BL-22 (Sub-temperate hilly regions of Northern India); UPB-110 (Southern Zone); Vardhan, JB-2, JB-3 (Entire Country)	50-110
Lucern	Anand-3 (Himachal Pradesh); CO-1 (Tamil Nadu and Karnataka); LLC-5 (Punjab); Chetak, LLC-3, Type-9 (Entire Country)	60-100

(Source: IGFRI 2001; Tripathi & Agrawal 2003)

They not only provide nutritious forage but also improve land fertility by adding nitrogen equivalence to the tune of 40-60 kg/ha (Rai *et al.*, 1980). A list of some promising pasture legumes is provided in Table-4 that could be incorporated for developing forage resource base for buffaloes under agroforestry.

Table-4: List of promising forage grasses for developing forage resource under agroforestry

Species	Region
Agrostis Sp.	1
Andropogon gaynus	3,7,10,12,20
Bothriochloa intermedia (Black Soils)	4,5,6,7,9,10,11, 12
Brachiaria brizantha	13
B. decumbanse	6,8,13,15,17
B. mutica	13,15,17
B. ruiziensis	8.20
Cenchrus ciliaris (Sandy Soils)	2,4,20
C. setigerus (Sandy Soils)	2,6
Chloris gayana (Salt Affected Lands)	2,5,9,18
Chrysopogon fulvus	3,4,7,10,12
Coix lacryma-joba	16
Cynodon dactylon	5,8,9,11,12,13, 18,19, 20
Dactylis glomerata	14
Dichanthium annulatum	3,4,5,6,7,8,9,10, 11, 18, 19
Festuca rubra	14
Lasiurus sindicus (Sandy Plains)	2
Lolium perenne	14
Panicum maximum	5,6,11,19

Panicum turgidum (Sand Dunes)	2
Paspalum notatum	13
Pennisetum clandestinum	16,19
Pennisteum pedicellatum	6,7,9,10,11,12, 18, 20
Pennisteum polystachyon	19,20
Poa alpine	1
Sehima nervosum	3,4,5,6,7,8,9,10, 11., 18, 19
Setaria sphacelata	11,19
Sporobolus marginatus (Salty Lands)	2
Tripsccum dactyloides	16
Tripsccum laxum	20
Tresetum spicatum	1
Urochloa mosambicensis	12,18

(Source: Shankar *et al.*, 1996)

1=Western Himalayas (cold and skeletal soils); 2=Western Plains and Kacch Peninsula; 3=Deccan Plateau; 4= Northern Plains (hot, semiarid); 5=Central (Malwa) Highlands, Gujarat Plains and Kathiawar Peninsula; 6=Deccan Plateau; 7= Telangana and Eastern Ghats; 8=Tamil Nadu Uplands and Karnataka; 9=Northern Plains (hot, subhumid); 10=Malwa, Bundelkhand and Satpura; 11=Chhatisgarh; 12=Eastern Plateau and Eastern Ghats; 13=Eastern Plains; 14=Western Himalayas (subhumid and podzolic soils); 15=Bengal and Assam Plains; 16=Eastern Himalayas; 17=North Eastern Hills; 18=Eastern Coastal Plains; 19=Western Ghats and Coastal Plains; 20= Islands of Andaman, Nicobar and Lakshdweep.

Table-5: List of promising forage grasses for developing forage resource under agroforestry

Species	Region
Atylosia scarabaeoides	10,12
Cassia rotundiflora	2
Centrosema pubescens	20
Clitoria ternatea	3,5,9,17,20
Desmanthus virgatus	4,5,8,9,10,11,13
Desmodium uncinatum	15,17
Desmodium heterophyllus	15,17
Lotus carniculatus	14
Macroptelium atropurpureum	4,7,9,10,12,20
Macrotyloma axillare	7,12
Medicago sativa	1
Pueraria phaseoloides	16
Pueraria thunbergiana	19
Stylosanthes guinensis	8,18,19,20
Stylosanthes hamata	3,4,5,6,8,9,10,11,19
Stylosanthes scabra	3,4,5,6,7,10,20
Trifolium pretense	14
Trifolium repens	14

1=Western Himalayas (cold and skeletal soils); 2=Western Plains and Kacch Peninsula; 3=Deccan Plateau; 4= Northern Plains (hot, semiarid); 5=Central (Malwa) Highlands, Gujarat Plains and Kathiawar Peninsula; 6=Deccan Plateau; 7= Telangana and Eastern Ghats; 8=Tamil Nadu Uplands and Karnataka; 9=Northern Plains (hot, subhumid); 10=Malwa, Bundelkhand and Satpura; 11=Chhatisgarh; 12=Eastern Plateau and Eastern Ghats; 13=Eastern Plains;

14=Western Himalayas (subhumid and podzolic soils); 15=Bengal and Assam Plains; 16=Eastern Himalayas; 17=North Eastern Hills; 18=Eastern Coastal Plains; 19=Western Ghats and Coastal Plains; 20= Islands of Andaman, Nicobar and Lakshdweep.

(Source: Shankar *et al.*, 1996)

MPTS : A number of multipurpose trees and shrubs (MPTS) yielding top feed are available for incorporation under agroforestry. A list of some promising MPTS is provided in Table-6 that could be incorporated for developing forage resource base for buffaloes under agroforestry. Fodder from the leaf or pod of trees/shrubs constitute an important source of livestock diet. This material is usually rich in proteins, vitamins and minerals like calcium. Many of these species have ability to fix nitrogen thus benefiting the grasses and other ground level crops directly or indirectly. Besides fodder, many of such resources may be used as firewood, charcoal, food (leaves and pods) and for medicinal purposes. They have also potential service roles like soil conservation and fertility enhancement (Roy and Pathak 1994).

It has been observed that leaf meal of *Grewia optiva* contained one-third of the DCP in ground nut cake and could replace completely on DCP basis in maintenance ration of large ruminants in Himachal Pradesh (Pachuri *et al.*, 1974). In Philippines, *Leucaena leucocephala* was given a great deal of attention for livestock industry. Small dairy farmers fed their large ruminants with 10-20 kg of fresh leaves of this species in combination with fresh grass fodder and obtained

Table-6: List of promising MPTS for developing forage resource under agroforestry

Species	Region	Species	Region
Acacia nilotica	2,3,4,6	Ficus nemoralis	15,16
Acacia tortilis	2	Ficus semicordiflora	16
Ailanthus excelsa	2	Gliricidia sepium	5,7,20
Ailanthus malabarica	8,18,19	Grewia optiva	14
Albizia amara	3,4,10	Grewia subinequalis	4,

Buffalo Production under Different Climatic Regions

Species	Region	Species	Region
Albizia lebbek	3,4,5,7,10	Hardwickia binata	10
Albizia procera	4,6	Holoptelia integrifolia	5.7
Albizia stipulata	9	Leucaena leucocephala	3,4,6.7,9,10,11,12,20
Annona squamosa	4,5,7,11	Moringa oleifera	6,11,12
Azadirachta indica	4,9	Morus alba	14,15,17
Bauhinia purpurea	6,20	Pithecellobium dulcee	5,20
Bauhinia variegata	6,11,13	Prosopis cineraria	2
Capparis decidua	2	Pterocarpus marsupium	6,13
Carissa carandus	4,8,10	Quercus incana	14
Celtis australis	14,16	Robinia pseudoacacia	9,14,17
Cordia gharaf	2	Salvadora oleoides	2
Dalbergia sissoo	1,13	Salvadora persica	2
Dichrostachys cinerea	2,4,10	Sesbania grandiflora	4
Erythrina variegata	8,18,19,20	Sesbania sesban	4,6
Fagus sylvatica	14	Tamarindus indica	3
Ficus carica	9	Trema tomentosa	20
Ficus hookeri	15,16	Ziziphus nummularia	2
Ficus racemosa	9	Zizyphus mauritiana	2

1=Western Himalayas (cold and skeletal soils); 2=Western Plains and Kacch Peninsula; 3=Deccan Plateau; 4= Northern Plains (hot, semiarid); 5=Central (Malwa) Highlands, Gujarat Plains and Kathiawar Peninsula; 6=Deccan Plateau; 7= Telangana and Eastern Ghats; 8=Tamil Nadu Uplands and Karnataka; 9=Northern Plains (hot, subhumid); 10=Malwa, Bundelkhand and Satpura; 11=Chhatisgarh; 12=Eastern Plateau and Eastern Ghats; 13=Eastern Plains; 14=Western Himalayas (subhumid and podzolic soils); 15=Bengal and Assam Plains; 16=Eastern Himalayas; 17=North Eastern Hills; 18=Eastern Coastal Plains; 19=Western Ghats and Coastal Plains; 20= Islands of Andaman, Nicobar and Lakshdweep.

(Source: Pathak and Roy 1992; Singh and Roy 1994)

around 7 kg of milk/day (Moog 1991). Another promising and extensively studied species is *Gliricidia sepium*. The leaves of this species are succulent but not very palatable when first introduced. However, livestock freely eat when they become accustomed to its taste. In Sri Lanka mixing of the leaves of this species in 50:50 ratio with the grass (*Brachiaria milliformis*) resulted in average live weight gain of 700 g/day in large ruminants (Liyange and Jayasudera 1988).

Agroforestry Systems

*Agrisilviculture:*In this technology, cultivation of forage crops is integrated in association with fodder trees/shrubs, both under irrigated and rainfed situations. Trees like *Leucaena leucocephala* and *Sesbania aegyptica* have been found to have positive associative effect with several forage crops *viz.*, hybrid napier, endull, oat and other cereal forages (Singh 1988). In another study with different forage crops like barley, oat, berseem, safflower, Chinese cabbage in association with three fodder trees *viz.*, *Acacia tortilis*, *Albizia lebbek*, *Leucaena leucocephala* it was found that different winter crops behaved differently with respect to seed and forage yield and even with same tree it was different over the seasons as radiation curtailment had differential influence in different seasons (Hazra 1985). Agrisilvicultural technology for fodder production in *terai* belt of Uttar Pradesh in irrigated conditions involving poplars has yielded 50 t fodder/ha in five to six months (Chaturvedi 1982).

The trees are one of the major components in such systems and require pruning each year to maintain availability of light for intercrops. Even due to pruning forage yield are obtained. In an experiment involving six year old fodder trees *Albizia procera* and *Azadirachta indica* (under rainfed conditions), additional forage yields of 1.63 and 0.37 t DM/ha were obtained (through pruning), respectively. Similarly, comparatively slow growing species like *Hardwickia binata*, *Anogeissus endulla* and *Anogeissus latifolia* at 11 years contribute 0.86, 0.48 and 0.42 t DM/ha, respectively through pruning (Rai and Misra 2003).

Alley Cropping: Alley cropping, also known as hedgerow intercropping, integrates the benefits of fallow period directly into the cropping period. In this system, trees are incorporated into traditional farming practices that have generally considered trees as hindrance to crop production. The idea is to maintain and improve soil fertility and crop yields with simultaneous forage production. It is quite a flexible technology that benefits crop and livestock activities and can, through a modification of tree management techniques, provide firewood for the household as well. It is a system that could be adapted to meet particular priorities of individual farmers.

The two most extensively tree species for alley cropping are *Leucaena leucocephala* and *Gliricidia sepium* on account of their fast growth and leaf production. The trees are pruned at regular intervals at about 0.6 m height and the pruning could be used as mulch or livestock feed. As the trees are not allowed to grow to full height, the crops are not shaded. During productive period the tree foliage contains more than 20 per cent crude protein (CP) and thus is a high quality supplementary feed for the ruminants or mulch to enhance soil fertility or both (Kang *et al.*, 1989). However, it is necessary to decide upon the quantity of leaf material to be utilized as mulch and fodder. A general recommendation is to utilize 25 per cent as fodder and remaining as mulch. Inclusion of animals to such systems was found to be profitable as net output of livestock producers increased by 25–30 per cent. Also, the animals yield substantial quantities of manure to the system. So the tree leaves consumed by the animals are in fact recycled in the system.

In a modified alley farming system, grasses may be planted between tree rows in place of food crops. A combination of *Leucaena leucocephala* and *Gliricidia sepium* with grasses like *Panicum maximum* yield over 20 t DM/ha/year. Cut and carry system may be followed for more intensive livestock operations like stall feed dairy units (Reynold 1991).

In humid zone, alley farming is more profitable than conventional crop fallow cycle, even though more labour is required to prune the trees regularly and spread the mulch (Sumberg *et al.*, 1987). In semiarid tropics, water competition between trees and crops could be dominant factor but availability of fodder during late part of dry season may increase attractiveness of legume forage to livestock owners (Singh *et al.*, 1991). Apart from these two species, the other potential species for alley cropping include *Calliandra callothyrsus, Flemingia congesta, Erythrina poeppigiana, Gmelina arborea, Inga edulis, Gliricidia maculata, Cajanus cajan* etc. (Kang *et al.*, 1989; Singh *et al.*, 1991).

Silvopastoral Systems

Fodder trees and shrubs can be incorporated on pasture lands in a variety of ways. In such systems, grown up trees/shrubs may be lopped for fodder and also for reducing shade effects during growing period of pastures. The livestock integration in such systems is very important and now there is a growing realization that such integration can, in fact, improve productivity per unit of land besides meeting feed requirements of livestock (Payne 1985). Three major ways of utilization of silvopasture systems are as under:

(i) Grazing and or browsing in natural forests

(ii) Grazing or harvesting pasture grown under planted trees

(iii) Browsing or harvesting of tree foliage

Management of both forage species and cattle is of paramount importance in fragile ecosystems of drier tropics. It is necessary to maintain equilibrium between fodder trees, ground cover and number and type of livestock. In wet African savanna a density of 100-400 trees/ha has been considered to be desirable. Such trees provide over 20 per cent of nutritional requirements of livestock. In higher tree densities forage plants more shade tolerant than others should be grown. Eriksen and Whitney (1977) reported that at low light intensity pasture grasses like *Panicum maximum, Brachiaria brizantha* and *Brachiaria milliformis* are more productive. The ranking of pasture legumes for shade tolerance was *Desmodium intortum, Macroptelium atropurpureum* and *Stylosanthes guianensis*. The shading tended to increase shoot/root ratio and mineral content of the pasture grasses. While practicing *in situ* grazing, it is always advisable to follow deferred rotational system of grazing so that possible effects of overgrazing are minimized. Otherwise compaction could be a problem, especially on heavier soils during wet conditions (Thomas 1978).

In India, semiarid regions lands (producing hardly up to 1 t/ha/yr) have been improved to produce > 10 t/ha/yr (10 year rotation) through such systems. Out of this total yield contribution of pasture grasses, top feed, lopped firewood, harvested firewood and minor timber were 45, 18, 15, 9 and 13 per cent, respectively. Besides yield improvement, quality of forage was better by 6-7 times. Apart from the products the gains to environment are enormous, reduction in the soil loss (1.3 t/ha/yr from17.8 t/ha/yr in bare situations) (Pathak & Roy 1994). In other regions, especially arid and hilly parts, such systems provided more forage yield/grazing availability for longer periods and of better quality over the existing situations (Shankarnarayan *et al.,* 1987; Sharma & Koranne 1988; Sharma *et al.,* 2001).

The management considerations for silvopastures are influenced by site conditions, inputs, demand of forest products, market prices etc. Trees have the potential to provide better returns (through biomass) only when they are managed effectively as per their growth and phonological attributes. Canopy management besides use of compatible shade tolerant grasses like *Panicum maximum*, is an important consideration besides other ones as practiced in pasture management systems. The canopy management options in relation to trees include coppicing, lopping, pollarding and pruning. Trees in the wood lots may be coppiced to produce firewood, while trees in fodder lots or hedge rows intercropping systems are often lopped to produce fodder and mulch and where necessary to reduce shade for the pasture component. Trees in pastures or on boundaries may be pollarded for the same purpose. Pruning is often done on the trees to increase flowering and fruiting, reducing shade and availability of some fodder.

Agrosilvopastoral Systems

Such systems integrate crops and pastures with livestock. Such systems can be used for food production and soil conservation besides providing fodder and firewood. Such systems hold promise in highland humid tropics. It may tree-livestock-crop mix around homestead, woody hedgerows for browse, green manure, soil conservation or an integrated production of pasture, crops, animals and wood (firewood and poles etc.). In Philippines, a system for small farmers known as Simple Agro-Livestock Technology (SALT-2), in which 40 per cent of farmers' land is devoted to agricultural crops, 40 per cent to livestock and 20 per cent to forestry has been developed. The contour lines are located on sloping fields. Food and cash crops are grown on the upper half of the farm so that soil loosened by cultivation is caught on the lower half which is devoted to permanent fodder crops. In order to minimize the disturbance of soil 75 per cent of the cropping area is put under permanent fodder crops and only 25 per cent to short term crops. This system has been found to protect and improve the soil besides contributing substantially to the self-sufficiency of the farm family (Tacio 1992).

In India, such agrosilvopastoral systems provide an opportunity in arable highland humid tropics for mixed tree-crop-livestock husbandry. During cropping seasons, the animals are fed with stored straw (hay), supplemented with grass and foliage of trees. In semiarid rainfed systems, incorporation of MPTS in forests and wastelands may supplement an additional resource of 4-5 t/ha/yr fodder and 6-8 t/ha/yr firewood at every fourth year (Singh 1992).

Plantation Systems

In these systems the main objective is to obtain commercial tree products like rubber, palm oil, cashew and other fruits like mango, tamarind, cocoa etc. However, these systems have potential for livestock integration as wide interspaces (10 x 10 m or 8 x 8 m) are available under such trees. In young stage of the tree, the light penetration is also quite adequate for many of the forage species. A great deal of work is available on integration of animals with coconut plantation, where growing pastures in between tree interspaces resulted in lower weeding cost, higher coconut production and additional income derived from animals (Reynolds 1988). In order to improve productivity and carrying capacity of such systems, establishment of improved pasture grasses and legumes may be undertaken. The degree of light interception varies with age of coconut, hence pasture thrives best under young trees (less than five years) or old trees (more than 20 years). Since, tropical C_4 grasses show a greater decrease in relative growth rate under shade than C_3 legumes, in many commercial plantations, fast growing leguminous cover crops reduce soil erosion, fix nitrogen, improve nutrient cycling and besides supplying forage (Nair 1984). However, a negative effect of pasture on coconut yield is expected in coconut plantations where rainfall is marginal and competition for moisture becomes critical (Plucknett 1979).

In a specialized plantation system used in Bali for vanilla production under coconuts, the vanilla orchard is supported and shaded by tree legumes like *Gliricidia sepium* or *Erythrina* species under lopping management for regulating shade levels and the tree leaves are used as byproduct to feed the livestock. The annual live weight gain up to 550 kg/ha are reported from such systems (Shelton *et al.* 1987).

The potential of livestock integration with rubber is comparatively low because young rubber trees may be damaged by animals and in mature trees latex tapping cups may be disturbed (Shelton *et al.*, 1987). In a long term study in rubber plantations of Sri Lanka, a decline (between 2-7 years) in rubber yield was detected when *Brachiaria brizantha* was grown in the interspaces (probably on account of competition). This effect was less pronounced with other grasses. The grass yield also declined on account of shading. However, this effect was partially mitigated by inclusion of herbaceous legumes (Waidyantha *et al.*, 1984). Experiments have also shown that productivity of livestock was moderate under young rubber (2-5 years) provided leguminous cover crops were grazed, but low under mature rubber where light levels have fallen to <20 per cent transmission of light. There is potential to increase productivity by using shade tolerant forage species and by altering the conventional rubber planting system to a hedgerow system (Gutteridge and Shelton 1994).

Other Systems

Boundary Planting: Farm boundaries, banks, river banks etc. may be suitably utilized for lean period fodder. Species like *Leucaena leucocephala* have the potential to provide 1 t of nutritious fodder per row km (at 2 m spacing) along farm roads when harvested at 3 years. In the second rotation the yield were over 2.5 times of the first rotation. Additionally the firewood yields were 11 t and 27 t/row km, respectively (Singh and Pathak 1992)

Live Hedges: These are lines of trees or shrubs planted on farm boundaries or on borders of home compounds, pastures or animal enclosures. Although main purpose of such systems is to control human and livestock movement, they provide nutritious fodder besides firewood. They also act a wind breaks and enrich soil depending on the species used. The fodder species selected for such purpose should be adaptable to local agro-climatic conditions, of fast growing nature, have high coppicing ability and should provide economic returns. Depending on the site conditions species like *Gliricidia sepium*, *Erythrina* species, *Zizyphus mauritiana*, *Prosopis juliflora*, *Dichrostachys cinerea*, *Acacia* species, *Albizia* species may be used (Westley 1990). In South America, it is a practice to harvest live fences of *Gliricidia sepium* and *Erythrina* species to provide fodder for livestock, sometimes meeting up to 25 per cent of total intake while thicker branches used for firewood (Budowski 1987).

Conclusion

Various agroforestry technologies *viz.*, agrisilviculture; alley cropping; silvopastoral system; agrisilvopastoral system; plantation systems; live fence system; shelterbelts and wind breaks; tree on farmers' bunds, around farm land and forest margins offer scope of overcoming the problems related to low quality and quantity of livestock feeds and seasonal deficits of green forage for feeding of buffaloes that constitute a very important livestock component in

Indian farming systems. Also such technologies have great potential in combating the challenge of land degradation. Livestock based agroforestry systems promote nutrient cycling for soil fertility enhancement and provide a channel for utilization of plant biomass through animal products into human nutrition.

Future Thrusts

1. Introduction and management of buffaloes in agroforestry systems in a way so that such systems require less human intervention are sustainable. More research is required on controlled grazing trials to recommend stocking rates for a given situation and selection of promising pasture combinations that can withstand grazing and lead to higher livestock productivity.

2. Emphasis of growing fodder trees and shrubs on wastelands through use of proper rhizobia and mycorrhizas and incorporation of tree leaves and pods into the diet of buffaloes for reducing the cost of supplementation and also increased productivity. More research is required on use of top feed in the diet of livestock with due consideration of palatability, nutritive value, anti-quality factors and management for reducing its effect for positive gains.

3. Emphasis on improvement of grazing areas through development of suitable agroforestry systems looking to the farmers' requirements, including forage for livestock. More research is required on forage yields of natural *versus* improved pasture grasses and legumes and their adequate proportion and design under silvopastures and other plantation systems.

4. Emphasis on promoting forage based agroforestry systems on arable lands. More research on economic analysis of such systems to determine the opportunity cost to the farmer of planting forage species *vis a vis* food crops. Similarly, more research on alley cropping

systems, especially in semiarid regions where demand for animal feed is very high. It is important to identify more number of hedge row species and also generating information on their cutting management and compatibility with other crops/crop mixtures.

5. Socio-economic aspects are extremely important for adoption of agroforestry systems. Proper understanding of social attitudes and aspirations along with technical feasibility of an agroforestry system on site considerations is required for positive gains in livestock productivity.

References

Akbar, M. A. and Gupta, P. C. 1985. Subabul (*Leucaena leucocephala*) as a source of protein supplement for buffalo calves. *India. J. Anim. Sci.*, **54**: 731-735.

Budowski, G. 1987. Live fence in tropical America – a widespread agroforestry practice. *In: Agroforestry – Realities and Possibilities* (ed. H. L. Gholz). Martinus, Nijhoff (the Netherlands): 169-178.

Chaturvedi, A. N. 1982. *Poplar Farming in Uttar Pradesh*. Govt. of Uttar Pradesh, Lucknow: pp 42.

Dharmraj, P., Rao, M. R. and Rao, V. P. 1985. Feeding *Subabul* leaf meal to lactating *Murrah* buffaloes. *India. J. Anim. Sci.*, **55**: 389-391.

Eriksen, F. and Whitney, A. S. 1977. Performance of tropical forage grasses and legumes under different light intensities. *Proc. Reg. Sem. Pasture Res. Dev. Solomon Island and Pacific Region*: 180-190.

Gopalakrishnan, C. A. and Lal, G. M. M. 2000. *Livestock and Poultry Enterprises for Rural Development*. Vikas Publishing House Pvt Ltd., New Delhi: pp 1096.

Gutteridge, R and Shelton, M. 1994. Animal production potential of agroforestry systems. *In: ACIAR Proc. On Agroforestry and Animal Production for Human welfare* (Eds J. W. Copland, A. Djajanegra and M. Sabrani). ACIAR, Canberra (Australia).

Hazra, C. R. 1985. Forage and seed yield of crops under agroforestry production system: Radiation and temperature relations. *J. Agron. & Crop Sci.*, **155**: 186-102.

Huxley, P. 1999. *Tropical Agroforestry*. Blackwell Science, London: 51-63.

IGFRI 2001. *Production Technology for Potential Kharif Forages*. Indian Grassland & Fodder Research Institute, Jhansi: pp 20.

Kang, B. T., Reynolds, L. and Atta-Krah, A. N. 1989. Alley farming. *Advances in Agronomy*, **43**: 315-359.

Kelley, T. G. and Rao, P. P. 1994. *Trends in Cultivated Green Fodder Crops in India*. ICAR & Agricultural University, Wageningen (Netherlands): 68.

Liyanage, L. V. and Jayasudera, H. P. S. 1988. *Gliricidia* as a multipurpose tree for coconut plantation. *Coconut Bull.*, **1**: 1-4.

Moog, F. A. 1991. Forage and legumes as protein components for pasture based systems. *In: Feeding Dairy Cows in the Tropics* (eds A. Speedy and R. Sansoucy). FAO, Rome: 142-148.

Nair, P. K. R. 1984. *Soil Productivity Aspects of Agroforestry*. ICRAF, Nairobi (Kenya): pp 85.

Nitis, I. M. 1997. Silvipastoral systems in tropical context. *In: Proc. Of the XVIII International Grassland Congress (Vol. III)*. Manitba (Canada): 9-17.

Pachuri, V. C., Pal, R. N. and Negi, S. S. 1974. A not on *Grewia optiva* leaf meal as DCP supplement in ruminant rations. *Indian J. Nutri. Dietet.* **11**: 33.

Pathak, P. S. and Roy, M. M. 1992. Fodder trees in agroforestry: Their selection and management. *Range Mgmt. & Agrof.* **13**: 63-87.

Pathak, P. S., and Roy, M. M. 1994. *Silvipastoral System of Production – a Research Bulletin*. IGFRI, Jhansi: 55 p.

Payne, W. J. A. 1985. A review of possibilities for integrating cattle and tree crop production systems in the tropics. *Forest Ecol. & Mgmt.* **12**: 1-36.

Plucknett, D. L. 1979. *Managing Pasture and Cattle under Coconuts.* Westview Tropical Agriculture Series 3. Boulder, Colorado (USA): pp 364.

Rai, P. and Misra, A. S. 2003. Agroforestry systems for sustainable livestock production in India. In: *Agroforestry: Potentials and Opportunities* (eds P. S. Pathak and Ram Newaj). Agrobios (India), Jodhpur: 281-303.

Rai, P., Kanodia, K. C., Patil, B. D., Velayudhan, K. C. and Agrawal, R. 1980. Nitrogen equivalence of range legumes introduced in natural grasslands. *Indian J. Range Mgmt.* **1**: 97-101.

Rajora, R. 1998. *Integrated Watershed Management.* Rawat Publications, Jaipur: 616 p.

Ranjhan, S. K. 1998. *Nutrient Requirements of Livestock and Poultry.* Indian Council of Agricultural Research, New Delhi: pp 72.

Ranjhan, S. K. 2003. Livestock and feed situation in India. *In: Sustainable Animal Production* (eds R. C. Jakhmola and R. K. Jain). Pointer Publishers, Jaipur: 1-24.

Reynolds, L. 1991. Livestock in agroforestry: A farming systems approach. *In: Biophysical Research for Asian Agroforestry* (eds M. E. Avery, M. G. R. Cannell and C. Ong). Oxford & IBH Publishing Co., New Delhi: 233-258.

Reynold, S. G. 1988. *Pasture and Cattle under Coconut.* FAO Plant Production and Protection Paper 91. FAO, Rome: pp 321.

Roy, M. M. and Pathak, P. S. 1994. Agroforestry interventions for livestock producers. *Range Mgmt. & Agrof.,* **15**: 217-227.

Shankar, V., Gupta, J. N. and Singh, J. P. 1996. Forage production technology for rangelands. *In: Transfer of Forage Production Technologies* (eds Bhag Mal and R. J. Haggar). IGFRI, Jhansi: 41-51.

Shankarnarayan, K. A, Harsh, L. N. and Katju, S. 1987. Agroforestry in arid zones of India. *Agroforestry Systems,* **5**: 69-88.

Sharma, B. K. and Koranne, K. D. 1988. Present status and management strategies for increasing biomass production in north-western Himalayan rangelands. *In: Rangeland Resources and Management* (eds P. Singh and P. S. Pathak). RMSI, Jhansi: 138-147.

Sharma, N. N., Singh, R. P., Singh, M. and Harsh, L. N. 2001. *Silvopasture Management in Hot Arid and Semiarid Ecosystems*. Agrotech Publishing Academy, Udaipur: pp 144.

Shelton, H. M., Humphreys, L. R. and Batello, C. 1987. Pastures in the plantations of Asia and the Pacific: Performance and Prospects. *Tropical Grasslands*, **21:** 159-168.

Singh, P. 1988. Forage production: Present status and future strategies. *In: Forage Production in India* (ed. Panjab Singh). RMSI, Jhansi: 1-10.

Singh, P. 1992. Agrosilvipasture systems in India. *In: Shrubs and Tree Fodders for Farm Animals* (ed. C. Devendra): IDRC, Ottawa (Canada): 183-195.

Singh, P. and Pathak, P. S. 1992. Revegetation of degraded lands. *In: Status of Indian Forestry – Problems and Prospects* (ed. P. K. Khosla). ISTS, Solan: 60-81.

Singh, P. and Roy, M. M. 1991. Forage production through agroforestry: Constraints and priorities. *Range Mgmt. & Agrof.* **12:** 169-178.

Singh, P. and Roy, M. M. 1994. Present and prospective uses of tree species in small farms of semiarid India. *Range Mgmt. & Agrof.* **15:** 175-185.

Singh, R. P., van den Beldt, R. J., Hocking, D., and Konwar, G. R. 1991. Alley farming in the semiarid regions of India. *In: Alley Farming in the Humid and Subhumid Tropics* (eds B. T. Kang and L. Reynolds). IDRC, Ottawa: 108-122.

Sumberg, J. E., McIntire J., Okali, C. and Atta-Krah, A. N. 1987. Economic analysis of alley farming with small ruminants. *ICLA Bulletin*, **28:** 2-6.

Tacio, H. D. 1992. Goat, trees and other things. *Agroforestry Today*, **4:** 12-18.

Thomas, D. 1978. Pastures and livestock under tree crops in humid tropics. *Tropical Agriculture (Trinidad)*: **55:** 39-44.

Tripathi, S. N. and Agrawal, R. K. 2003. Intensive fodder production system. *In: Sustainable Animal Production* (eds R. C. Jakhmola and R. K. Jain). Pointer Publishers, Jaipur: 120-151.

Waidyanatha, U. P., de Wijesinghe, D. S. and Strauss, R. 1984. Zero grazed pastures under immature *Hevea* rubber: Productivity of some grasses and grass-legume mixtures and their competition with *Hevea*. *Tropical Grasslands*, **18**: 21-26.

Westley, S. B. 1990. Live fences – an agroforestry technology. *Agroforestry Today*, **2**: 11.

CHAPTER 23

Status of Reproductive Problems in Indian Buffaloes and Strategies for their Management

Satish Kumar

Indian Veterinary Research Institute, Izatnagar-241 322 (U.P.)

Contribution of livestock to Agriculture GDP has been constantly rising during last few decades. The livestock sector growth during last few decades has been able to record an annual steady growth of more than 5%. The buffaloes are the main stay of Indian dairy industry contributing over 50% of the total milk produced in the country besides being used for meat and draft purposes. The importance of buffalo in Indian subcontinent is well recognized targeting the country's big challenge of food production during the coming decades. This thrifty, versatile, adaptable and productive domestic animal has drawn national, international and multinational attention in recent past. The buffaloes are commonly grouped into two categories i.e. river and swamp. The world buffalo population is about 150 million out of which India alone contribute approximately 76 million. There are about 26 descript breeds available in our country. Besides a huge population of unknown non-descript buffaloes which constitutes about 3/4 th of the total bovine population. The worlds best dairy breeds are originated from this country and their superior germ plasm has been distributed to several countries of the world.

Considering its unique potential, a great deal of attention has been paid in the last three world buffalo congresses to exploit this species for human welfare. Poor knowledge of buffalo reproduction limited the species in to casual attention since its inception. However, a proper care and management offers a new dimension for increasing its productivity. The study, therefore, appeared to be necessary to understand the reproduction for optimal production and the extent of loss occurring due to infertility problems prevailed in the field situation. Although the information on the descript breeds of buffaloes in India are available from the farm conditions but very less information is available on the overall reproductive status of rural buffaloes which constitute to national economy in a big way and also if positively exploited it could generate a heavy production potential in rural condition. Hence, it is the right time to look into the details of the reproductive status of our buffaloes spread through out the length and breadth of our country.

Reproductive efficiency can be defined as activity of an individual to produce young ones. In order to obtain maximum output the main interest of a breeder is to achieve more young ones in life time, reduced mortality and healthy and superior young ones. There is a great potential for increasing fertility and productivity of farm animals. According to a survey fertility / productivity can be enhanced to 100% with a three pronged approach;

- 30% by improving reproductive management
- 40% by use of modern reproductive tools and
- 30% by control of reproductive disorders

Low reproductive efficiency of buffaloes remains an economic problem specially to buffalo rich countries and its incidence is higher in our country. Climatic stress, nutritional deficiencies, improper management and lack of disease prevention are some of the main attributing factors. The major reproductive problems with buffaloes are; delayed age at maturity, anestrus both in heifers and post-partum, repeat breeding and certain obstetrical problems such as genital prolapse, torsion and retention of placenta and endometritis.

Broadly the problems of infertility can be grouped as:

- Congenital / hereditary causes – ovarian hypoplasia, aplasia and malformations of organs etc.

- ◉ Acquired environmental causes – stress on reproduction due to somatic diseases such as Brucellosis, Tuberculosis etc.; Puereperal infections and coital infections, deficiencies like over and under feeding, defective fodder composition, deficiencies in minerals, vitamins and trace elements and climate management factors.

- ◉ A combination of both the above 1 and 2 factors.

The functional forms of infertility are the most commonest and important in buffaloes. This arises due to impaired physiological process of reproduction. Management, nutrition, endocrine system and heredity may impair the physiological process of female infertility. The impaired physiological functions are; absence / faulty manifestation of heat, abnormal cyclic periodicity, delayed / failure of ovulation, fertilization failure, gestational errors embryonic mortality, mummification, abortion etc. The entire functional form of infertility caused due to above factors may further be classified as: anoestrus, subestrus or silent heat or quiet ovulation, ovulatory dysfunction, delayed ovulation, cystic ovaries, failure of fertilization and dysfunction of fallopian tube, infertility associated with death or loss of conceptus.

In a herd or in an individual, various types of infertility problems are encountered. These infertility problems are manifested by following functional aberrations:

1. Ovarian dysfunctions

2. Fertilization failure

3. Early embryonic / fetal wastage and

4. The disorders during / after parturition.

The main problems of reproduction causing infertility in buffaloes are:

- ◉ Late maturity in heifers.

- ◉ Long interval between calving and first post partum heat.

- ◉ Repeat breeding in both heifers and cows.

- ⊙ Low conception rate and

- ⊙ Management problems of heat detection.

Incidence of Various Types of Reproductive Disorders in Buffaloes

Majority of reproductive disorders in buffaloes are due to either of ovarian disorders, disorders of oviducts, disorders, of uterus or disorders of cervix, vagina or vulva. Out of these disorders the main problems arise due to the ovarian problem, which leads to the conditions of various types of anestrum. The other important disorders in female buffaloes are the repeat breeding.

Ovarian disorders: The disorders of ovarian in nature leads to the anestrum, subestrum or silent estrus or leading to other conditions like; ovarian hypo-plasia, bursal adhesion, ovarian cysts and tumours. In a report from rural buffaloes Kumar and Agarwal (1986) reported that 47.40% bovines have some or the other type of ovarian abnormalities leading to infertility. Conducting a survey of Indian bovines Lagerlof (1955) found subfunctional ovaries to the extent of 30.7% in rural animals.

Pubertal anestrus : Underfeeding and stress of summer having its detrimental effect on the anterior pituitary to cause reduced output of gonadotrophins and subsequent release of estrogen / progesterone. The treatment of this condition is done using GnRH, PMSG, PRID, Norgestomate. The pubertal anestrus is a common condition in our Indian buffaloes.

Post-partum anestrus: Resumption of ovarian activity is delayed in buffaloes. The predisposing factor being the reduction in post-calving weight of buffalo due to negative energy balance, suckling, low plane of nutrition especially during last trimester of pregnancy etc. Since role of nutrition is the main cause hence improving the feeding may correct this situation. The onset of post partum estrus is a physiological phenomenon of immense economic importance in buffaloes. It is unduly delayed in those calving in winter and early summer. Efforts have been made to treat post-partum anestrus in buffaloes using several hormonal and non-hormonal preparations.

Post-partum supplementary feeding and treatment with norgestomate+ PMSG and many other combinations are useful.

Nutrition since plays an important role pre-partum feeding is more important than post-partum feeding. During summer poor reproductive performance is attributed to meager availability of green fodder, this further restricts the onset of estrus. It has been found that if buffaloes are fed good pasture with supplemented extra concentrate diet they can be bred round the year.

True Anestrum: Absence of periodic manifestation of estrus is termed as anestrum. On gynecological examination, ovaries of such animals are smooth and quiescent and do not reveal presence of palpable follicle or corpus luteum. The reason being insufficient releases of production of gonadotrophins to cause folliculogenesis along with other associated factors like nonavailability of green, plane of nutrition etc. and is common during summer months. Such ovaries are usually termed as inactive ovaries. This condition may result due to hypoplasia of ovaries, infantile genitalia, malnutrition and poor body growth in the heifers. In the post-partum animals it may be due to summer stress and early embryonic death. The other management factors like lactational stress, suckling negative energy balance and heat stress are also the contributing factors. During summer high ambient temperature, relative humidity and poor feed supply have been identified as the cause for anestrum in summers in this species.

Incidence of anestrum has been reported higher in buffaloes ranging from 14 to 45% (Luktuke and Sharma, 1978; Singh et al., 1979, Pandit et al., 1982). Reported extent of true anestrus 9-23% from abattoir material (Samad et al. 1984) and 56.4% records of sub-fertile buffaloes (Rao and Sreemannarayana, 1982). In a study conducted on 1600 buffaloes Kumar et al. (1988) reported 20% buffaloes to be non-pregnant with subactive ovaries, 24.31% buffaloes with true anestrus, 2% cases of hypoplasia of ovaries, 3.2% hypoplastic / atrophied uterus and 6.68% under developed / infantile genitalia. They concluded that underdeveloped and infantile genitalia are the major problem in rural buffaloes and the reason for this has been defined as the low level of nutrition during growing phase of heifers.

Kumar and Kumar (1993) examined 721 rural buffaloes from Tarai region of U.P and reported 70% buffaloes more prone to some type of anestrus problem. The number of cases of anestrus due to physiological nature (either or both of ovaries subactive) was 67.54%, true anestrus (both ovaries inactive) was 25.10% and anestrus due to infantalism was approximately 7.35%. It is apparent that a great proportion of (67.54%) buffaloes considered to be anestrus were not true anestrus, since they had subactive ovaries of normal texture and size carrying small to medium sized follicle or CL indicating its active functionality and normal cyclicity. The management and treatment of this condition includes:

⊙   Protection against heat

⊙   Initiation of ovarian activity using hormones like-GnRH, PRID alone or in combination with GnRH or PMSG, norgestomate ear implant, Synchromate B etc. have been found effective not only in inducing estrus but in resumption of subsequent cyclicity.

Subestrus: Subestrus or silent estrus is another important ovarian dysfunction in buffaloes where behavioral signs of estrus are not manifested. Such cases are usually reported from the field by the livestock owners. Gynaecological examination of such animals usually reveal presence of a palpable CL and characteristic cyclic changes in the genital tract, which shows that probably the animal is normally cyclic but the heat manifestations are not prominent to be observed by the owners. Usually in buffaloes the first heat after attaining puberty or after calving are not exactly preceded by behavioral sign of estrus leading to truly silent heat.

The incidence reported for this condition is 10-15 in organized farm to 34-36% in villages (Sharma *et al.*, 1967). Occurrence of subestrus or silent estrus in buffaloes have been reported to vary from 6 to 30% (Agarwal, 1978.; Chauhan and Singh, 1979).

Since nutritional factor is the most crucial predisposing factor the management of this condition applies to regression of CL of previous cycle- PG is the drug of choice and has been found quite effective, besides improving the nutrition and management during the above mentioned phases of the reproductive phases. Many hormonal preparations viz. estrogen, progesterone, gonadotrophins, prostaglandins alone or in combinations and certain non-hormonal preparations such as Prajana, aloes compound, clomiphene citrate, Janova and many others have been tried with varying degree of success in such cases.

Ovarian Hypo-plasia: As reported earlier this is a congenital abnormality. The affected ovaries are very small, hard with reduced number of primordial follicle. The hypo-plasia may be partial or total, unilateral or bilateral. Buffaloes having such type of ovaries do not come into heat and remain anestrus. Treatment in such conditions are not successful and these animals are to be culled. Occurrence of ovarian hypo-plasia in the buffaloes are less reported but the available literature reveals that it ranges from 0.24% to 2.0% in Indian and Egyptian buffaloes (Rao and Sreemanarayana, 1982; El Khouly, 1985; Kumar and Agarwal., 1986). In a field study from Ludhiana, Singh *et al.* (1986) reported the hypo-plastic / atrophied ovaries to the tune of 6.9% in the rural buffaloes and the condition of cystic ovarian degeneration was found to be 2.3%.

Ovarian cysts: It is one of the common condition in cattle but its incidence in the buffaloes are less. Presence of single or multiple persistent fluid filled structures on the ovary are characteristics of the ovarian cysts. The cyst may be follicular or luteal in nature. The follicular cyst is an ovulatory follicle that persist on the ovary for a longer period and is usually characterized by persistent estrus also known as the nymphomania or anestrus.

On the other hand the luteal cyst is also anovulatory in nature. The follicle gets partially / totally lutinized and persist for a longer period and are usually characterized by anestrus. Clinically lower percentage of these cysts are reported in buffaloes in the range of 0.9% to 2.0% (Dessouky and Juma, 1973; Rao and Sreemanarayana, 1982; Kumar and Agarwal., 1986). Bhattacharya (1954) reported 5.17% unilateral cystic ovaries in buffalo heifers and 3.31% in buffalo cows. The incidence of bilateral cystic ovaries was 1.65% in buffalo cows. The incidence of ovarian cysts in buffaloes from other parts of the country has been reported in the range of 1.03 to 2.4% (Sharma et al., 1967; Dwivedi, 1968).

The management of these conditions is usually attempted by either manual rupture or use of hormones such as LH, Progesterone, GnRH and Prostaglandin. Since delayed ovulation, anovulation and cystic ovaries are always associated with lack of adequate LH surge the practices of using LH for such conditions are in vogue. Beta-carotene is known to reduce the incidence of ovarian cysts. The predisposing factors from management point of view are as follows and proper attention is mandatory in order to take care of the following:

- ◉ Genetic selection for high yielder

- ◉ Larger group of animals and stressful conditions

- ◉ Parity of the animals-Increased parity higher risk

- ◉ High occurrence at the time of peak yield

- ◉ Post-parturition uterine infections

Ovario-bursal adhesion: This is the condition where an adhesion develops between ovary and ovarian bursa, even some time the bursa may encapsulate the ovary. This is the condition arising mostly due to faulty handling of ovary or due to infection. The incidence of this condition in buffalo from the abattoir material revealed its presence to the tune of 10.9% (Dobson and Kamonapatana, 1986). Incidence of ovario-bursal adhesion in clinical surveys are reported to the tune of 0.8 to 2.0% (Rao and Sreemanarayana, 1982; Ahmed, 1986).

Adhesion of ovarian bursa has been frequently observed and the per cent incidence reported was 10.29% by Sharma et al. (1968), besides the hydroponics bursa (1.4%), lymphoid growth in bursa (1.3%) and bursal cysts 2.47% cases. Dwivedi and Singh (1971) reported 6.26% chronic adhesions and 2.8 and 12.8% of unilateral and bilateral adhesion from the infertile cases. From field condition Rao and Keshavamurthy (1971) reported only 0.09% bursal adhesions. Kumar (1977) recorded 0.26 and 1.55% adhesions in heifers and cows, respectively. Agarwal (1978), however, recorded 4.07% adhesions in the buffalo cows.

Ovarian tumours: Very less incidence of ovarian tumour in buffalo are reported and this constitute of Granulosa cell tumour 0.2 to o.24%; dysgerminosa 1.9%; Ovarian teratoma / Dermoid cysts-0.17 to 1.9% (Dwivedi and Singh, 1971; El Khouly. 1985; Ahmed, 1986).

Oviductal disorders: The major abnormalities of the oviductal part in buffaloes are salpingitis, hydrosalpinx, pyosalpinx and adhesions of the salpinx. These abnormalities form a important cause for the infertility in buffaloes. Dobson and Kamonapatana (1986) reported the incidence of hydrosalpinx to the tune of 3.1% in the buffaloes. Rao and Sreemananarayana (1982), Ahmed (1986) reported that oviductal abnormalities formed 1.5 to 5.9% of the total genital tract abnormalities in the buffaloes. One of the most common affection of the oviduct the inflammation of salpinx (salinities) has been reported in the range of 0.8 to 4.5%. Besides these the other affections of the oviduct like pyosalpinx is also reported. Although the detection of affections of fallopian tube is a difficult and experienced process its treatment usually requires a great skill.

The incidence of salpingitis was reported much higher in buffalo cows than in buffalo heifers (Bhattacharya, 1954). Sharma *et al.* (1968) observed 2.0, 0.6, 1.03 and 0.48% cases of salpingitis, pyosalpinx, and hydrosalpinx in local and graded Murrah buffaloes. Dwivedi and Singh (1971) reported about 2.9% cases of slaughter house material to have hydrosalpinx. Chatterjee (1977) reported 3.2% chronic salpingitis and 1.6% hydrosalpinx in repeat breeder buffaloes.

Affections of the Tubular Genitalia/Uterus : Uterine infections have been reported to be one of the main cause of infertility in buffaloes. These uterine abnormalities include endometritis, perimetritis, pyometra, mucometra, uterine abscess and tumor. several morphological and functional disorders of fallopian tube, uterus, vagina and vulva have been reported in buffaloes during last few years. Among all these affections of the uterus, endometritis and pyometra are the most common conditions encountered under field. Abattoir survey and clinical investigations revealed higher incidence of endometritis in buffaloes than cows (Dobson and

Kamonapatana 1986). The incidence of endometritis has been reported 4.5 to 25% (El Hariri *et al.* 1980; Vale *et al.*, 1984) in abattoir and 2.4 to 20 % in clinical survey (Rao and Sreemananarayana, 1982). Incidence of pyometra in Indian buffaloes has been reported from 0.5% to 2.5% (Agarwal, 1978). Further the buffaloes as been reported to be more resistant to Brucellosis, campylobacteriosis and Tuberculosis (Kulshreshta *et al.* 1978; Shalaby, 1986).

Uterine tumours (0.3 to 0.72%), paraovarian cysts (0.4 to 2.1%) and uterus unicornis conditions have also been reported in buffaloes (Agarwal, 1978; Samad *et al.*, 1994). Besides cases of double external os in cervix and uterus bicornis have also been reported. Sharma *et al.* (1967) reported pyometra, mucometra, atrophied uterus, external uterine cysts, hydrometra, lymphoid growth and internal cysts to the tune of 9.2, 4.9, 3.1, 2.9, 2.6 0.8 and 0.025% from the abattoir material of 3684 samples of buffalo genital organs. Kodagali *et al.* (1973) reported 32.19% incidence of endometritis.

The incidence of cervicitis was observed in 6.90 and 8.96% of buffalo heifers and buffalo cows (Bhattacharya, 1954). In another study Kodagali (1968) reported 2.5 and 0.75 incidence in Surti buffaloes. Kumar (1977) reported 0.26 and 0.48 % cervical malformation and 1.55 and 0.54% enlargement of cervix in heifers and adult buffaloes. Agarwal (1978) recorded 0.9% cases of cervicitis in buffalo cows from the rural area.

For the treatment of endometritis estrogen alone or in combination with oxytocin have been used but because of possibility of development of ovarian cysts this therapy does not hold good. Hence prostaglandin has widely been used for regression of CL and evacuation of pus from the uterus. Parentral and intrauterine antibiotic therapy has also proved beneficial for treatment. The recent approach for endometritis now a days are the use of immunostimulators such as *E. Coli* lipopolysaccharide (LPS) and lower dosage of Levamisol.

Repeat breeding: The condition of repeat breeding in buffaloes has been reported in the range of 6 to 11% (Bansal *et al.*, 1978; Narsimha Rao and Kotayya,1980). Under field condition the incidence is too high and it has been reported to the tune of 20 to 25% (Agarwal,

1978, Kumar, 1987).

For the settlement of repeat breeding condition in buffaloes intra-uterine antibiotic treatment, use of certain immunomodulators, GnRH, LH and other hormones have been tried. In fact the condition of repeat breeding arises due to many reasons and before start of any treatment regime the responsible cause for this needs to be ascertained and this can only provide the best results.

It is well known that the managemental factors play an important role towards repeat breeding problem. Hence animal should be properly fed with balanced ration and they should be kept healthy and free from diseases by employing a routine deworming and vaccination programme. All the animals should be observed for estrus at least twice daily to ensure insemination at an appropriate time. Besides the semen and the hygienic conditions should be very proper.

Obstetrical problems: The cases of fetal mummification, maceration, uterine torsion, dystocia, cervical vaginal prolapse, retention of placenta and other obstetrical problems are also reported in buffaloes. Goswami and Nair

(1968) reported 6 cases of twinning comprising of 3 heterosexual twins, 1 male twin and 2 female twins out of approx. 8000 calving from different farms. A case of extrauterine pregnancy in buffalo is also reported (Gupta and Singh, 1980). The incidence of abortion (1.5 to 2.5%) and dystocia (1 to 2%) in buffaloes are reported (Kaikini et al., 1976; Patel and Kodagali, 1983). The cases of retained placenta, metritis, prolapse, cervicitis/ vaginitis, adhesion from field buffaloes are reported to the tune of 11.49, 18.69, 6.68, 3.25 and 5.66% cases (Kumar, 1988). In a survey from the rural buffaloes Kumar (1987b) reported that the occurrence of retained placenta and metritis is affected by parity level, season of calving, age of dam and farmer's land holding capacity as well.

The incidence of retained placenta in buffaloes have been reported low as compared to cows (1 to 3 vs. 5 to 10%) (Devraj and Rai, 1979, Pandey et al., 1981; Agarwal et al., 1984). Conducting a survey of the post-partum reproductive disorders in rural buffaloes

Kumar (1988) reported maximum occurrence of retained placenta (43.12), followed by endometritis (34.12%; including pyometra, mucometra, metritis etc.), uterine prolapse (6.12%), vagino-cervical prolapse (5.21%) , cervicitis (2,84%), vaginitis (2.84%) and adhesion of organs (5.68%); based on the total observations of 221 cases of ailments. Further higher incidence of retained placenta have been reported during the rainy season (Kumar, 1992, Bhalaru, 1983). The treatment of retained placenta is accomplished either by manual removal 24 to 48 hr. after calving or/and using intra-uterine and parental administration of antibiotics.

Attainment of Reproductive Management

Genetic make up, environmental condition, nutritional status and management are the four key factors which determine the performance of livestock production and reproduction. Among all these factors probably the nutritional factors are the most important one, as it can influence the effects of other factors. Majority of the problems of reproduction in livestock faced in Indian conditions are nutritional in origin. Late maturity, delay in postpartum estrus, repeat breeding and anestrus particularly during summer are affected by adverse climate, poor nutrition and lack of proper management. Ample scope exist for increasing the reproductive efficiency by modification in the traditional methods of breeding, feeding, management and diseases control.

Generally functional form of infertility tends to occur in individual animal of a herd. However, when the number of affected animals in a herd / locality becomes larger it reflects some other problem-specially that of nutritional and / or environmental origin. Most of the functional aberrations are the reflections of the endocrinological abnormality which many times becomes difficult to specify even with current methods of hormonal assay particularly when single spot samples of blood or milk are investigated. In modern dairy farming, stress is another factor that contributes to endocrinological imbalances mainly mediated through hypothalamus (opiod-peptide-encephalin-bradykinin substances-P etc.) are known to be produced during stress which suppress Gn RH release and / or LH surge.

Strategies: Majority of buffaloes are kept in small numbers by the livestock owners and are accustomed to individual attention, being groomed and managed and hand fed. In hot climate there must be allowance for shade or wallowing or a routine or sluicing or spraying/ plastering of mud protects against biting flies and rays of summer.

Excessive stress of milk production results in negative energy balance and due to it about 25% cows in European countries are culled due to reproductive regress, 35 % of cows have cystic ovaries and only 1% endometritis. similar analysis of subfertility in buffaloes revealed a long post-partum interval to fertile heat and incidence of endometritis can be as high as 20% in Indian situation. Suckling is also present for longer time in buffaloes and this also affects the reproductive efficiency.

Land holding capacity in the farmers owning the buffaloes in rural pockets also had a marked effect on the incidence of anestrus. Kumar (1987) reported that maximum number of anestrus animals were owned by marginal farmers (land holding capacity up to 1 hectare), followed by Large farmers (More than 2 hectares) and small and land less farmers (less than 1 hectare or nil). This probably occurs as a result of non or less availability of green fodders with various levels of farm holdings.

The literature on the reproductive status of almost all the zones and regions / parts of the country are available but the entire information is widely spread. There is a great need for pooling all these information's at one place so that a national status of the problems related with buffalo reproduction can be had and further programmes for improving the production through reproduction at national level can be achieved.

Target and interference level: To address the infertility problems associated with a dairy farm one should, through proper record keeping, define and set the normal expectations/ targets of fertility of the breedable females. The causes for deviations from expectations become the most important matter of concern. The parameters of reproductive efficiency help us to understand the severity of problems with an individual animal and / or with a herd as a whole and the farmer / manager can decide when to intervene to correct the problem. In the following table it has been tried to give a target and interference level required to be achieved for the cattle and buffalo herd:

Index	Cattle		Buffalo	
	Target level	Interference level	Target level	Interference level
Age at puberty (months)	18	20	21	24
Age at first conception (months)	21	24	32	36
Mean calving to first service interval (days)	65	70-75	75	90
Mean calving to conception interval (days)	85	95-100	90	120
First service submission rate (%)	80	70	70	55
Reproductive efficiency (%)	46	35	40	30
Overall pregnancy rate (%)	58-60	50	50	45
Animal served that conceive(%)	95	90	80	70
Annual culling rate(%)	<10	15	<10	15

The following aspects if cared judiciously the problems of reproductive disorders and the fertility in the buffaloes can be handled successfully:

◉ Optimum feeding – if proper feeding is provided from the very beginning the age at maturity can be reduced down.

◉ Heat detection– Since heat detection itself is a problem if proper care is exercised for heat detection by employing the scientific methods and timely observations this can be taken care of.

◉ Post-partum rest– proper post-partum rest of 60 days as

recommended in past may be provided, although this is not a problem because already the rural buffaloes evince their first heat after a long period.

◉ Proper management during and after calving

◉ Feeding green fodder

◉ Routine examination of the breedable stock

◉ Periodic disease testing programme

◉ Calf-hood vaccination

◉ Keeping the animals in a positive energy balance

◉ Protection from heat stress during summer

Nutritional Interventions

The problem infertility both in buffaloes of rural areas are delayed puberty, anoestrus and increased calving interval (of non-infectious origin) are due to deficiencies of either energy or protein or vitamins and minerals.

For investigating nutritional factors of infertility one should address:

◉ Evaluation of feeding values of available forages and concentrates

◉ Feed intake feeding and managemental practices

◉ Body condition score

◉ Other supportive environmental factors

◉ Metabolic profile of animal need to be ascertained

◉ Soil and plant contents of major and trace minerals need to be ascertained.

Methods to Overcome the Deficiency of Minerals

◉ Since common feeds and fodders are deficient in the sodium its deficiency can be overcome by including common salt in the daily ration. Supplementation of about 30 gm of common salt per 100 kg of ration will maintain normal health and production.

⦿ For the correction of the Cu deficiency any of the four forms of the copper supplementation (injectables, mineral or salt mixtures, copper given as drench and bolus containing copper oxide needles) can be used.

⦿ Zinc deficiency can be prevented by supplementation of zinc salts in the mineral mixture. Incorporation of 50 mg. Zn / kg. DM is normally recommended.

⦿ The problem of Se deficiency can be overcome by supplying Se in the diet. A level of 0.2 to 0.3 ppm Se in the diet is recommended. This can be achieved by injections, supplement in the feed, as free choice salt mixture, oral administration in the form of pellet, drenching of Se salt etc.

⦿ The most common way to combat the Mn deficiency is to supplement it through feed supplement in the mineral mixture, in the form of Manganese sulfate and Manganese oxide (approx. 3-4 gm as daily supplement).

⦿ The deficiency of cobalt can be prevented by supplementing either Vita. B_{12} or cobalt in the diet , injection, pellets or other routes.

⦿ In the cattle the most suitable method for overcoming deficiency of iron is by supplying 2-4 gm. ferrous sulfate for 2 weeks orally.

⦿ Iodine should be given to dairy animals along with common salt. Potassium iodide (PI) or calcium iodide is commonly used at the dose rate of 200 mg of PI per kg of salt. Besides intramuscular injection, application of tincture of iodine are the other routes.

⦿ The deficiency of vitamin A can be corrected by supplementing feed with provitamin A rich substances.

Management Tips for Obtaining Better Economic Returns from Buffaloes

⦿ Ensure the bull/ semen used is fertile.

⦿ Ensure that the time and method of insemination is correct.

⦿ If the infertility is found to be associated with nutritional deficiencies (i.e. delayed puberty, anestrus) improved feeding specially energy intake supplemented with minerals and vitamins for 1 to 2 months is advisable. Then only the ailing animals failing to respond to above regime should be considered for hormonal therapy.

◉ If the infertility problem is of infectious type always sample the tissues / secretions and seek for laboratory diagnosis. Identify the cause and take necessary therapeutic, preventive and control measures to the disease.

◉ Soil-plant -animal relationship is very important to address the nutritional form of infertility of cattle and buffaloes of rural areas. Metabolic profiles of sampled animals from a locality help to understand the depth of the problem. Further information on soil and roughage composition becomes necessary.

◉ Investigation of individual animal should involve.

◉ History, fertility records, calving problems etc.

◉ Clinical examination and diagnostic tests.

◉ Evaluate at least once or twice annually the dairy herd fertility status from the available records for the parameters of reproductive efficiency.

◉ Do not hesitate to cull long standing problematic animals.

Certain Important Points to Note

◉ It is always better to manage the fertility rather than to treat the infertility cases. Hence, attempts should be made to educate the farmers about the fertility management and if done properly there will be less cases for treatment

◉ For successful fertility of animals the body condition is more important than the body weight. Hence proper care for feeding and management should be taken for all types of animals.

◉ For successful reproduction certain vitamins (A, D and E) and minerals (calcium, phosphorous, selenium and chromium) are mandatory in the diet of mature animals.

◉ Vitamin A -35,000 to 45,000 I.U. per day should be provided to buffaloes. If deficiency of this vitamin is suspected then the dose level per day can be increased to 1,00,000 I.U. per day.

⊙ Vitamin E-1000 I.U. per day during the last 40 days of gestation should be provided to animals.

It is evident from the above discussion that considering the extent and nature of infertility problem the drugs should be chosen and used for its effective treatment. Besides treatment management is a very important aspect for fertility improvement. Some of the points for better management are as follows:

⊙ Regular, systematic and well-planned sexual health control programme with periodic gynaeco-clinical examination and treatment of detected cases be implemented with proper recording of individual case history.

⊙ Regular monitoring of field cases is a must.

⊙ Optimum management practices should be followed as it go a long way in reducing stress conditions and help in maintaining good fertility.

⊙ Hygienic breeding, prophylaxis, heat detection, periodic gynecological check up should be strictly followed.

Managemental Considerations

Malnutrition and excessive negative energy balance are the main cause of delayed ovarian function after calving. Mediated by a complex mechanism the negative energy balance results in insufficiency of LH and GnRH and consequently post-partum anoestrus. This condition is more prevalent in our field animals. Besides this the commonly occurring disorders during the post-partum period are; Retained placenta, puerperal endometritis, delayed involution of uterus and delayed post-partum ovarian activity.

Following are some of the management and therapeutic measures to be adopted for their corrections:

⊙ For retained placenta management: Oxytocin, ergot derivatives, beta adrenergic antagonists have been used along with prostaglandins. The results with prostaglandin are superior to other preparations.

⊙ For post-partum endometritis: It occurs within a few days of parturition because of exposure and as a complication to retained placenta, dystocia, and stillbirth. The line of treatment is suggestive for use of prostaglandin, as this induces luteolysis, stimulates uterine defense mechanism and also increases myometrial contractions, which in turn helps in evacuation.

⊙ For abnormal/ delayed ovulation: Involution of uterus in normal course is completed by 25-35 days of calving, up to 20 days post-partum prostaglandin is released and beyond 20 days injection of prostaglandin helps in involution of uterus.

⊙ For post-partum anoestrus: Delay in uterine involution and ovarian cyclicity is often associated with post-partum anoestrus. Prostaglandins alone or in combination with GnRH improves post-partum ovarian activity.

Approaches for Correction of Certain Reproductive Ailments

Condition of subestrus is a great problem in our livestock population. This condition is more common in buffaloes ranging from 10% in organized farm to 40-50% in village conditions. The failure to show oestrus symptoms may be due to lack of secretion of estradiol by the mature follicle or may be due to need for a higher threshold. If the CL of the previous breeding season has not regressed completely, the progesterone withdrawal is prevented, which results in non-occurrence of tonic LH surge and the animal become subestrus. For the treatment of this condition the prostaglandins (PGF_2 alpha) is the drug of choice which will cause the regression of CL. i/m injection of 30 mg. Prostaglandin in presence of a well developed and mature CL results in induction of estrus within 3-4 days. The other condition like anovulation (although less occurrence reported) is recognized as evincing normal cyclic changes, estrus symptoms, maturation of follicle but no ovulation due to lack of LH/ insufficiency of LH. This condition can be treated using injections of either GnRH (Busereline-20 microgram or Fertireline-250 microgram) or LH or hCG (1500-2500 I.U. per animal i/ m).

In certain cases of delayed ovulation (due to hereditary, nutritional or hormonal origin) due to aging of gametes fertilization failure or early embryonic mortality may result leading to infertility.

Since it is very difficult to ascertain the cases of delayed ovulation, the following practices can be adopted for improving the results of conception:

◉ The animals may be inseminated twice at 24 hr. interval

◉ Inseminate at 12 hr. interval

◉ Injection of 1000-2000 I.U. –hCG or LH

◉ Injection of 100 mg of GnRH –6 hr. before or at the time of insemination

Other condition namely early embryonic mortality the embryo dies at an early age and depending upon the period the animal may show either normal cycle length or increased cycle length. The reason being that the early lysis of the CL leads to progesterone insufficiency. The other reason could also be due to any defect in the maternal recognition of pregnancy or any autolysis signal from conceptus. The following practices can be adopted:

◉ Use of progesterone –100 mg i/ m daily from day 10-20 of cycle.

◉ Insertion of PRID for a week

◉ Administration of hCG 1000-1500 I.U. i/m on day zero of the cycle

◉ GnRH or hCG therapy.

The conditions which are most encountered in the field are animals which have no palpable structures on the ovary and the ovaries are smooth most of the time referred to as true anestrus. This condition is due to insufficient gonadotrophin release to cause follicular development, maturation and ovulation. This condition can be diagnosed per rectal or by estimation of progesterone 11 days apart. For correction of this condition following practices can be adopted:

◉ Supplementation of mineral mixture

◉ Hormonal therapy

◉ Use of herbal drugs

⊙ Use of progesterone-P4 –Nogestomate ear implant

⊙ Administration of GnRH, FSH, PMSG or P4.

Management of Post-Partum Anestrum in Buffaloes

1. Use of PGF_2 alpha

⊙ *7-14 days post-partum* – PGF_2 alpha is released from the cotyledons in cows around parturition. Since it plays an important role in the process of parturition and post-partum involution of the uterus, PGF_2 alpha may be used for expulsion of exudates as well as increasing the phagocytosis, resulting in early occurrence of cyclicity.

⊙ *14-28 days post-partum* – In cases of endometritis with presence of CL PGF_2 alpha acts by regression of CL, development of new follicle and exhibit estrus signs. The Graffian follicle produces estrogen which is responsible for increased uterine contraction and increased phagocytosis.

2. Use of GnRH

⊙ *7-14 days post-partum* – Causes first emergence of follicular wave and facilitates the resumption of ovarian cyclicity.

⊙ *At about 30 days post-partum* – One month after calving some animals have inactive ovaries. Injection of GnRH alone or in combination with progesterone induces follicular growth, maturation and ovulation.

3. Use of Prostaglandin

⊙ After rectal palpation inject PGF_2 alpha to those cows, which have the CL, express estrus within 3-5 days.

⊙ Observe all the cows in estrus for 7 days period and perform AI. in estrus cows. The rest of cows should be inject PGF_2 alpha on the following day and perform AI twice at fixed time.

◉ All the cows are injected on day of treatment and second injection is repeated 11 days later. AI is performed 72 hr later.

◉ The other method is to inject PGF$_2$ alpha in all cows and inseminate them, which come in heat. For rest inject PGF$_2$ alpha again on 1 day and inseminate.

4. Use of Progestrone

These are placed either in vagina or as ear implant s/c for 8-10 days. Cows express heat 48 hr after removal.

Management of Cystic Ovarian Degeneration (COD)

Cysts can either be follicular or luteal.

Follicular – Causes repeat breeding, due to insufficient LH.

Luteal – Causes anestrus, due to continuous presence of progesterone.

Depending upon the cyst the treatment is advisable.

◉ Use GnRH followed by PGF$_2$ alpha 9-11 days after has been found most successful.

◉ Administration of GnRH 12-14 days post-partum reduces incidence of ovarian cysts, thus reduction in inter-calving period can be obtained.

◉ Use of progesterone 200 mg for about 8-1 days has negative feed back on anterior pituitary which releases LH after withdrawl leading to either leutinization or ovulation of the cysts.

Management of Cycling Non-Breeding Animals

◉ For the cases of anovulation: - LH, hCG and GnRH at 6-8 hr before AI or time of AI.

◉ GnRH is also used on day 11 or 12 (mid cycle) to improve conception rate by reducing chances of early embryonic mortality.

◉ Use of GnRH on day 5-6th after AI causes ovulation of first wave dominant follicle with the formation of accessory CL – more progesterone – implantation – maintenance of pregnancy.

Management of other Conditions

⊙ Injection of PGF$_2$ alpha for Pyometra – A second injection may be required in some cases that can be given 12-14 days apart.

⊙ Use of PGF$_2$ alpha in cases of RP.

⊙ For the cases of maceration and mummification – use of estrogen.

Management Tips to Overcome Cystic Ovarian Problems

1. *Monitoring the estrus cycle in following type of animals*

⊙ Older animals – 5/6th lactation

⊙ Animals in early lactation

⊙ Animals under lactation stress

⊙ Animals being fed diets of high levels of estrogens.

2. *Per-rectal examination of susceptible animals at regular intervals for the presence of*

⊙ Developing / regressing CL on the ovary

⊙ Developing / mature follicle on the ovary

⊙ Tonicity of the uterine horns

Management and Treatment of Infertility in Buffaloes

While dealing infertility problem the veterinarian has two tasks to perform

1. To investigate and determine the cause of infertility.

2. To assist in maintenance of optimal fertility.

To solve the reproductive problem the veterinarian relies more on his palpating skill, knowledge of reproductive physiology, pathology and sound management practice. Accordingly the drugs are selected for treatment. Besides the following are to be well taken care of:

⊙ Deworming at regular intervals.

⊙ Regular examination of the reproductive tract for the

cervicitis and metritis.

◉ Regular examination of the ovaries for the presence of cysts, etc.

◉ Feeding the animals properly i.e.

    a. Maintaining energy balance in the ration

    b. Maintaining protein balance in the ration

    c. Maintaining vitamin and mineral balance in the ration

Therapy for Functional Disorders of Gonads

Functional anestrus due to ovarian inactivity is considered to be most expensive and frustrating problem associated with productivity in India. Treatment combination involving short term use of progesterone and gonadotrophin or progesterone and prostaglandin have produced better synchrony of estrus response and higher fertility. The administration can be done by way of feeding, daily i/m injections or inserting long active depot preparation under the skin or in vagina of buffalo.

Prid is a silastic coil impregnated with 1.55 gm progesterone having gelatin capsule of 10 mg extradiol benzoate attached to inner surface. On removal of coil on 10th day AI is done at 48-72 hr. Ear implanted by special syringe s/c on the outer portion of ear for 10 days. At the time of implant 10 mg estradiol valerate and 3 mg norgestomate are injected i/m.

Gondadotrophins

◉ *PMSG:* The delay to first post-partum estrus is the most vital factor responsible for reproductive insufficiency of dairy animals. PMSG will stimulate follicular growth in the ovaries producing endogenous estrogenous estrogen which will expert positive feed back on the anterior pituitary function and in term the ovarian cyclicity in early post-partum period is obtained. PMSG can also be used for cases of delayed sexual maturity.

◉ *HCG:* In the condition of delayed ovulation this is used. Delayed ovulation is one of the major cause of repeat breeding and hence this is a drug of choice for treatment of repeat breeders.

- **GnRH:** Administration of GnRH stimulates the release of endogenous LH and FSH and ovulation can be induced and normal estrus cycle initiated on day 13 post-partum. Treatment of dairy cows with GnRH also reduces the incidence of ovarian cysts.

Clomiphence citrate: It has action to stimulate hypo-thalamo-pituitary axis to release gondadotrophin. A non-hormonal combination of two isomers of Triphenyl Ethylene compound (Clomiphene) each containing Cis-clomiphene citrate 180 mg and trans clomiphene citrate 120 mg dissolved in 500 ml water and given daily as a drench for a period of 5 days result in induction of estrus.

Prostaglandins: This is a drug of choice for correction of many types of infertility problems, like sychronization of estrus, removal of fetus due to defective gestation, uterine infections and other conditions, like silent heat, subestrus condition etc.

The commonly used drugs for treatment of functional disorders are:

- Clomiphene citrate
- Fertivete
- Prajana therapy
- Aloes compound
- Indigenous/herbal preparations
- PMSG
- Progesterone – MAP, CAP, MGA, PRID etc.
- GnRH
- Prostaglandin
- Antibiotics for infectious condition

For Anovular Heats   –   1. GnRH

                                2. hCG/LH treatment

For delayed ovulation – GnRH or LH treatment at the time of AI

For cystic ovarian degeneration – hCG and GnRH 3000 – 4500 I.U. of hCG is recommended

For Persistence corpus luteum – Prostaglandins can be used. Some commonly available prostaglandins are: -

- ◉   Lutase – Hoechst
- ◉   Dinoprost – Upjohn
- ◉   Dinofertin – Alved
- ◉   Estrumate – ICL – 80996
- ◉   Synchromate
- ◉   Cloprostenol

References

Agarwal, S.K. 1978. M.V.SC. Thesis , I.V.R.I., Izatnagar

Ahmed, Y.F. 1986. Ph. D. Theses, Cairo University, Egypt.

Bansal, R.S., Gupta, S.K. and Dugwekar, Y.G. 1978. J. Remout & Vety. Corp. **17**: 9.

Bhalaru, S.S., Tiwana, M.S. and Dhillon, J.S.1983. Trop. Vert. Anim. Sci. Res. **1**: 81.

Bhattacharya, P. 1954. Current Science: **23**: 335.

Chatterjee, S.K. 1977. M.V.Sc. Theses abstract, Pantnagar J. Res. **2**: 265.

Chauhan, F. S. and Singh, , M. 1979. Indian Vet. J. **60**: 665.

Chauhan, F. S. Takkar, O.P.and Singh, M. and Tiwana, M.S. 1981. Indian J. Anim. Reprod. **1**: 31.

Chauhan, F. S. and Takkar, O.P.1983. Indian Vet. J. **15**: 6.

Dessouky, F and Juma. K.H. 1973. Indian J. Anim. Sci. **43**: 187.

Devraj. M and Rai. A.V. 1979. Mysore. Agri. Sci. **13**: 62.

Dobson , H. and Kamonapatana, M. 1986. J. Reprod. Fert. **77**: 1.

Dwivedi, J.N. and Singh, C.M.1971. Indian J. Anim. Health **10**: 27.

El Hariri, M.N. Omar, M.A.and Shalash, M.R. 1980. Proc. 9th Indian Congr. Animal Reprod. & A.I. Madrid. Vol.IV P.754.

El Khouly, A.B. 1985. Ph.D. Theses, Cairo University, Egypt.

Goswami, S.B. and Nair, A.P. 1968. Indian J. Dairy Science. **21**: 50.

Gupta, S.K. and Singh, Surjit. 1980. Indian Vet. J. **57**: 79.

Kaikini, A.S. Kadu, M.S. Bhandari, R.M. and Belorkar, P.M. 1976. Indian J. Anim. Sci. **46**: 19.

Kaikini, A.S. and Patil, R.K. 1978. Indian J. Anim. Sci. 48: 411.

Kaikini, A.S.and Deshmukh, M.J. 1984.. Indian J. Anim. Reprod. **5**: 30.

Kodagali, S.B. 1967. Indian Vet. J. **44**: 773.

Kodagali, S.B. 1968. Indian J. Vety. Science. & A.H. **11**: 286.

Kodagali, S.B. and Bhavsar, B.K. 1973. Gujrat College of Vety. Sci. and A.H. Magazine **6**: 73.

Kulshreshta, R.C. Kaira, D.S. and Kapur, M.R. 1978. Indian J.Publ. Hlth. **22**: 232.

Kumar., M. 1977. M.V.Sc. Theses. Rohilkhand University, Bareilly

Kumar, S. 1992. Bovine Reproduction: A Rural Experience. Published From Indian Veterinary Research Institute, Izatnagar.

Kumar, S. and Agarwal, S.K. 1986a. Indian Vet. Med. J. **10**: 11.

Kumar, S. Agarwal, S. K. and Purbey, L. N.1988. Indian Vet. Med. J. **12**: 125.

Kumar, S. 1987a. Dairy Guide. **1987**: 44.

Kumar, S. 1986b. Indian Dairy man **38**: 10.

Kumar, S. and Kumar, H. 1993. Indian J. Dairy Sci. **46**: 80.

Kumar, S. 1987b. Livestock Advisor.12: 41.

Kumar, S. 1988b. Livestock Advisor. 13: 47.

Luktuke, S.N. and Sharma, S.S.1978. Indian Vet. J. **55**: 940.

Luktuke, S.N. 1977. Indian Dairyman. **29**:281.

Narsimha Rao, A.V. and Suryanarayana, Murthy, T. 1978. Indian Vet.**55**: 1003

Narsimha Rao, A.V. and Kotayya, K. 1980. Indian J. Anim. Health. **19**: 21.

Narsimha Rao, A.V. Sreemanarayana, O. and Rao.K.D. 1985. Anim. Reprod. Sci. **8**: 129.

Pandey, M.D. 1979. F A O Animal Production and Health Paper No.13, Rome, p.52.

Pandey, S.K. Pandit, R.K. and Chaudhary, R.A. 1981. Livestock Advisor. **6**: 43.

Pandit, R.K. Gupta, S.K. and Raman S.R. 1982a. Livestock Advisor. **7**: 51.

Pandit, R.K. Gupta, S.K. and Raman S.R. 1982b. Indian Vet. J. **59**: 854.

Pandit, R.K. Gupta, S.K. and Raman S.R. 1982c. Indian Vet. J. **59**: 975.

Patel, D.M. and Kodagali, S.B. 1983. Indian J. Anim. Reprod. **4**: 99.

Rao, A.V.N. and Sreemannarayana, O. 1982. Theriogenology. **18**: 403.

Rao, A.V.N. and Keshavamurthy, A. 1971. Indian Vet. J. **48**: 1007.

Samad, H.A. Ali, C.S. Ahmed, K.M.Z. and Rehman, M.1984. Proc. 10th Int. Congr. Anim. rep. & A.I. Urbana. Vol. XIV-25.

Shalaby, M.N.H. 1986. Ph. D. Theses. Cairo University, Egypt.

Sharma, O.P. Bhalla, R.C. and Soni, B.K. 1967. Indian J. Animal Health. **6**: 21.

Sharma, R.D. and Gupta, S.K. 1978. Indian J. Dairy Sci. **31**: 86.

Singh, B. Singh, K.P.Parihar, N.S. Bansal, M.P. and Singh, C.M. 1980. Vet. Bulletin. **51**: 3454.

Singh, C.P.S. Singh, S.K. and Singh, B. 1981. Indian Vet. J. **58**:909.

Singh, N. Chauhan, F.S. and Singh, M. 1979. Indian J. Dairy Sci. **32:** 134.

Singh, S.S., Prasad, B. Sharma, S.N. and Ram Kumar, R. 1986. Indian Vet. J. **63:** 693.

Vale, W.G. Ohasi, O.M. Sousa, J.S. and Vale-Filho. 1984. Proc. 10th Int. Congr. Anim. rep. & A.I. Urbana. Vol. IV-465.

CHAPTER 24

Advances in Reproductive Management for Sustainable Buffalo Production

O.P. Dhanda

CCS HAU, Hisar-125 004 (Haryana)

Poverty and food security are closely related. Generally poor people and that too, in rural areas, are devoid of proper quality and quantity of food. In India, about 33% population is below poverty line and under-nourished (World Bank, 2001). Under such a scenario, livestock, in general, and dairy animals, in particular, offer immense potential towards alleviating the problem of food insecurity. It is now realized that crop sector alone will not be able to provide sufficient livelihood to rural poor as the land per capita availability has declined rapidly due to increase in population. During the last decade, livestock sector grew at the rate of 3.9% as against 2.2% of crop sector. These figures indicate almost double the rate of growth for livestock sector. In addition, a number of studies (Singh *et al.*, 2000; Bhowmick and Sharma, 2002) have shown that poor households depend considerably on livestock for their survival. Presently livestock contributes 25% towards agricultural domestic gross product and is an importance source of employment.

Out of livestock, buffalo forms the backbone of India's dairy industry by producing around 56% of the total milk in the country in spite of their number being 1/3rd to that of cows. On an average, Indian buffalo yields around 500 litres of milk per lactation in comparison to about 190 litres by non-descipt cows. Buffalo is also capable of utilizing low-grade roughages more efficiently in terms of feed conversion efficiency against cows and can sustain in very harsh agro-climatic conditions. In a recent report (NABARD, 1997) it has been shown that the rate of economic return was higher from buffaloes than cows. There is a continuous increase in demand for buffalo milk and milk products even in southern states like Kerala where consumers used to favour cow milk until recently.

However, per unit productivity from almost all livestock species including buffaloes is very low in our country in comparison to those in advanced ones although we derive proud in claiming number one in milk production in the world. It needs serious consideration. Dairy industry must operate an economically efficient production system, which, in turn, depends upon high level of reproductive efficiency. Failure to achieve this aim has major implications for future improvement of economic traits. According to a survey, reproductive efficiency in cattle and buffaloes can be improved by 30% through reproductive management without any extra economic inputs. Hence, this paper endeavours to identify fertility parameters critical for assessing the reproductive status of a dairy buffalo herd and also suggest suitable and cheaper measures for enhancing reproductive efficiency of herds in the light of recent research findings and experiences of a number of workers so as to make the profession as more rewarding.

Measures of Reproductive Efficiency

In order to determine the effect of management on reproductive status of the herd, it is essential to establish some benchmarks for measuring reproductive efficiency. Until two decades ago, the reproductive efficiency was used to be determined by following parameters:

1. *Traditional Approach*

i) **Services per Conception:** It is measured on a herd basis by dividing total services by number of pregnancy. Services per conception are a valid measurement for a single herd or individual female. However, on a herd basis, unidentified sterile females will make calculations less meaningful. Optimally, it should not be more than 2.0 in any case. Otherwise, it will denote reduced reproductive efficiency of the herd.

ii)   Calving Rate: It is calculated by dividing the total number of females bred by number that calved. It is also expressed as per cent calf crop. At least 80% of the breedable population must calve yearly for efficient reproductive efficiency.

iii)   Non-Return Rates: A non-return rate is the percentage of females that do not return to estrus or secure a second service within designated period of time. A time interval of 28-35 days after A.I. or service is generally used to evaluate fertility status of the herd in the shortest possible time as information on pregnancy is usually not available at this time. Non-return rates are always higher than actual pregnancy rates because of some open females.

2.   Endocrine Fertility Parameters

Recently, with advent of Radio Isotopic Assey (R.I.A.) techniques, some more parameters based on endocrine profiles which can provide precise and pertinent information about the reproductive status of the herds have been analyzed. These are as follows:

i)   Pregnancy to First A.I.: Conception or pregnancy as a result of first insemination is considered more meaningful for determining reproductive efficiency of a cow/buffalo. Calculations are made after including subsequent inseminations. The pregnancy is diagnosed from milk progesterone profiles using ELISA/RIA techniques. Around 55% and 45% conception in cows and buffaloes, respectively, as a result of first A.I. is considered as satisfactory level.

ii)   Post-Partum Commencement of Luteal Activity (CLA): According to Darwash *et al.* (1997), a period of luteal activity is defined as the occurrence of two or more consecutive milk progesterone concentration of more than 3 ng/ml. In other words, the day on which the female achieves a definite level of progesterone (3 mg/ml in cows) may be taken as post-partum period. They further reported that cows starting to cycle early had a reduced interval to conception, higher pregnancy rates and fewer services per conception than those with prolonged unovulatory periods.

---

iii) **Luteal Phase:** Following ovulation, the period of time for which a corpus luteum secretes progesterone (> 3 mg/ml) is referred as luteal phase. The length of luteal phase is measured from time of first elevated milk progesterone and the final consecutive milk progesterone level. Measurement of luteal phase length provides an indication about uterine environment.

iv) **Inter-Ovulatory Interval:** This is defined as the interval between milk progesterone rise from the corpus luteum of one oestrus cycle and the milk progesterone rise from the corpus luteum of the next oestrus cycle. This is considered to be an objective assessment and measurement of oestrus cycle length. In case of cows and buffaloes, the oestrus cycle length is 21 days. Pregnancy rates are higher for cows with normal interovulatory interval (18-24 days) and lower for cows with interovulatory intervals of 4 to 17 days and beyond 24 days (Royal et al., 2000). Hence, IOI or oestrus cycle length is of great value in analyzing the reproductive status of the herd.

v) **Inter-luteal Interval:** The interval in days between the death of one corpus luteum and the rise of the next is known as inter-luteal interval and generally measured from the first milk progesterone level following luteolysis and the last consecutive milk progesterone profile. This allows indirect detection of delayed ovulation and luteanization. Shorter the ILI, higher are pregnancy rates (Royal et al. 2000).

Reproductive Management of Females

Most of the farmers attend to the fertility problems only at the time of their occurrence. It is very essential to know as to what can or cannot be achieved in terms of reproductive efficiency of a milch herd (Sprecher *et al.* 1995). Several reports have shown that well supervised reproductive management programme can improve the fertility. The various falls of this programme for both female as well as males are detailed in subsequent discussions.

1. First Time Breeding

The main concern relative to when young females should be inseminated or bred for the first time is the weight and age of the animal. This period is known as sexual maturity. Most of farmers confuse with first time expression of oestrus (puberty) with sexual maturity. Both age and weight at puberty are influenced by genetic factors and vary between the breeds and within the breed of a species (Table-1). Although large number of follicles may develop in very young calves, the oocytes they contain lack developmental competence until animals are 6-8 months old (Duby *et al.* 1996). Studies in U.S.A. in late 1980s also demonstrated that pregnancy rate in heifers bred at pubertal estrus was about 20% lower than heifers breed at third estrus. In such situations, uterine environment has been observed to play an important role. Bovine reproductive system continues to grow and mature beyond the first heat (puberty). For practical consideration, heifers should be bred after passing at least 2-3 estrus cycles for higher conception rate. The data in Table-2 shows the desired weight at puberty and first breeding for different livestock species and breeds.

Table-1: Species and Breed differences in age and weight at puberty.

Animals	Age (months)	Weight (kg)
Sow	4-7	68-90
Ewe	7-10	27-34
Mare	15-24	(Varies with mature size of breed)
Cow	8-13	160-270
Jersey	8	160
Guernsey	11	200
Holstein	11	270
Ayrshire	13	240
Reverine buffalo	24	240
Egyptian buffalo	17	270
Swamp buffalo	36	250-275

2. Oestrus Detection

Oestrus detection is the single most important problem for making A.I. programme as successful. Failure to detect oestrus or erroneous diagnosis of estrus results in huge economic loss. Various methods are employed to find the female in heat. Before discussing the various options for detecting estrus in dairy animals, it is pertinent to understand their heat characteristics. A female particularly cow in heat usually will stand when mounted by a male or female companion. Other heat symptoms would include clean mucus discharge from vulva, tail raising frequent urination, licking, rubbing against other colleagues, swelling and reddening of vulva, restlessness, temporary decline in milk. However, these symptoms are less prominent in buffaloes as a result of which these animals are known as shy breeders.

i) **Use of Teasers:** Vasectomised bulls are used as teasers in detection of estrus by parading at least two times in the morning and evening. Early morning check for oestrus detection is most important as more females come in estrus between 2.00 a.m. and 5.00 a.m. than any other similar period during the day. Sometimes a chin ball marker is attached to the bull as to identify estrus females.

ii) **Heat Mount Detectors:** Several types of heat mount detectors are available to record that a female animal has been mounted repeatedly. Devices such as Kamar heat detector are glued to the hair over the midline just in front of the tail head. Pressure from a mounting animal squeezes dye from a reservoir so that a colour change is visible to the observer. These females are identified to be in estrus. An error rate of 10-12% has been observed in this technique (Britt, 1987). However, Cavalieri and Fitzpatride (1995) found Kamar heat detectors as more efficient for estrus detection than other methods like chin-ball, tail paint or visual observations.

iii) **Vaginal Probes:** There has been a number of reports providing evidence of a change in electrical resistance of the fluids in the vagina at the time of estrus and various workers have suggested to use this as basis for detecting the females in estrus (Lewis *et al.*, 1989, Kitwood *et al.* 1993). Detection in vaginal mucus is known to be markedly influenced by ovarian steroids. Resistance falls markedly with lowest values with highest peripheral estradiol levels. Changes in vaginal resistance have been shown to be correlated with pre-ovulatory surge of LH. Results by Canfield and Butler (1989) in Holstein Friesian cows suggested that estimation of vaginal resistance predicted LH surge and could be employed as basis of A.I.

iv) Progesterone Profile: Since progesterone is secreted from corpus luteum, its determination in body fluids such as plasma and milk is a good marker for determining the functional status of ovaries and serve as a valuable diagnostic tool for confirmation of accurate estrus. This technique has been successfully employed for diagnosis of pregnancy at 20-24 days post A.I., silent heat, cystic ovaries disorders, parturition induction etc. in bovines (Kaul and Prakash, 1994). This aid is being used commercially in U.S.A. and other developed countries in large herds for estrus detection. The daily checks can show the reduction in progesterone level falls and a female could be served on day third of low progesterone level.Claycomb et al. (1995) have advocated a new rapid progesterone ELISA assay for use with a biosensor to measure hormone concentrations each time a cow/buffalo is milked as a part of automated system. It is cost oriented but such a system could be extremely helpful in providing progesterone profiles of post partum females as well as detecting estrus. A typical progesterone level can assist in identifying sub-fertile dairy animals (Darwash and Laming, 1995).

v) Use of Pedometer: Some of the research workers have measured increase in physical movement and activity as an aid for the females to be in estrus with the help of pedometers. Peter and Bosu (1986) found the device to be most useful in cows maintained in loose housing system. The other data dealing with activity profile of cows indicated rapid increase in activity until the peak of oestrus followed by a sudden decline after estrus (Arney et al., 1994). However, the device has not been found suitable and convenient due to their frequent losses.

vi) Body Temperature: A rise in vaginal or milk temperature during estrus has been reported by various workers (Rajamahendren et al., 1989). Onset of pre-ovulatory temperature increase may be considered as reliable indicator of LH surge (Clapper *et al.*, 1990; Mosher *et al.*, 1990). Likewise Kyle and Kennedy (1994) demonstrated that continuous monitoring of vaginal temperature could serve as a very sensitive and accurate means of predicting oestrus during breeding season. However, method could be unreliable in case of individual cow and buffalo because of high level of false detections (25%) as rise in temperature is dependent on a number of factors.

vii) **Visual Observations:** It is essential that a farmer or manager should take out sometime to observe dairy animals for estrus. Observations should be made by regular visit to the farm for at least half an hour. Especially in the late evening or early morning. Estrous symptoms in buffaloes are much less obvious than in cattle and silent estrus is a major problem in buffalo breeding (Jainudeen *et al.*, 1986). The incidence of silent heat is higher in herds using AI rather than natural service and this may indicate that the problem lies with heat detection rather than the animal itself. Symptoms such as swollen vulva, mucous discharge and frequent urination may not be reliable indicators of estrous detection in this species. An evaluation of three methods of estrus detection in buffaloes by Alonso *et al.* (1992) suggested the use of a vasectomised bull and twice daily observations as the most successful tool for heat detection for buffaloes. Our experience with buffalo heifers has been very effective when a combination of visual observation and use of teaser bull for heat detection were employed (Saini *et al.*, 1998).

Season is known to have marked influence on reproductive behaviour of buffaloes. Female buffaloes are not as sexually active in summer as during winter. Barkawi *et al.* (1993) observed pronounced estrus behaviour in Egyptian riverine buffaloes in cold season in comparison to the hot season. Sheokand *et al.* (1983) and Banerjee et al. (1989) drew the attention to a peculiar phenomenon in dairy buffaloes which show temporary engorgement of teats for a period of 4 to 5 days prior to occurrence of estrus 8 days later.

3. *Time of Breeding/A.I.*

Proper timing of A.I. or natural service is very essential for optimum pregnancy/conception rate. The length of sperm life in female reproductive tract is 24 hrs. Sperm capacitation takes about 6-8 hrs. The ovum remains viable up to 8-12 hrs with a better chance of survival when capacitated sperm comes in contact with it soon after ovulation. Therefore, timings of A.I. will depend upon time of ovulation and duration of estrus. In cows and buffaloes, the best conception rates are obtained when females are inseminated from middle until the end of estrus in view of the fact that ovulation occurs 11 to 17 hrs and 15 to 18 hrs after end of estrus in Indian cows and buffaloes (Raut and Kadu, 1990). The conception rate for those inseminates performed near the beginning of estrus is considerably lower indicating sperm have lost viability prior to time of ovulation. A thumb rule to follow in a practical management situation is for cows first observed in estrus in the morning to be bred late same day. Those females first observed in the afternoon should be bred early next day. This is known as a.m./p.m. rule. However, subsequent studies have shown that once daily insemination carried out between 8.00 and 11.00 hrs could provide satisfactory pregnancy results with no difference in calving rates from traditional a.m./p.m. system (Nebel et al. 1994).

4.   Post-Partum Breeding

Sexual rest after parturition is also very critical for subsequent pregnancy and varies in different species. Uterine involution includes the return of uterus to the pelvic area, return to non-pregnant size and normal uterine tone. The average time required for the process in cow and buffaloes is 45 days. Histological studies have shown that further 15 days are required before endometrium is normal. However, with the advent of RIA procedures, much more accurate picture of hormonal changes occurring when the cows return to full ovarian cyclicity after calving has been built up. There are sizable number of females which ovulate within 3 weeks after calving although this interval can be influenced by factors such as nutritional status, body condition, age of animal, season, milking frequency (Hanzen, 1986).

According to Peters and Lamming (1986) as occurring in the normal cow after calving included the following (i) some GnRH is secreted immediately after calving but not in sufficient amounts to cause gonadotrophin release, (ii) plasma FSH concentrations rise rapidly after parturition, stimulating follicular development; (iii) there is gradual increase in the frequency of LH pulses and in plasma LH concentrations; (iv) gonadotrophin secretion stimulates follicular growth and the production of oestradiol and perhaps inhibin and (v) concurrent with such endocrine changes, there is a gradual recovery of the positive feedback mechanism so that ovarian inactivity by the secretion of hormones.

The pre-mature death of the first corpus luteum is believed to be the result of pre-mature release of PGF2a by the uterus (Lishman and Inskeep, 1991). Many authors have reported that first ovarian cycle after calving was 6-7 days shorter than the average (Guven and Bolukbasi, 1989). The interval from calving to detection of first dominant follicle averaged 11 days and that to first ovulation 26 days (Kamimura et al. 1994).

A number of factors are known to affect the post-partum periods and is higher in tropical breeds as compared to the temperate ones. A study by D'Occhio et al. (1990) in Australia examined whether patterns of gonadotrophin secretion for zebu and taurine cattle differed during the postpartum period. It was found that around day 30 postpartum, taurine cattle had higher plasma LH concentrations than zebu cows; this difference appeared to increase as the postpartum period progressed. Such findings, according to these authors, could be taken as further evidence that, in a comparable reproductive state, zebu cattle have a lower capacity for LH secretion than European breeds.

Likewise, buffaloes also take long time to resume cyclicity after calving (75 days). However, long post-partum period in buffaloes may not be endocrine problems of this species and has been erroneously attributed (Jainudeen, 1986). By estimating progesterone concentration, Usmani et al. (1985) and Lohan et al.(2000) showed that the general pattern of uterine involution and return to ovarian cyclicity in buffaloes was similar to those in cows. In a study (Vale, 1990), the effect of management on the duration of post-partum interval, indicated that the first ovulation was delayed until 102 days after calving under traditional conditions but occurred at 30 days under improved nutritional conditions.

5. Use of Bio-stimulation

Bio-stimulation is used for stimulatory effect of animal on oestrus and ovulation through genital stimulation, pheromones and other external cues (Chenowith, 1983). In domestic animals, the priming pheromones from the male have been found to induce puberty, oestrus, synchronization of oestrus, reduce silent heat cases and improve the ovulation rate (Burns and Spitzer, 1992). The Bio-stimulation technique using a male has been explored as very effective management tool to modulate the reproductive tract activity in sheep and goats (Rosa et al., 2000; Vatiz et al., 2002).

The magnitude of response to male exposure depends on a number of factors. One of the factors is to rejoin previously isolated anoestrus female with sexually vigorous males before the start of normal breeding season (Singh et al., 1987). An isolation period of about two months is commonly accepted requirement to respond well to pheromones. The observations suggest that olfactory stimuli produced by the male are responsible for induction and synchronization of estrus in females. The role of priming pheromones in cattle reproduction is not as clearly defined as that in other species. However, Roberson et al. (1991) observed the effect of bio-stimulation on reproduction in cattle. Since the buffaloes are anoestrus for some part of the year, the introduction of buffalo bulls may be helpful in induction of estrus in buffaloes during non-breeding season and also decrease the rate of silent estrus.

6. Summer stress Management

Harsh climate of subtropical regions like India severely affects the reproduction in dairy cattle. This effect gets aggravated in buffaloes, which necessitates the modifications in the management of species during summer. Roy et al. (1968) were able to show a near uniform breeding activity by buffaloes during breeding and non-breeding seasons by protecting these animals against hot wind. Subsequently, the beneficial effect of modified management on breeding have been studied by Sastry *et al.*, 1973; Thomas *et al.*, 1975 and Sastry *et al.*, 1991). Recently Saini et al. (1998) showed that about 93% buffalo heifers came in fertile estrus by splashing 5 times water on their body as against 17% in the control group.

7. Identification of Problematic Females

Early detection of reproductive problems is very essential as it is much easier to correct them if identified at early stage so as to save time and money. Before one can identify these problems, it is necessary to know some basic standards for comparison and the same can be done with the help of records of reproductive status about individual female. The following criteria may be applied in dairy herds:

1. At least 90% of the cycling cows/buffaloes should be detected for estrus after 40 days post-partum.

2. Between 50 to 55% of the cows/buffaloes should become pregnant to first A.I. If the animals do not settle up to three inseminations, they are classified as problematic ones.

3. The average number of services per conception should not be three in any case. Those requiring more need to be checked.

4. At any given time, there should not be more than 10% problem cows/buffaloes under efficient system of management.

5. Estrus cycle length normally should be 21 days. Short cycles are an indication of ovarian dysfunction and long cycles are an indication of embryonic loss.

6. Continuous estrus for more than 24 hrs in a female is an indication of follicular cysts.

7. The number of cow/buffaloes leaving the herd due to breeding failure should not exceed 5% per year.

If more than 10% of the herd is affected by a combination of above-mentioned symptoms, the herd suffers from reproductive problems and an immediate action is warranted to correct the situation. Prevention is far superior than cure.

Management of Bulls

It is important to consider bull reproductive soundness regardless of whether the male is used in natural service or A.I. Evidence of bull fertility differences has been available for many years. The effect of nutrition on reproductive efficiency of bulls before and after puberty has been reviewed by Brown (1994). Reproductive functions appear to be more susceptible to restrictions of dietary energy in growing bulls via hypothalamic – hypophyseal axis. However, many reports have shown that onset of puberty in bulls appears to be more influenced by body weight than the age which in turn are likely to be influenced by feeding. Scrotal circumference of testes is an important component while selecting a breeding bull for breeding soundness. It is indicator of testis size and highly correlated with sperm production and quality (Brinks, 1994). Therefore, it will be pertinent if scrotal circumference of testis is also considered as criterion while selecting a breeding bull in addition to semen tests like volume, sperm number and motility score. With advent of a number of laboratory tests for semen analysis, it has become easy to predict the fertility of the bulls to be put in the herd for breeding. However, it has been observed that semen with normal semen index may not always yield acceptable conception rates if semen contain abnormally higher number of sperms with damaged acrosome. It will be useful if test for intact acrosome is also included in above-mentioned tests.

Generally two types of cares are important while maintaining the bull for optimal seminal quality. The first one is nutrition. Many reproduction related problems are due to either underfeeding or overfeeding. The later may be more serious and results in excessive fat which is responsible for low conception, abortion, dystokia and more retention of placental cases, reduced libido. However, underfeeding is more common which reduces total sperm output and libido in bulls. The mature bulls should be restricted to maintenance diet along with regular exercise.

Second factor which influence semen and its characteristics is the climatic environment. It is well recognized that higher temperature and humidity are associated with marked seasonal decline in semen quality of bulls. Sengupta *et al.* (1968) studied physical characteristics and metabolic attributes of Murrah buffalo bulls and reported superiority of all parameters in spring season.

The effect of season on fertility of frozen buffalo semen as examined by Heuer *et al.* (1987) showed adverse effect of summer heat on quality of fresh semen with damaged acrosomes. Bahga and Khokar (1991) obtained lowest sperm motility in summer and highest in winter. Semen quality with acceptable level of sperm motility has been shown to be maintained by adopting modified management in summer season. The classical work by Roy *et al.* (1968) at Mathura demonstrated that good semen quality can be ensured during severe summer in Indian conditions provided the bulls are protected against hot wind and temperature. The effect of water sprinkling during summer months on semen quality of Holstein bulls was studied by Salah *et al.* (1992) and obtained increased sperm motility and decreased incidence of dead and abnormal spermatozoa.

In conclusion, a combination of traditional and modern tools like progesterone estimation may be employed to assess the reproductive efficiency of a dairy herd. Appropriate steps as mentioned in the paper be taken so that dairy production can become economically viable and sustainable.

References

Arney, D.R.; Kitwood, S.E. and Phillips, C.J.C. (1994). The increase in activities during oestrus in dairy cows. Applied Animal Behaviour Science, **40**: 211-236.

Banerjee, A.K.; Chaudhary, R.R and Bandopadhaya, S.K. (1989). Temporary Engorgement of teats (TET) – its relationship with occurrence of estrus in buffaloes. Indian J. Ani. Reprod. **10**: 166-169.

Barkawi, A.K.; Bedeir, L.H. and El-Wardani, M.A. (1993). Sexual behaviour of Egyptian buffaloes in post-partum period. Buffalo Journal, **9:** 225-218.

Bhowmick, B.C. and Sharma, A.K. (2002). Livestock in Assam. In Livestock in different farming systems.

Brinks, J.S. (1994). Relationship of scrotal circumference to puberty and subsequent reproductive performance in male and female offspring. In Fields, M.J. and Sand, R.S. (eds.) Factors Affecting Calf Crop. CRC Inc. Boca Raton, Florida, pp. 363-370.

Britt, J.H. (1987). Detection of oestrus in cattle. The Veterinary Annual Issue, **27:** 74-80.

Brown, B.W. (1994). A review of nutritional influences on reproduction in boars, bulls and rams. Reproduction, Nutrition, Development, **34:** 89-114.

Burns, P.D. and Spitzer, J.C. (1992). Influence of biostimulation on reproduction in post-partum beef cows. Journal of Animal Science, **70:** 358-362.

Canfield, R.W. and Butler, W.R. (1989). Accuracy of predicting the LH surge and optimal insemination time in Holstein heifers using a vaginal resistance probe. Theriogenology, **31:** 835-842.

Cavalieri, J. and Fitzpatrick, L.A. (1995). Oestrus detection techniques and insemination strategies in Bos indicus heifers synchronized with norgestomet-oestrediol. Australian Veterinary Journal, **72:** 177-182.

Chenowith, P.J. (1983). Reproductive management procedures in control of breeding. Aust. J. Animal Prod. **15:** 28-33.

Clapper, J.A.; Ottobre, J.S.; Ottobre, A.C. and Zartman, D.L. (1990). Estrual rise in body temperature in the bovine. Temporal relationships with serum patterns of reproductive hormones. Animal Reproduction Science, **23:** 89-97.

Claycomb, R.; Delwiche, M.; Munro, C. and Bon Durant, R. (1995). Rapid enzyme linked immunosorbent assay for on line measurement of bovine progesterone during milking. Biology of Reproduction, **52** (Suppl. 1): 107.

Darwash, A.O. and Lamming, G.E. (1995). To define and quantify atypical ovarian function in untreated post-partum cows. Biology of Reproduction **52** (Suppl. 1): 72.

Darwash, A.O.; Lamming, G.E. and Woolliams, J.A. (1997). Estimation of genetic variation in the interval from calving to post-partum ovulation of dairy cows. Journal of Dairy Science, **80**: 1227-1234.

Duby, R.T., Damiani,; Looney, C.R.; Fissore, R.A. and Robl, J.M. (1996). Prepubertal calves as oocyte donors: promises and problems. Theriogenology, **45**: 121-130.

Guven, B. abnd Bolukbasi, F. (1989). Determination of milk progesterone levels in cows during the post-partum period by microtitration plate enzyme immunoassay. Veteriner Facultesi Dergisi (Ankara Universitesi), **26**: 565-582.

Hanzen, C. (1986). Endocrine regulation of post-partum ovarian activity in cattle: a review. Reproduction, Nutrition, Development, **26**: 1219-1239.

Heuer, C.; Tahir, M.N. and Amjad, H. (1987). Effect of season on fertility of frozen buffalo semen. Animal Reproduction Science, **13**: 15-21.

Jainudeen, M.R. (1986). Reproduction in the water buffalo. *In:* Morrow, D.A. (ed.) Current Therapy in Theriogenology. W.B. Saunders, Philadelphia, pp. 443-449.

Kamimura, S.; Samshima, H.; Enomoto, S. and Hamana, K. (1994). Turnover of ovulatory and non-ovulatory dominant follicles in post-partum Japanese Black cows. Journal of Reprod. and Dev. **40**: 171-176.

Kaul, V. and Parkash, B.S. (1994). Application of milk progesterone estimation for determining the incidence of false heat detection and ovulation failures in Lobu and crossbred cattle and Murrah buffaloes. Ind. J. Anim. Sci. **64**: 1054-57.

Kitwood, S.E.; Phillips, C.J.C. and Weise, M. (1993). Use of a vaginal mucus impedance meter to detect estrus in the cow. Theriogenology, **40**: 559-569.

Kyle, B.L. and Kennedy, A.D. (1994). The use of pedometers to detect

---

estrus in beef cows. Journal of Animal Science, **77** (Suppl. 1): 292.

Lewis, G.S.; Aizinbud, E. and Lehrer, A.R. (1989). Changes in electrical resistance of vulvar tissue in Holstein cows during ovarian cycles and after treatment with prostaglandin F2a. Animal Reproduction Science, **18**: 183-197.

Lishman, A.W. and Inskeep, E.K. (1991). Deficiencies in luteal function during reinitiation of cyclic breeding activity in beef cows and in ewes. South African Journal of Animal Science, **21**: 59-76.

Lohan, I.S.; Malik, R.K.; Saini, M.S.; Dhanda, O.P. and Singh, Baljit. (2000). Uterine and ovarian changes during early post-partum period in Murrah buffaloes. Buffalo Bullten, **19**: 20-23.

Mosher, M.D.; Ottobre, J.S.; Haibel, G.K. and Zartman, D.L. (1990). Estrual rise in body temperature in the bovine. The temporal relationship with ovulation. Animal Reproduction Science, **23**: 99-108.

Nebel, R.L.; Walker, W.L., McGillard, M.L.; Allen, C.H. and Heekman, G.S. (1994). Timing of artificial insemination of dairy cows: fixed time once daily versus morning and afternoon. Journal of Dairy Science, **77**: 3185-3191.

Peter, A.T. and Bosu, W.T.K. (1986). Post-partum ovarian activity in dairy cows: correlation between Behavioural estrus, pedometer measurements and ovulation. Theriogenology, **26**: 111-115.

Rajamahendran, R.; Robinson, J.; Desbottes, S and Walton, J.S. (1989). Temporal relationships among estrus, body temperatue, milk yield, progesterone and luteinizing hormone levels, and olvulation in dairy cows. Theriogenology, **31**: 1173-1182.

Raut, N.V. and Kadu, M.S. (1990). Observations on ovulation and its association with fertility in Nagpuri buffaloes. Indian Vet. Journal, **67**: 130-132.

Roberson, M.S.; Wolfe, M.W.; Stumpf, T.T.; Werth, L.A.; Cupp, A.S.; Kojima, N.D.; Wolfe, P.L.; Kittok, R.J.; Kinder, J.E. (1991). Influence of growth rate and exposure to bulls on age at puberty in beef heifers. J. Anim. Sci. **69**: 292-298.

Roy, A.; Raizada, B.C.; Tewari, R.B.L.; Pandey, M.; Yadav, P.C. and Sangupta, B.P. (1968). Effect of management on the fertility of buffalo cows bred during summer. Indian Journal of Veterinary Science, **38**: 554-560.

Royal, M.D.; Darwash, A.O.; Flint, A.P.F.; Webb, R.; Wooliams, J.A. and Lamming, G.E. (2000). Declining fertility in dairy cattle: Changes in traditional and endocrine parameters of fertility. Animal Science, **70**: 487-501.

Saini, M.S., Dhanda, O.P., Singh, N. Georgie, G.C. (1998). The effect of improved management on reproductive performance of pubertal buffalo heifers during summer. Indian Journal of Dairy Sci. **51**: 250-253.

Sastry, N.S.R.; Juneja, I.J.; Yadav, R.S.; Gupta, L.R.; Thomas, C.K. and Tripathi, V.N. (1981). Effect of provision during summer of extra shelter and sprinkling water on young buffalo heifers on their later productive and reproductive functions. Indian Veterinary Journal, **58**: 753-754.

Sastry, N.S.R.; Thomas, C.K.; Tripathi, V.N.; Pal, R.N. and Gupta, L.R. (1973). Effect of shelter and water sprinkling during summer and autumn on some physiological reactions of Murrah heifers. Indian Journal of Animal Sciences, **43**: 95-99.

Sheokand, B.S.; Kapoor, C.M. and Dhanda, O.P. (1983). A peculiar phenomenon among buffaloes. Farmers' Journal **11**: 56-58.

Singh, Khub, Dad Kumar, P. (1997). Annual Report of CIRG, Makhdoom, Mathura (India).

Sprecher, D.J.; Farmer, J.A.; Nebel, R.L. and Mather, E.C. (1995). The educational implications of reproductive problems identified during investigations at Michigan Dairy Farms. Theriogenology, **43**: 373-380.

Thomas, C.K.; Tripathi, V.N. and Sastry, N.S.R.; Pal, R.N. and Gupta, L.R. (1972). Effect of shelter and water sprinkling on buffaloes: Growth rate. Indian Journal of Animal Sciences, **42**: 745-749.

Usmani, R.H.; Ahmad, M.; Inskeep, E.K.; Dailey, R.A.; Lewis, P.E. and Lewis, G.S. (1985). Uterine involution and post-partum activity in Nilli-Ravi buffaloes. Theriogenology, **24**: 435-448.

Vale, W.G. (1990). The water buffalo in Latin America. *In:* Livestock Reproduction in Latin America. FAO/IAEA. Pp. 199-200.

Vatiz, F.G.; Moreno, S., Daure, G.; Vielma, J.; Chemineau, P.; Poindron, Malpaux, B.; Delgadillo, J.A. (2002). Male effect in seasonality anovulatory lactating goats depends on the presence of sexually active bucks but not estrus females. Anim. Reproduction Science. **72:** 197-207.

CHAPTER 25

Management of Buffalo Diseases with Special Emphasis on the Diseases of Economic Importance

M.C. Sharma

Indian Veterinary Research Institute, Izatnagar-234 122 (UP)

The domestic buffalo (*Bubalus bubalis)* is closely related to in phylogeny to cattle and have broadly been known as batter buffaloes and swamp buffaloes. These are the triple purpose farm animals and are the major source of milk, meat and tractile energy for most of the developing countries where these are reared. These are mainly available in the Asian subcontinent since about 97% of total buffalo population exists here due to which these are referred as Asian Animal.

Of total buffalo population of 152 million, India possesses 83.5 million (53%) while China and Pakistan have 22.8 and 20.2 million buffaloes, respectively. Out of total buffalo milk production of 54.9 million tons of world, India produces 34.8 million tons while share of China and Pakistan is 2.2 and 14.8 million tons, respectively. In India, it contributes about 52% of total milk production and average milk production of buffalo is high as compared to average milk production by a cow. The annual growth rate of buffaloes is 1.7% in India, which is the home of the best Murrah and Nili-Ravi breeds of buffaloes.

Buffalo health about four decades back was a neglected subject, but looking to the potential of the animal, it has gained priority and research has focussed on buffalo health problems. There are ongoing studies to determine the susceptibility of the animal to various diseases, and its relation to age, environmental and management factors etc. Buffalo health is a very vast topic, but some of the economically important non-infectious and infectious problems have been discussed.

Viral Diseases

Foot and Mouth Disease

Foot and mouth disease (FMD) or aphthous fever is a contagious acute viral disease characterized by fever and formation of vesicles in the mouth and on feet. FMD is caused by an aphthovirus (family picornaviridae), which occurs in seven major serotypes A, O, C, SAT 1, SAT 2, SAT 3 and Asia 1. Of these, the predominant ones in India are, the serotypes O, A, Asia 1 and C. The morbidity rate is up to 100 per cent.

Clinical findings: In adult buffaloes, the clinical findings are mild in the majority of cases and they recover rapidly. Lesions occur mainly in the mouth and foot lesions are rare and mild in nature. Small vesicle formation is noticed on buccal mucosa, dental pad and tongue while sometimes these may be observed in interdigital space also.

Diagnosis: Serological methods which can be used for diagnosis include CFT, ELISA, plaque reduction assay, virus neutralization, radial immunodiffusion and the virus infection associated antigen test. ELISA has been found to be more sensitive for demonstration of antibody levels than microserum-neutralization test.

Treatment and control: Treatment is directed to the use of mild disinfectants and protective emollients on the lesions, administration of antibiotics to prevent secondary bacterial infection and the use of NSAIDS like flunixin meglumine to reduce inflammation. Vaccination should be done regularly with multivalent vaccines against the serotypes prevalent in that area. Vaccination twice in a year is recommended. Presently tetravalent oil adjuvant binary ethyleneimine inactivated vaccine is available which can be given @ 3 ml im or sc in buffaloes. This vaccine is to be repeated after 44-48 weeks and can be used in endemic areas as well as in containing the spread of disease during outbreak. In calves it is given at 1 month, 4 months and 1 year of age and then repeated after every 6 weeks.

Ephemeral Fever

The disease is also referred as three days sickness and characterized by hyperpyrexia, lameness and muscular stiffness. The ephemeral fever virus belongs to the family rhabdovirus. It primarily affects the adult buffaloes and calves below 6 months of age are not affected by it.

Clinical findings: The clinical signs are sudden rise in body temperature (41° C), anorexia and reduction in milk yield. Shivering, stiffness and clonic muscular movements, nasal discharge and increased heart rate are also noticed. Lameness becomes very prominent and typical posture of laminitis may be observed. Sometimes abortions in pregnant buffaloes may also occur.

Diagnosis: The cases can be easily diagnosed by clinical findings and can be confirmed by blood analysis, which reveals leukocytosis, neutrophilia, lymphopenia and increased fibrinogen. The disease can also be confirmed by agar gel precipitation, complement fixation, ELISA and fluorescent antibody tests

Treatment and control: Usually the clinical findings disappear in 3 days so supportive treatment is recommended for affected animals. Paracetamol and phenyl butazone are given by parental route. To prevent secondary bacterial complications, broad-spectrum antibiotics like streptopenicillin or tetracyclines should be used.

Rinderpest

Rinderpest also known as cattle plague is an acute highly contagious viral disease of ruminants. Rinderpest virus, which is a RNA virus, is a morbilli virus belonging to family paramyxoviridae causes the disease. A national eradication programme is been actively implemented in India and certain zones have already been declared as Rinderpest free.

Clinical findings: There is sudden onset with high fever, anorexia, rough hair coat and depression. Mucosal lesions appear 2-3 days after the onset of fever. Diarrhoea, often haemorrhagic, sets in after the fever subsides. These animals become prostrate 7-8 days after the onset of these clinical findings with temperature becoming subnormal a few hours prior to death. Abortions in the pregnant animals have also been noticed.

Diagnosis: Serological techniques that can be used include, AGID (serum must be taken 3-5 days after the onset of fever for accuracy), CIEP, CFT, FAT, ELISA, immunoperoxidase, serum neutralization and the use of specific cDNA probes.

Treatment and control: Disease can be controlled easily as virus can not survive outside the host for longer period and various strains of virus have immunological identity. It is effected by good quarantine measures and the culling and slaughter of all infected and in contact animals.

Rabies

Rabies also known as lyssa, is a highly fatal viral infection caused by rhabdovirus (genus lyssavirus). It is a truly neurotropic virus and causes lesion only in the nervous tissue. The source of infection is always an infected animal and the spread is invariably through a bite.

Clinical findings: The affected animals either show paralytic or furious form. There is drooling of saliva, eructation, grinding of teeth, continuous tail movements, anorexia, and stiffness of hind limbs paralysis and recumbency. These recumbent animals die in 2-3 days in paralytic form. However in furious form, the animals become alert but tense and hypersensitive. They show sexual excitement, apparent inability to swallow, violent ramming of head on fixed objects and loud bellowing. Such clinical findings persist for 36-48 hr and then animal collapses and die.

Diagnosis: Impression smears prepared from brain can be tested by FAT for confirmation. CFT or ELISA can also confirm it. Animal inoculation test using mice is also of greater importance. Differential diagnosis with sub acute lead poisoning, polioencephalomalacia, listeriosis, lactation tetany and deficiency of vitamin A.

Treatment and control: Immediately after exposure the wound should be irrigated with soap solution and water. Control is effected by the destruction of wild fauna in around animal holdings and the vaccination of all domestic cats and dogs being maintained at the premises. Live and inactivated vaccines both of chick embryo origin and tissue culture origin are available and can be used.

Malignant Catarrhal Fever

It is a highly fatal acute disease of buffaloes characterized by gastroenteritis, lymphnode enlargement and encephalitis. There are two forms of the disease, one of which is caused by herpesvirus-1 (AHV-1) which is a herpes virus and transmitted from blue wildebeest. The other form is associated with the domestic sheep and is caused by ovine herpes virus-2 (OHV-2).

Clinical findings: In head and eye form, the course of disease is comparatively longer and buffaloes are suddenly infected and reveal high fever (41-42°C), anorexia, rapid pulse rate, reduction in milk yield, profuse purulent nasal discharge and severe dyspnoea. Discrete local areas of necrosis are seen on the hard palate, gums and gingivae.

Diagnosis : The disease can be confirmed by polymerase chain reaction test. It should be differentiated from mucosal disease, rinderpest, infectious bovine rhinotracheitis, and viral encephalitis or haemorrhagic septicemia.

Treatment and control: Though there is no specific treatment, but use of non-steroid anti-inflammatory drugs may be given to relieve from discomfort. Supportive treatment given in the form of broad-spectrum antibiotics and fluid therapy are also of some value. For the control of disease, buffaloes should be isolated from sheep flock.

Buffalo Pox

It is a mild viral disease of buffaloes and is characterized by the presence of typical pox lesions on the teats and udder. The disease is caused by buffalo pox virus of the genus orthopox of the family pox viridae. Morbidity up to 70% has been reported but mortality rates are comparatively low.

Clinical findings: The affected buffaloes suffer from fever (40° C), anorexia, dullness, depression and congestion of conjunctivae. These animals show typical pock lesions mainly involving teats, udder and medial aspect of thighs, there is development of mastitis due to secondary bacterial infection in teat.

Diagnosis: The cases can be detected by the clinical findings and can be confirmed by isolation of the virus from the lesions present on the teats or udder. Precipitation, complement fixation and neutralization can also confirm it.

Treatment and control: There is no specific treatment. However to check the secondary bacterial contamination, antibiotic therapy with broad-spectrum antibiotic must be initiated as early as possible. Care must also be taken to prevent the occurrence of mastitis in affected animals. Presently no suitable vaccine is available against the disease so adopting strict hygienic measures can only control it.

Bacterial Diseases

Black Quarter

Black quarter or emphysematous gangrene or black leg, is caused by *Clostridium chauvoei,* a gram positive, spore forming, rod. The disease is a soil borne infection and spreads mainly through intestinal mucosa after the ingestion of contaminated feed.

Clinical findings: There is marked swelling in the muscles of shoulder, hip, chest, back or flank region. The swelling that is hot and painful to touch initially becomes very extensive, cold and painless later on. On palpation of swollen area, crepitating sounds can be listened. The affected area later on, turns black and dry and froth mixed foul smelling fluid is released by puncturing it. There is high fever (41°C), depression, anorexia, and ruminal stasis with increased heart and pulse rates. In terminal stage, diaphragm, tongue and heart muscles are also involved and the animal dies within 12-36 h after appearance of clinical findings.

Diagnosis: In typical cases of black quarter, a definitive diagnosis can be made on the clinical signs and the necropsy findings. Isolation of the causal agents and serological tests are also helpful in the confirmation of the disease. The disease must be differentiated from anthrax, lightening strike, lactation tetany, bacillary haemoglobinuria, haemorrhagic septicemia and acute lead poisoning.

Treatment and control: The severe cases must be treated immediately with BQ antiserum and high doses of antibiotics. The antiserum can be given @ 200-400 ml iv and repeated after 24 h. Penicillin @ 1000 units/kg b wt is highly effective. Treatment should commence with crystalline penicillin iv followed by long acting preparations for 5-7 days, some of which must be injected into the affected tissue Ciprofloxacin and norfloxacin are also effective. Vaccination of calves at 3 weeks of age is recommended when the incidence of disease is very high.

Anthrax

Anthrax or spleenic fever may be peracute or acute characterized by sudden death with the exudation of dark coloured blood from the natural orifices. Anthrax is caused by *Bacillus anthracis*, a gram-positive spore. Anthrax infected carcasses are never opened. The disease is of zoonotic importance as human beings suffer from "Wool sorter's disease". In recent past it has been used as weapon by the terrorist.

Clinical findings: In buffaloes, the disease is manifested in peracute or acute forms. The infected animals die suddenly within two hours without showing any clinical findings, but escape of blood from nostrils, anus and mouth occurs. Before death, fever, muscle tremors, dyspnoea and congestion of the mucosa may be observed with collapse following terminal convulsions. Pregnant animals abort, and milch animals show reduction in milk yield and milk is usually blood tinged. There may be diarrhoea, dysentery, and edema of the tongue, throat, sternum and perineum.

Diagnosis: Peripheral blood or edema fluid smears will reveal the organism, in positive cases. For ascoli test, a small piece of the muzzle or ear should be collected for preparing antigen for conducting the precipitation test. The disease should be differentiated from peractue black quarter, lead poisoning, acute leptospirosis and bacillary haemoglobinuria.

Treatment and control: Anti-anthrax serum @ 100-150 ml iv can be given to valuable animals in conjunction with antibiotics. Streptomycin @8-10 g/day in two doses im has been found to be

---

much effective. Oxytetracycline @ 5 mg/kg b wt per day parentally is also more effective. It is desirable to prolong the treatment up to a minimum of 5 days to prevent relapse of the disease.

Haemorrhagic Septicemia (Pasteurellosis)

Haemorrhagic septicemia also known as shipping fever or locally in India as galghontu and ghurrha. *Pasteurella multocida* type 2B is mainly responsible to cause the disease in buffaloes. The causal organism is a small gram negative, bipolar, coccobaccilus. The tonsillar region of the nasopharynx has been identified to be the main portal of entry of the organism whether through inhalation and ingestion. Carriers include cattle and buffaloes while reservoir hosts include pigs, sheep, goat and horses.

Clinical findings: The disease is characterized by sudden rise in temperature up to 42°C, profuse salivation, severe depression, sub mucosal petichae and death in about 24 h. In the common throat form, there is a hot painful swelling of the throat, brisket or perineum and severe dyspnoea may occur. The animals die as a result of respiratory distress.

Diagnosis: It is made based on clinical findings, demonstration of the causal organism in smears of blood / edema fluid made immediately after death, since the organism disappears from dead animals fast. It is necessary to differentiate the disease from anthrax, leptospirosis and black quarter.

Treatment and control: Sulphadimidine is highly effective and earlier used to be the drug of choice and given at the dose rate of 0.5-1.0 gm/kg b wt sc for 3-4 days. Combination of trimithoprim and sulphamethroxazole is highly effective when used @ 3-5 ml/ 50 kg b wt im for 4-5 days. Oil adjuvant vaccines are favoured for systematic vaccination programmes.

Tuberculosis

Tuberculosis is characterized by the progressive development of tubercles in various organs of the body. The causative organism in buffaloes is *Mycobacterium bovis*. The infected animals serve as the main source of infection to other animals since the organisms are excreted in the sputum, faeces, milk, urine, vaginal and uterine discharges and open peripheral lymphnodes. The organisms enter via inhalation or ingestion but inhalation is the common route of spread

Clinical findings: General malaise is the first observable sign in affected buffaloes. In the pulmonary form, low grade fever, inappetence, chronic dry husky coughing, progressive weakness and dryness of skin are observed. In later stages pleurisy and dyspnoea are also noticed. In the intestinal form, there is persistent diarrhoea. The affected mammary glands are painlessly enlarged and the supramammary lymph node may also be enlarged. The milk becomes watery and large numbers of organisms are present in it.

Diagnosis: The cases can be detected on the basis of clinical findings and post-mortem lesions. Delayed hypersensitivity reactions like the single intradermal test (SID), short thermal test, Stormont test and the comparative test with tuberculin of various origins, are used for diagnosis. The indirect hemagglutination test is reported to be a very sensitive test to detect early and advances cases of tuberculosis, in which the tuberculin test failed.

Treatment and control: Animals may be treated with a combination of 2.5 gm streptomycin im, 1.5 gm rifampin and 2.0 gm isoniazid given orally daily for 4-6 months has been reported highly effective. Effective eradication of the disease is based on removal of infected animals, prevention of spread of infection and avoidance of further introduction of disease.

Brucellosis

The disease also known as contagious abortion or Bang's disease and is caused by *Brucella abortus* that are small gram-negative, non-motile, non-sporulating coccobacillus organisms. The organism can also penetrate intact skin or mucous membranes. Congenital infection can also occur but is seen in calves. The disease is of zoonotic importance as dairy workers, veterinarians or butchers may pick up infection and suffer from undulant fever.

Clinical findings: Highly susceptible pregnant animals suffer from abortions after 6 months, retained placenta and catarrhal metritis. After one or two abortions the animal may give birth to full term calves. In herds there is usually a 'storm' of abortions. In bulls there is epididymitis and orchitis involving one or both scrotal sacs. In mild cases, sinovitis and painful swelling of affected joints are noticed.

Diagnosis: The cases can be diagnosed by history of abortions in late pregnancy and can be confirmed by isolation of the organisms from uterine discharges of an infected animal or the stomach contents and heart blood of the aborted foetus. Ziehl Neelsens staining is specific for confirmation of the organisms Serological tests include the rose Bengal test for herd testing and the standard tube agglutination test, spot agglutination test, ELISA, and CFT in individual cases.

Treatment and control: Chloramphenicol or combination of long acting oxytetracycline and streptomycin are useful in its treatment. Calfhood vaccination with *Br. Abortus* strain 19 vaccine, between 4-8 months is carried out. The only other vaccine with some accuracy is Strain 45/20 in adjuvant. Greatest care must be taken in handling and disposal of aborted foetus, foetal membranes and uterine discharges etc. as it may serve as source of infection to other animals and human beings.

Johne's disease

It is also known as paratuberculosis and is caused by *Mycobacterium paratuberculosis*. Economic losses mainly occur due to ill health, poor growth and reduced productivity and working efficiency.

Clinical findings: In affected buffaloes reduced working efficiency and productivity, submandibular edema and progressive weight loss are noticed in spite of normal appetite. There is chronic diarrhoea and faeces resemble pea soup like but without any odour. The disease runs a protracted course and terminates in death.

Diagnosis: Demonstration of the organisms by microscopic examination of the faeces or of the rectal mucosa also helps in its detection. A CFT having a sensitivity of 90% and specificity of 70% and an AGID that has 96% sensitivity and 94% specificity are the most widely used modes of diagnosis of this disease

Treatment and control: Streptomycin has the maximum activity against the organism. A combination of dihydro streptomycin, rifampin and isoniazid has been found effective when used for three months. Eradication of infected animals and carriers and segregation and proper feces disposal can help to control the disease. A vaccine of live bacilli suspended in lanolin can be employed in calves less than one month of age.

Leptospirosis

It is a disease of zoonotic importance and is characterized by interstitial nephritis, hemolytic anaemia and abortions. The pathogenic leptospires are classified into one species *Leptospira interrogans*. The organism gains entry from contaminated pasture and water by infected urine, aborted fetuses, and infected uterine discharges.

Clinical findings: Buffaloes suffer from acute, sub-acute or chronic form of disease. In acute form, fever, anorexia, acute haemolytic anaemia, haemoglobinuria, jaundice, petechial haemorrhages on mucosae, dyspnoea, abortions and blood in milk are noticed. In sub-acute form, moderate fever, anorexia, dyspnoea and haemoglobinuria are observed. Abortions occur after 3-4 months of infection.

Diagnosis: Acute and convalescent sera taken 7-10 days apart should be submitted from each clinically affected animal or from those with a history of abortion. Of all the laboratory tests the examination of urine samples for the organism probably offers the most profitable opportunity of demonstrating the presence of infection. Of the serological tests, the microscopic agglutination test (MAT) is the most commonly used one. Other tests include ELISA and FAT.

Treatment and control: Streptomycin @12-15 mg/kg b wt twice daily for 3 days is effective when given as soon as the signs appear. Elimination of infection in carriers can be effected by a single dose of streptomycin @ 25 mg/kg b wt. Animals which are severely affected with hemolytic anemia can be given blood transfusion @ 5-10 lit/450 kg b wt). Vaccination with formalin inactivated bacterin with either aluminum hydroxide or Freund's complete adjuvant can be used. A live vaccine prepared from avirulent *L. interrogans* serovars *pomona* gives good immunity and is quite safe.

Listeriosis

It is an infectious disease of some zoonotic importance and is caused by *Listeria monocytogenes*. Virulent strains can be identified by their ability to multiply in macrophages and monocytes and their ability to produce a hemolysin, listeriolysin 'O'.

Clinical findings: Among buffaloes, it produces two types of manifestations listerial encephalitis or listerial abortions. The course of listerial encephalitis in adults is 1-2 weeks while in calves, deaths occur in 3-4 days. The affected buffaloes develop cranial nerve dysfunction and press their head onto fixed objects and there is unilateral facial paralysis. Mostly sporadic abortions occur due to the disease in buffaloes during the late third of pregnancy. Retention of after births is commonly noticed and in such cases temperature rises up to 40.5° C. In the calves, septicemic listeriosis may occur in which case, emaciation, depression, fever and diarrhoea are noticed. In 3-7 days old calf, corneal opacity, nystagmus and dyspnoea have been observed.

Diagnosis: Examination of CSF for presence of inflammatory cells and increased protein content are also helpful in the diagnosis.

Treatment and control: Though the organism are resistant to many drugs, but these are sensitive to chlortetracycline when given @ 10 mg/kg b wt iv for 5 days. Penicillin when used @ 44000 units/kg b wt for 10-14 days has also been proved effective. The feeding of excess/ spoilt silage must be avoided in these areas. Killed vaccines can be used to control the disease.

Tail necrosis

The drying of tail due to gangrenous necrosis is quite common in buffaloes. This condition is caused by *Corynebacterium bovis*. It begins with a swelling and inflammation at the tip of the tail, which gradually extends to the entire tail if not checked by amputation. The primary lesion becomes necrosed and a part of the tail sloughs off. The whole of the tail may also be lost as it is a chronic disease. Treatment usually consists of amputation of the affected part. Gangrenous syndrome involving tail, ear tips and extremities has been reported to be caused by mycotoxins.

Vibriosis

The disease vibriosis in buffaloes may be widespread as it has been recorded in India. It is caused by *Campylobacter fetus (Vibrio fetus)*. Infection is spread by the use of infected bulls for artificial insemination / natural service. The disease causes abortion between the 4-6 months of gestation. Isolating the organism from the uterine exudates and stomach contents of aborted fetus can make diagnosis. A mucous agglutination test has been developed and is used as a reliable diagnostic tool. Adoption of good hygienic measures and screening of bulls can help to control the disease.

Calf Diarrhoea

It is commonly seen in newly born calves and is recognized as one of the major causes of neonatal fatality. The bacteria usually found associated with this condition include strains of *E. coli, S. typhimurium, S. dublin, S. newport, S. bovis morbificans, S. weltevreden,* etc. Maximum deaths were seen in third week of age and in summer months while there was no sex difference.

Clinical findings: The affected calves suffer from profuse diarrhoea, sometimes with dysentery, straining during defaecation, cyanotic mucous membranes, depression, weakness, and incoordination of gait and severe dehydration and lymphopenia, neutrophilia and hyperproteinemia.

Diagnosis: Rotavirus induced calf diarrhoea can be detected by dot immunobinding assay. Blood examination may also help in its diagnosis.

Treatment and control: The management of calf diarrhoea is based on alteration of diet, replacement of lost fluid and electrolytes, use of antibiotics and use of intestinal protactants. Effective control can be achieved by reduction of the exposure of the neonate to infectious agents, providing adequate colostrum and increased non-specific resistance, and increasing specific resistance by vaccination of the dam or the neonate.

Mastitis

It is referred as the inflammation of mammary gland resulting in physical, chemical or bacteriological changes in the milk of lactating animals. It is caused by Streptococcus including *Str. agalactiae, Str. dysglactiae, Str. faecalis,* and *Str. uberis*. The main sources of infection are the infected udder and the environment, with infection taking place through milkers' hands, contaminated litter or milking machines.

Clinical findings: In acute inflammatory reaction, significant alterations in milk quality occur as it contains large number of clots or flakes and there is change in its colour which may be even blood mixed sometimes. Udder reveals hot and painful diffuse swelling. In sub-clinical form, apparent clinical findings are not seen and there is increase in leukocyte count in the milk. In coliform mastitis, the buffaloes reveal sudden onset of pyrexia, muscular tremors of head and upper hind legs, rumen atony and diffuse local swelling of the udder with straw-coloured watery non-odourous flaky secretion

Diagnosis: Periodical examination of udder, strip cup testing, use of indirect tests like whiteside and California mastitis tests give an indication of the presence of disease, N-acetyl-beta-D-glucosaminidase and lactate dehydogenase in milk is reliable tests for detecting the disease.

Treatment and control: Broad-spectrum antibiotics like cephalosporins or penicillin G with other drugs may be infused. The cases of coliform mastitis can be successfully treated with parenteral injection of norfloxacin, frequent stripping of affected quarter and oral administration of sodium salicylate and potassium nitrate.

Contagious bovine pleuropneumonia (CBPP)

It is a highly contagious and septicemic disease characterized by high rise of body temperature and respiratory grunts. The disease is caused by *Mycoplasma mycoides* var. *mycoides* which are pleomorphic, gram negative and highly fragile organisms.

Clinical findings: There is a sudden rise of temperature, up to 40° C, anorexia, absence of ruminal movement, dullness, depression and a fall in milk yield. Coughing is observed first on exercise and then even at rest. There is pain in chest region. The respiration becomes shallow and accompanied by grunting which is pronounced during expiration. Edema of the dewlap and throat is also seen.

Diagnosis: Isolation of organisms from serous fluid in thoracic cavity and identification of the organism can confirm this. Complement fixation and serum agglutination tests are highly useful in its diagnosis.

Treatment and control: Sulphadimidine and organic arsenicals have been used extensively and appear to reduce the mortality rate.
Good hygiene and sanitation practices and removal of sources of infection effect control. Vaccination by a live attenuated vaccine administered at the tip of the tail can be given to calves after the age of 2 months.

Metabolic and Deficiency Diseases

Parturient paresis

It is also referred as hypocalcemia or milk fever and characterized by general muscular weakness, depression and low blood calcium level. Deficiency of calcium and phosphorous may occur as a result of defective calcium absorption from intestine during parturition, insufficient mobilization of calcium from skeletal reserve sources during terminal stage of gestation to meet out the demand or due to excess drainage of calcium in cholestrum.

Clinical findings: The disease is commonly seen in high milk producing dairy buffaloes in third to sixth lactation. Disease is noticed within 72 h after parturition. The affected buffaloes initially reveal excitement and tetany. They become off feed, disinclined to move and hypersensitive. Such animals fall easily and remain in sternal recumbency in semi-conscious or unconscious stage. Afterwards, the animal goes in lateral recumbency and reveals flaccidity of muscles, marked depression, bloat, impalpable pulse and weak and inaudible heart sound.

Diagnosis: Blood examination confirms the disease as it reveals eosinopenia, lymphopenia, neutropenia, hypocalcemia, and hypophosphatemia and increased activity of AST and creatine phosphokinase enzymes. Urine can be examined for presence of calcium, which also helps in the confirmation of disease. The disease should be differentiated from hypomagnesemia, ketosis, Downer's syndrome and ephemeral fever.

Treatment: A 25% solution of calcium boro gluconate given 300-350 ml iv and 250-300 ml sc provides immediate response. In reoccurrence of disease, solution containing 20.8% calcium gluconate, 4.4% boric acid, 5% magnesium hypophosphite and 20% dextrose should be given @ 350 ml iv and 200 ml sc after 24 h. Mild cases may be treated by giving calcium boro gluconate orally @ 60-100 gm daily for 3-4 days. Maintaining proper calcium and phosphorous ratio in diet (2.3:1) can prevent the disease.

Hypomagnesemia

It is also referred as grass tetany, grass staggers or lactation tetany and characterized by clonic and tonic convulsions and muscular spasm. The disease occurs due to deficiency of magnesium in the body. High producing buffaloes excrete more magnesium in milk and if it is not replaced by dietary source, hypomagnesemia develops. The calves maintained on magnesium deficient milk or milk replacers, also suffer from it.

Clinical findings: In adults, it occurs commonly in second or third lactation. Magnesium deficiency alters impulse transmission and sensitivity of motor end plate resulting in muscular irritability. The animals have reduced appetite and milk yield. Pulse rate becomes very high and in terminal stage convulsion disappears, pulse becomes impalpable and mucous membranes look cyanotic.

Diagnosis: It can be confirmed by measuring the level of magnesium in serum, cerebrospinal fluid or urine. The blood calcium level is also reduced in affected buffaloes. Hypomagnesemia should be differentiated from nervous form of ketosis, acute lead toxicity or rabies.

Treatment: The animals respond quickly if calcium and magnesium salts are given. A mixture containing 25% calcium boro gluconate and 5% magnesium hypophosphite is used @ 500 ml iv followed by 200-ml sc after 12 h. For the control of disease, diets should be supplemented with magnesium carbonate, magnesium oxide or magnesium sulphate or salt licks containing magnesium salts.

Phosphorous deficiency haemoglobinuria

It is also known as haemoglobinuria and characterized by intravascular haemolysis, haemoglobinuria and anaemia. The disease is related to the deficiency of phosphorous and can occur due to feeding of low phosphorous containing feeds for long periods. Molybdenum and copper present in some feeds compete with the absorption of phosphorus and thereby the disease may occur.

Clinical findings: One of the most consistent findings in affected buffaloes is the coffee-coloured urination. These buffaloes have almost normal appetite but their milk yields are reduced significantly and have constipation with hard and black tinged faeces. In terminal stage, jaundice and jugular pulsation are also recorded. All these clinical findings are mostly observed after 4 months of pregnancy.

Diagnosis: Serum inorganic phosphorous is significantly decreased while calcium remains normal or slightly decreased. Chemical analysis of urine reveals presence of albumin and haemoglobin while ketone bodies, glucose, bile and erythrocytes are absent. The disease should be differentiated from leptospirosis, bacillary haemoglobinuria, babesiosis and chronic copper poisoning.

Treatment: 120 gm sodium acid phosphate can be dissolved in 600 ml water and half of it is given iv and half sc. Phosphorous deficiency can also be overcome by oral administration of 125 gm bone meal or dicalcium phosphate twice daily for 5 days or 60 gm monobasic sodium phosphate twice daily for 3 days.

Ketosis

It is also referred as acetonemia and characterized by hypoglycemia, ketonuria and moderate loss of appetite and milk yield. Main cause of disease is hypoglycemia, which is produced as a result of impaired carbohydrate metabolism. During early stage of lactation if buffaloes are provided low energy ration or given inadequate exercise, they suffer from disease. It is commonly recorded in high producing dairy buffaloes in early and at peak lactation.

Clinical findings: Due to hypoglycemia, liver glycogen level is also reduced and ketonemia develops. Wasting and nervous forms of ketosis as recorded in cows, are seen in buffaloes also. Wasting form of ketosis is more common and animals show gradual decrease in feed intake and fall in milk production. There is excessive urination of watery consistency and urine and milk have ketotic smell. Buffaloes reveal excess salivation, false chewing movements, hyperasthesia and abnormal movements.

Diagnosis: Ketone bodies in urine can be detected by Rothera's test. Blood examination reveals eosinophilia, lymphocytosis, and neutropenia, reduction in glucose and calcium and increase in ketone bodies. The disease should be differentiated from parturient paresis, abomasal displacement, TRP, pyelonephritis, metritis or rabies.

Treatment: The drugs that maintain blood glucose level, are highly effective. For it 500 ml of 50% glucose solution can be given by slow iv injection and repeated daily for 3-5 days. To enhance gluconeogenesis, dexamethasone can be given @ 15-25 mg im once and may be repeated after 24 h if necessary. Sodium propionate given @ 120 gm in 250 ml water twice daily for one week also helps in recovery of the case.

Parasitic Diseases

Trypanosomosis

It is also referred as surra and results into development of intermittent fever, loss of condition and weakness of hindquarter. The disease is caused by *Trypanosoma evansi* in most of the countries.

Clinical findings: The affected animals reveal hyperthermia or intermittent fever (up to 41^0 C), emaciation, conjunctivitis, anaemia, nasal discharge, loss of milk, dyspnoea and weakness of muscles of back and hind-quarter resulting in difficulty in movement and staggering gait. In the buffaloes suffering from acute form of the disease, intermittent fever, anaemia, edema of the legs and other lower parts of the body, inappetence, loss of condition and weakness.

Diagnosis: The disease, can be confirmed by seeing the organisms in blood smears and by animal inoculation test by injecting blood from suspected buffalo into rat or mice iv. Serological tests like capillary agglutination, indirect fluorescent antibody and ELISA also confirm the disease. Of late PCR has also been used for its confirmation.

Treatment and control: The cases can be treated either with suramin (3 mg/kg), quinapyramine sulphate (5 mg/kg) or samorin (0.25 mg/kg b wt). Diminazine aceturate given @ 10-15 mg/kg b wt and isometamidium @ 0.5 mg/kg b wt also effective. Presently quinapyramine prosalt containing quinapyramine methyl sulphate and chloride salts is commonly used and is very effective when given @ 2.5 gm im as single dose. Supportive treatment given in the form of haematinics and liver tonics help in early recovery of the cases. The disease can be controlled by checking the vector population and by preventing the buffaloes from getting bitten by *Tabanus* or *Stomoxys*.

Babesiosis

The disease is also known as red water disease and caused by *Babesia* parasite. The affected buffaloes revealed anaemia, anorexia, dullness, depression, disinclination to move, severe haemoglobinuria and rise in body temperature up to 41°C. Dyspnoea, reduced milk yield and low PCV, Hb, TEC and TLC are also noticed. Blood smear examination and serological tests like card agglutination, slide agglutination, ELISA, indirect fluorescent antibody and passive haemagglutination help in confirmation of the disease. Imidocarb has also been found highly effective when given @ 1 mg/kg b wt sc. As the disease is transmitted by ticks, control of tick population can help in controlling the disease.

Theileriosis

The disease is caused by *Theileria* parasite, which has many species. These buffaloes revealed marked fever and leukopenia and thrombocytopenia before death, which occurred by day 16. There is high fever (41.5ºC), dullness, emaciation and ocular discharge. Use of berenil and tetracycline has been reported to have some efficacy. Buparvaquone which is highly effective against theileriasis in cattle, can be used in buffaloes also @ 2.5 mg/kg b wt im. Supportive treatments given in the form of liver tonics and fluid administration are advantageous.

References

Dhanda, M.R. 1977. Diseases caused by bacteria and fungi. *In:* Handbook of Animal Husbandry, I.C.A.R., New Delhi, p. 788.

Joshi, H.C; Sharma, M.C. & Kumar, Mahesh. (2002). "Handbook of Animal Husbandry". Chapter 21 page 816-877.

Kumar, S and Sharma, M.C. (1993). Buffalo Journal **1**: 69-73

Pathak, N.N. and Sharma, M.C. 1988. Buffalo Production and Health. I.C.A.R., New Delhi, pp. 156-184.

Sharma, M.C., Gupta, O.P., Dwivedi, S.K. and Das, S.C. 1981. Indian Vet. J. **58**: 241-242.

Sharma, M.C., Hung, N.N. and Vuc, N.V. 1982 b.Buffalo Breeding Res. Cen. Song Be Annual. **5**: 33-35.

Sharma, M.C., Pathak, N.N., Hung, N.N., Nhi, D.L. and Vuc, N.V. 1985. In: Veterinary Viral Diseases. Antony, J. Dhella Porta (editor), Academic Press, Sydney, pp. 302-303.

Sharma, M.C. and Kumar, M (2003) Infectious Diseases of Buffaloes. IInd Asian Buffalo Congress, New Delhi 25-28 Feb

Tsai, S.J., Hutchinson, L.J. and Zarkower, D. 1989. Canadian J. Vet. Res. **53**: 405-409.

Wiyono, A. and Damayanti, R. 1999. J. Ilmu-Ternak Vet. **4**: 264-272.

CHAPTER 26

Health Barriers to Buffalo Productivity in Different Agro-Climatic Regions and their Amelioration

D. Swarup

Indian Veterinary Research Institute, Izatnagar-243 122 (UP)

India is a vast country with continental geographical attributes. It has 16 agro-climatic zones, and a rich heritage of genetic resources of flora and fauna in form of 150,000 identified living species. There are 45,000 different plant species and 15,000 medicinal plants. A huge livestock population consisting of 209.5 million cattle and 91.8 million buffaloes dwell in the country. Buffalo (*Bubalus bubalis*), is an incredible multi-purpose farm animal providing high quality milk, meat and draught power. Archeological evidences in form of fossils recovered from the gravels of Narmada and the topmost beds of Siwaliks, suggest that the Indian buffalo is the lineal descendant of the gigantic *Bubalus palaeindicus* of the Pliocene. According to Cockrill (1982) all the buffaloes of river group, including Egyptian and Mediterranean breeds, have their origin in the Indian subcontinent. During Harappan civilization the Indus Valley was one of the centres, if not the sole, of domestication of buffalo in India (Randhawa, 1980). Murrah and Nilli Ravi breeds of riverine buffaloes may produce lactation yield of up to 2,000-2,400 kg with 7% butter fat, which is more than double that of cattle (Chantalakhana, 1988). Currently, 56.84% (9.3772 million) of the total world buffalo population dwells in India, contributing nearly 63% (3.9 million MT) of the total world buffalo milk and 46.41% (1.421 million MT) of the total world buffalo meat production. In terms of buffalo production and population, India enjoys the apex place in the world (FAO 2001). It is estimated that in India, 39 million breedable buffaloes contributes nearly 55% of total milk produced in the country. The sturdy animal holds greatest potential for sustainable livestock production, especially in the Indian subcontinent. However, number of factors including chiefly the health barriers, inadequate nutrition and poor reproductive performance limit the production potential of buffalo population in India and other countries.

Health Barriers to Productivity

Production potential of buffaloes is hampered by host of prevailing and emerging diseases causing inimical impact on buffalo health. Most of the diseases that occur in cattle also inflict harmful effects on buffaloes. But, there appears to some degree of difference in patho-physiological and clinico-epidemilogical profile, particularly in relation to severity and manifestations between the two species. For example, diseases such as HS, fasciolosis, ascariosis, filariosis trypanosomiasis, rumen indigestion, diaphragmatic hernia and milk fever are more frequently seen in buffaloes than cattle. Besides these conditions, MD, Acute undifferentiated diarrhea (ADD), Buffalo pox, Ephemeral fever, Infectious bovine rhino-tracheitis, Bovine viral diarrhea, Brucellosis, JD, Tuberculosis, mastitis and ketosis are important diseases causing huge economic losses in the form of qualitative and quantitative drop in milk and meat yield. Outbreak toxic conditions including nitrate/nitrite, cyanide, lead, fluoride pesticides and phyto toxins have also been recorded with increasing frequency in buffaloes limiting bubalian production in India.

Infectious Diseases and their Management

Foot and mouth Disease (FMD): FMD (Aphthus fever) is a highly contagious endemic disease in India, and assumes great economic significance, particularly in crossbred cattle and high yielding buffaloes. Economic losses are attributable to direct losses on account of reduction in milk yield, quality and quantity of meat, work capacity, and growth. Saxena (1994) estimated a milk loss of about 3,508 million liter due to FMD in early 90's. This was about 6.5% of total milk output of the country and valued to about Rs12, 520 million in terms of foreign exchange and Rs 16, 500-18,730 million in terms of domestic economic surplus losts on the basis of 1990 prices. Indirect losses due to FMD occur in form of reduction in fertility, and loss of semen quality in breeding bulls. On per sick animal basis maximum milk loses have been estimated for crossbred cattle followed by buffaloes. Considering both milk and non-milk losses, annual loss per buffalo averaged Rs 250, which was higher than the average per animal loss of Rs 125 occurring due to FMD in 5 categories of cattle and buffalo (Saxena, 1994).

There are conflicting reports on the susceptibility of buffaloes to FMD. Some workers have indicated low susceptibility, whereas others have found that both cattle and buffaloes are equally susceptible. Dutta *et al.* (1983) recorded 70 outbreaks of FMD in buffaloes involving mainly Asia -1, and O types of virus and in some cases type C virus. However, of late type C has not been reported from the country since 1996 (Singh, 2003). In 1990, the annual incidence of disease was 23% (Saxena, 1994). Highest incidence was recorded in indigenous cattle and draught animals (29%) followed by dairy cattle (25%), young animals (23%) and buffaloes (20%). Emergence of deadly duo of FMD and HS during 2001-2002 caused heavy mortality in buffaloes. It was possibly an outcome of unrestricted animal movement and improper immunization. However it has further underlined the significance of FMD as a health barrier to buffalo production.

Control of FMD, depends mainly on prophylactic immunization of susceptible animal population and assumes greater economic importance. It is indicated that control of FMD could lead to at least 5% annual increase in milk production and present level of export of about 80,000 tonnes of buffalo meat could be enhanced by 3-5 times as the same produce would attract better price (Bhat and Taneja, 2001). For effective control, vaccination at regular intervals is the only choice. But there is a need to develop a highly effective vaccine with known types using biotechnological tools and to evolve cold chain delivery system that is effective up to the village level. Further research to evaluate efficacy of FMD vaccine, developed from cattle strains, in buffaloes is also warranted.

Infectious bovine rhinotracheitis (IBR): IBR is usually a mild but highly infectious disease of cattle and buffaloes caused by Bovine Herpes virus-1. It is manifested by fever (40-42°C), drastic fall in milk yield, rhinitis, serous nasal and ocular discharge, salivation, conjunctivitis, dyspnea and abortion. Mortality is low (<1%) but considerable economic losses occur due to abortion. The disease is mostly confined to cattle, but aborting buffaloes have shown high antibody titers to IBR virus (Singh *et al.*, 1983) and 40% of the aborted buffaloes in Punjab tested positive for IBR-IVP virus (Gill *et al.*, 1987).

It is possible to establish an IBR sero-negative herd by careful planning and health management and some countries with low prevalence of the disease have achieved the IBR free status by adopting test and slaughter policy. Monitoring and surveillance should be taken up to find seropositive buffalo. The positive animals have to be eliminated and only seronegative animals are introduced in the herd. A national programme adapted by Switzerland in 1983 was based on annual serological testing of national herd, restriction on the trade with seropositive animals and stepwise elimination of these animals. Priority of eradication was given to breeding animals and by 1987 the breeding stock was virtually free from the disease (Ackerman *et al.*, 1990). This success story can be implemented in India as well. Serological testing of breeding bulls and semen for virus needs to be taken up on priority. Control of IBR by development of immunity following natural exposure and vaccination is an alternative to the eradication programme. Both modified live-virus and inactivated vaccine that can be used intramuscularly or intranasally are available for cattle (Radostits *et al.*, 2000). However, there is a need to find a highly immunogenic vaccine for buffalo.

Acute undifferentiated diarrhoea (ADD) of newborn calves: Perinatal mortality in buffalo calves is a major barrier to buffalo health and production. ADD, which occurs in calves under 30 days of age, is one of the principal causes for it. ADD is a multi-etiological syndrome commonly involving enterotoxigenic *E.coli* (ENTEC), rotavirus, corona virus, *Cryptosporidium* spp, and *Salmonella* spp. The disease is clinically characterized by acute profuse diarrhoea, progressive dehydration, acidosis and death in a few days. The incidence of disease depends upon several interrelated epidemiological determinants or risk factors such as colostral immunity to calves, overcrowding, parity of dam, quality of diet, meteorological factors and general care provided to calves. Low levels of serum immunoglobulin render calves highly susceptible to calf scour and deprivation or inadequate feeding of colostrum to newly born calves is a major contributory factor for it. Studies have shown that mortality among buffalo calves is directly related to the season and in India highest mortality, particularly in 2-4 weeks old calves occur during July to September (Rajya, 1988).

Management of ADD and reducing calf mortality is multi-pronged approach comprising case management by using specific therapy and correction of fluid and electrolyte losses; hygienic calf rearing, adequate feeding of colostrum; reduction of degree of exposure of newborn to infectious agents; protection to newborn during inclement weather and vaccination of the dam or newborn to enhance specific resistance. Survey conducted at IVRI; show that inadequate feeding of colostrum to buffalo calves is a big contributing factor to calve mortality. As such farmers should be educated that colostrum is essential for reducing calf mortality and the new born should get colostrum @ 50ml/kg (1/20th of BW) within first 2 hr of their life. It has also been observed that calf mortality is likely to be lower when the same person feeds calf regularly and preferably. Proper housing and ventilation are important to alleviate stress. Vaccination of pregnant buffaloes for specific pathogens, particularly those causing diarrhoea may be considerably beneficial to reduce calf mortality.

Haemorrhagic septicemia (HS): The recent emergence of deadly duo of FMD and HS in many parts of India has caused much concern to buffalo keeping. Buffaloes are highly susceptible to HS with overall mean case fatality rate nearly 3 times as high as in cattle. Retrospective analysis of disease occurrence in cattle and buffaloes in India during 1974-1986 showed that HS (*Pasturella multocida* infection) was responsible for the highest mortality and second highest morbidity in comparison to other major infections such as RP, FMD, anthrax, and BQ (Dutta *et al.*, 1990). The disease is also prevalent in other countries of Asia including Thailand, Myanmar, Sri Lanka, Philippines, Malaya and Pakistan.

Sudden onset of fever, profuse salivation, submucosal petechiae, depression and death in 50-100% animals within 24 hr are the characteristic symptoms of HS. Warm painful swelling about throat, dewlap, brisket or perineum and severe dyspnea due to obstruction of airways occur in animals surviving beyond this period. The disease can effectively controlled by vaccination using oil adjuvant vaccine. Immunity lasts for about 12 months. However, outbreaks have been recorded following improper vaccination. In case of outbreak, immediate vaccination of all animals in the area using ring method should be taken up irrespective of previous vaccination history. Various types of vaccines viz. double emulsion vaccine and adjuvant bactrim have been developed for more effective control of HS various types of vaccines. Treatment of affected animals at early stage using sulfa drugs and third generation antibiotics is also effective in reducing mortality.

Brucellosis: Brucellosis is widespread and is a major economic concern to dairy cattle and buffaloes in most countries of the world. FAO, OIE and WHO consider it as the most widespread zoonosis all over the world. The infection not only cause abortion late in pregnancy, but also is also subsequently induce high infertility rate in affected females and varying degrees of sterility in males. In India, *Brucella* is reported to be the most important cause of abortion in dairy buffaloes (Das *et al.*, 1990) and 1.8% of the 7,153 buffaloes from 23 states of the country tested positive for Brucella antibodies (Isloor *et al.*, 1997).

Control and eradication programme for brucellosis have several components notably including test and reduction of reservoir of infection, quarantine, deportation, vaccination, education and guidelines. The programme must be indigenous for a given country and area. Cooperation is required at all levels of government beginning from Village Panchayat to Central Government departments.

Identification of brucella positive villages using Milk Ring ELISA followed by screening of individual animals by Rose Bengal Plate test has been suggested to assess the status of disease (Singh, 20030. Vaccination with *Br. abortus* strain-19 live vaccine is a valuable control tool. It protects uninfected animals living in a contaminated environment and can thus be useful to lay groundwork for eradication by gradual disposal of infected animals. Controlled movement of animals from one area to another and production of quality semen from certified bulls are other important aspects of brucellosis control.

Mastitis: Mastitis is an important barrier to buffalo production. It is one of the most important diseases confronting dairy industry in India and elsewhere. Besides making milk unhygienic and unfit for human consumption, mastitis causes tremendous losses in form of reduced milk yield, and cost of discarded milk, treatment and labour. Even in highly developed countries, mastitis incidence is nearly 10%. A recent report from India, involving 10,891 buffaloes, indicated that 12.28% had clinical mastitis (Shinde *et al.,* 2001). Various species of staphylococcus and streptococcus are the major cause of mastitis. Other organisms incriminated with mastitis are *E.coli, Pseudomonas, Klebsiella,* and *Corynebacterium.*

Management of mastitis is a big challenge and is one of the foremost goals of dairymen and veterinarians. As milk is an excellent medium for growth of several pathogens, a number of food-borne pathogens are transmitted to man through contaminated milk. International food laws are in place with global agreement on Sanitary and Phytosanitary (SPS) Regulations and Technical Barriers to Trade (TBT). In the overall management of food quality and safety, the HACCP system and GMPs are the two guiding principles and to follow them, particularly in terms of quality milk production, control of mastitis becomes imperative. Recommended mastitis control programme for an organized dairy herd is aimed at reducing the duration of infection, reducing new infection rate and monitoring the infection rate for assessment of herd's mastitis status. However, most farmers in India maintain small herds and periodic monitoring is not available to them, the best option for mastitis management is its timely treatment. Use of effective antibiotics at early stage, dry cow therapy and provision of diet containing vitamin E and selenium are good options that can be adopted even by the small dairy farmers. Researches are also in progress on evaluation of bacteriocins, and herbal homeopathic drugs and other immunomodulatory agents as an alternative to antibiotic therapy.

Trypanosomiasis and other Haemoprotozoan Infections

Haemoprotozoan diseases: Infections such as Babesia, Theileria and Trypanosoma are widely prevalent in dairy cattle and buffaloes. Trypanosormiasis (Surra) caused by Trypanosoma evansi is of particular significance in terms of buffalo production. The disease increases significantly during rainy season, when the population of transmitting flies is at rise. The economic losses due to disease occur in form of mortality and the cost of treatment. In some countries Surra was responsible for huge financial losses. In Indonesia, it was ranked third most important disease of livestock accounting for estimated more than US dollar 20 million in 1984. Although case fatality rate in cattle and buffaloes is lower than horses and camel, long duration of disease coupled with treatment failure and responsible for significant losses to cattle and buffalo production. The severity of disease is exacerbated by stress from adverse climatic conditions, stress and concurrent diseases. Besides fever, progressive anemia and nervous signs, irregular estrus, abortion and still birth in females and poor quality semen in bulls are also seen in Surra.

Timely detection and treatment is important steps for limiting losses due to trypanosomiasis as there is no vaccine against the disease. Quinapyramine sulfate (curative) and Quinapyramine prosalt (both curatively and prophylactic) are the commonly used drugs for management of Surra. However, treatment is not always effective owing to low tryponocidal activity of the drug against *T. evansi*. Further more, the drugs generally fail to cross blood brain barrier to reach parasite in CSF (Radiostits *et al.*, 2000). The other measures for control of Surra include prophylactic treatment of susceptible animals and their protection from biting flies prior to rainy season.

Helminthosis: Helminthic infections, particularly fasciolosis, amphistomiosis, ascariosis, filariosis and hookworm infestation and parasitic gastro-enteritis cause substantial losses to buffalo production. Buffaloes, owing to their propensity to seek water and waterlogged land, expose themselves more frequently to snail borne helminthic infections. As such, in terms of economic losses, these helminthic infections, such as fasciolosis caused by *F. gigantica,* are undoubtedly, the most important disease with which buffalo industry has to contend. The losses due to fasciolosis are occasioned much less by fatalities than from reduction in milk yield, condemnation of liver, infertility, poor body growth, and higher susceptibility to other infections. Incidence of fasciolosis is more during rainy period and spring months (Swarup and Pachauri 1987).

Approaches to control fasciolosis include strategic deworming as preventive measure to limit incidence in endemic area; reduction of pasture contamination with metacercaria; reduction in snail population by chemical, biological and mechanical methods. At least 6 flukicidal compounds *viz.* triclabendazole, albendazole, oxyclozanide, closantel, clorsulon, nitroxynil are available that can be used therapeutically for treating the disease or prophylactically for preventing outbreaks. Closantel and clorsulan bind to plasma proteins and erythrocytic membrane, respectively and thus provide extended period of protection (Radostits *et al.*, 2000). Pasture contamination with metacercariae can also be reduced by deworming the animals, just prior and after the monsoon, to eliminate adult worms from the bile ducts. Chemical control of snail population by spraying pasture with copper sulfate and sodium pentachlorophenate has been widely used in western countries. However, it may not be very much fruitful in counties like India, where pastureland is not well defined and no systematic grazing is practiced. Further, these inorganic chemicals may be hazardous to human health. Duck rearing has also been proposed as an alternative method for eco-friendly snail control, but its effeteness has not been tested so far. More selective newer molluscicides such as n-trityl morpholine have recently been introduced. Vaccines against *Fasciola* are also under development and one, which has been developed from fluke, derived catepsin-L-protinase and haemoglobin molecule, has given 72% protection against *F. hepatica.*

Ascariosis is another helminthic infection having great economic impact on buffalo production. The disease caused by *Toxocara vitulorum* is responsible for heavy neonatal calf loss. The life cycle of *T. vitulorum* is more complex than *Ascaris.suum* and the infection is also transferred through colostrums. The incidence of infection is relatively higher in buffalo calves and many times the entire lot of newborn may be affected by the parasitism. The clinical findings include poor coat, diarrhoea, constipation followed by diarrhoea, anorexia, anemia, pot-bellied condition, stunted growth and occasionally colic.

Management of ascariasis is based on hygienic measurements and anthelmintic treatment to new born at the age of 10-16 days. Pyrental is the drug of choice for buffalo calves and can be administered orally @ 250 mg for a calf. Levamisol, febental, oxfendazole, fenbendazole and piperazine are also effective against ascariosis in buffalo.

Non-Infectious Conditions and their Amelioration

Metabolic and deficiency diseases: The metabolic diseases such as milk fever, ketosis and hypomagnesaemia and micro-mineral deficiency due to copper, selenium, zinc, etc. cause significant production loss to buffalo industry. Metabolic diseases are more important in dairy cows and buffaloes with higher incidences in high yielder and in animals fed inadequate diet. Milk fever occurs most commonly in high producing buffaloes in 3rd parity or older. The case fatality if low, but drastic reduction in milk yield and culmination of non-responsive cases into Downers cow syndrome results in substantial financial losses. Disease occurs around 48 hr of parturition and manifested clinically by anorexia, ruminal atony, inactivity, circulatory collapse, dry muzzle, hypothermia, lateral recumbency and tachycardia. Untreated animals may die within few to several hours.

Calcium borogluconate intravenously along with oral calcium chloride is the treatment of choice. If given at proper dose and time the response to therapy is very high. Insufflations of udder with air may be practiced in animals, which do not recover completely to repeated calcium injection. Dietary management during pre-partum period, calcium gel dosing using 49% calcium chloride 12 hour before parturition, and administration of vitamin D and its metabolites are recommended procedure for prophylaxis of milk fever.

Ketosis, particularly sub-clinical form is an important disease of high yielding dairy cows and buffaloes. Impaired glucose metabolism due to improper energy supply leads to excess production of ketone bodies and onset of clinical or sub-clinical form of disease manifested by sudden drop in milk yield, anorexia and wasting condition. Ketosis is reported to occur more frequently during 3rd lactation and in buffaloes of 9-10 years of age. Intravenous injection of 50% dextrose and together with subcutaneous insulin (0.5 iu/kg b.wt.) and oral feeding of jaggery achieved 100% cure in clinical cases of ketosis in buffaloes (Ahuja, 2003).

Toxic conditions: Toxicity due to fluoride, lead, cyanide, nitrate and nitrites and pesticides has significant bearing on buffalo health in India. Outbreaks of lead toxicosis, fluorosis, pesticide and nitrate and nitrite toxicity have been noted from different parts of the country, particularly due to pollution of environment and indiscriminate use of agro-chemicals. Fluorosis, which occurs due to ingestion of constant or increasing amount of fluoride / fluorine by animals, is of particular interest. It is stated that of all the pollutants, fluoride pollution has caused much damage to dairy animals. In India incidences of fluorosis in buffaloes have been encountered due to excess fluoride in natural water in Unnao (UP), phosphate fertilizer industry in Udaipur (Rajasthan), and due to pollution from brick work in Ghaziabad (UP). The affected animals suffer from inappetance, reduced milk yield, skeletal deformity and lameness. They become prone to infectious diseases (Swarup and Dwivedi, 2002). Outbreaks of lead toxicosis, causing heavy mortality have been reported in buffalo population in Punjab, Maharasthra, Rajasthan and Delhi. In one of this recurrence of lead poisoning due to pollution has caused much serious impact leading to drastic depletion of buffalo population in some villages.

Amelioration of toxicity depends on removal of the source of poison and restricting ingestion of contaminated forage and water by animals. There are no effective ways to cure fluorosis once the skeletal signs appear. However lead toxicities can be treated effectively by giving Ca- EDTA together with thiamine (Swarup and Upadhyaya, 1991). Other toxic conditions can be treated by using specific antidotal therapy.

References

Ackerman, M, Weber, HP and Wyler, R (1990). Aspects of infectious bovine rhinotracheitis eradication programmes in fattening cattle. *Preventive Veterinary Medicine* **9**: *121*

Ahuja, A (20003). Ketosis in buffalo *(Bubaus bubalsi)-* clinical case report. *Intas polivet* **4(2)**: *292*

Chantalakhana, C (1988). Role of buffalo in rural economy in the world.). 2nd Buffalo Congress-1988. Invited Papers and Special Lectures. *Proc. Vol-II (Pt.I): 111.*

Cockrill, WR (1982). The water buffalo: A review. *British Veterinary Journal* **137**: *8.*

Das, VM, Paranjape, VL and Corbel, RC (1987). Investigation of brucellosis associated abortion in dairy buffaloes and cows in Bombay. *Indian Journal of Animal Sciences* **60**: *1193.*

Dutta, J, Rathore, BS, Mullick, SG, Singh, R and Sharma, GC (1990). Epidemiological studies on occurrence of haemorrhagic septicemia in India. *Indian Veterinary Journal* **67**: *893.*

FAO (2001). FAO Database, 2001. http://www.apps.fao.org

Gill, BS, Sharma, DR, Kwatra, MS, Kumar, A and Gill, JS (1987). Seroprevalence of infectious bovine rhinotracheitis in the Punjab state. *Journal of Research- PAU* **24**: *317.*

Isloor, S, Renukaradhya, GJ and Rajshekhar, M (199&). A serological survey of bovine brucellosis in India. *Revue Scientifique et Technique- Office Interantioal des Epizooties* **17**: *781.*

Radostits, OM, Gay, CC, Blood, DC and Hinchcliff, KW (2000). *Veterinary Medicine.* 9th edn. WB Saunders

Mohanty, PK and Rai, A. (1988). Immunogenicity of cell culture vaccines against buffalo pox. 2nd Buffalo Congress-1988. India *Proc. Vol IV pp 1-4.*

Rajya, BS (1988). Major constraint in buffalo improvement: A need for coordinated approach in combating diarroheal disease (CDD). 2nd Buffalo Congress-1988. India. Invited Papers and Special Lectures. *Proc. Vol-II(Pt.II): 399.*

Randhawa, M.S.A *History of Agriculture in India. Vol I.* Indian Council of Agricultural Research, New Delhi.

Saxena, R (1994). *Working Paper.* Institute of Rural Management, Anand. *No 57: 27pp; 60: 20pp; 62: 15pp*

Shinde, SS, Kulkarni, GB, Gangane, GR and Degloorkar, NM (2001). Incidence of mastitis in buffaloes in Parbhani district, Maharashtra. *Mastitis* **2:** 35.

Singh, BK, Ramakant and Tongaokar,SS (1983). Adaptation of infectious bovine rhinotracheitis in Madindarvy bovine kidney cell line and testing of buffalo sera for neutralizing antibodies. *Indian Journal of Comparative Microbiology Immunology and Infectious Diseases* **4:** 6

Singh, D.K. (2003). Prevention and control of infectious diseases of buffaloes. *Intas polivet* **4(2):** *292*

Swarup D and Dwivedi, SK (2002). *Environmental Pollution and Effects of Lead and Fluoride on Animal Health.* Indian Council of Agricultural Research, New Delhi.

Swarup D and Pachauri SP (1987). Epidemiological studies on fascioliasis due to *Fasciola gigantica* in buffalo in India. *Buffalo Bulletin* **6:** 4

Swarup D and Upadhyay, AK (1991). Chemoprophylactic efficacy of thiamine hydrochloride in experimental lead toxicosis in Calves. *Indian Journal of Animal Sciences.* **61:** 1170.

CHAPTER 27

Feeding Management of Buffaloes with Special Emphasis on Body Condition Score

Shiv Prasad

Dairy Cattle Breeding Division, NDRI, Karnal - 132 001 (Haryana)

The management of the dairy farming enterprise is the pivot around which the sustainability and the profitability of the dairy farming business revolve. The efficient management of the animals essentially involves the two aspects namely (a) improving the breed and breeding efficiency of the animals and (b) providing optimum environment to the animals to express their productivity potential optimally by strategic application of modern technological interventions in feeding, housing and preventive health care (optimum management strategy). The optimum management strategy would include adoption of proper breeding plans, provide them optimum feeding, housing and health care so that they are able to realize their genetic potential optimally. In addition, the available management tools must also be used strategically and optimally to provide optimum environmental conditions to improve the productive performance of the animals.

A large majority of dairy animals in our country are grossly under nourished/mal nourished, as a result their growth, milk production, reproduction and health is adversely affected. As a result of generations of under nourishment the general body size of the animal is also coming down. To overcome this problem it is essential to manage the feeding regime optimally. Whereas under feeding will not allow the true potential of the animal to be realised, the over feeding will effect the economic efficiency of the enterprise. Over conditioned animals apart from lowering the efficiency of production also make them vulnerable for a number of diseases and metabolic disorders. Most common among these problems is fatty liver syndrome, which mainly results from the excessive mobilisation of fatty tissue and subsequent infiltration and deposition of this excessive fat in liver. This problem is commonly observed in those cows/ buffaloes, which are over conditioned at calving and are subsequently subjected to poor nutritional regime post calving or that remain off feed due to complications arising at the time of calving.

It is therefore utmost necessary that the feeding management of buffaloes is regulated efficiently at every stage of production cycle. Close monitoring of body condition provides valuable guidelines in this regard.

Feeding Management During Dry Period

The dry period is a time of rest and rejuvenation for the animal but it is also a time when important changes will take place that will affect future production levels. The mammary tissue involutes, regenerates, and prepares itself for colostrum production and a subsequent lactation. Foetal growth intensifies even at the expense of body maintenance of the animal. Feeding programmes should be designed to maintain body condition, even if the animal is over conditioned. Thin animals should only be allowed to gain not more than 35-40 kg during the dry period. Over conditioned buffaloes should not be allowed to lose or gain body condition as this may lead to fatty liver syndrome and or ketosis.

Feeding Management of Early Lactation

The best approach of feeding during early lactation would be to feed and manage them separately. This ration should be low in fat in order to stimulate rumen microbial protein. One of the greatest problems facing the early lactation animal is energy intake. Increasing concentrates rapidly after calving is necessary but care should be taken to insure adequate fiber intake so as to prevent rumen acidosis. Feeding high levels of crude protein at this stage could prove beneficial in stimulating higher dry matter intake, less post calving weight loss, and higher peak milk production. Another advantage of a separate fresh buffaloes group is that it allows for better observation of the health of the fresh buffaloes.

A major goal of proper nutrition during the dry period and early lactation period is to prevent metabolic problems during the peripartum period. By adhering to sound principles of nutrients management during the peripartum period, the risk of most metabolic diseases is minimized.

The feeding and management of dairy buffaloes during the entire period of lactation is vital for harvesting the optimum milk production from the dairy animal. Proper management of the dairy animal during first few days after calving and during early lactation is of particular importance. The following management principles must be observed:

◉ Feeding management during early postpartum must focus on attaining higher peak milk production and better persistency. This could be achieved by:

    i. Feeding the animal with higher energy diets and

    ii. Maximising dry matter intake

◉ Monitor the body weight and condition regularly during early stages of lactation. It must be ensured that the animal does not lose excessive condition during this phase as this may result in fat infiltration of liver also called 'fatty liver syndrome'

◉ Attempt must be made to return the animal in the positive energy balance soon as the long phase of negative energy balance results in poor persistency of milk production and lower reproductive efficiency. This can also be achieved by improving the quality as well as quantity of the feed.

◉ After the peak milk production has been achieved the feeding must be based on the level of milk production.

◉ The milk is most economically produced from fodders. All attempts must therefore be made to ensure supply of green fodders/silage/hay round the year. Concentrates should also be supplemented whenever necessary and depending on the level of production.

- In case of dairy buffaloes producing higher quantities of milk (>16-17 litres/days), no suitable combination of concentrates and fodders (even at high intake levels) can sustain this level of production without the mobilisation of body reserves. Such buffaloes can also be supplemented with oils/fats in their diets at 300 g per day level.

- Moderate levels of milk can be sustained on a suitable combination of green and dry fodders supplemented with desired amounts of concentrates. While feeding a mixture of straw and green fodders, it will be desirable if 1 Kg of straw is mixed with every 4-5 Kg of chaffed green fodder for each 100-Kg body weight.

- If plenty of quality green fodder is not available and the ration is based on low quality straws/stovers then additional concentrate feeding is required.

- For optimum results the protein requirement of total ration should be adjusted at 13-14 per cent level. Leguminous fodder (like berseem, lucerne) contain about 12-14 per cent crude protein, non-leguminous fodder (like maize, sorghum, oats and grasses etc) contain about 7-8 per cent protein. Straws like wheat and paddy straws contain only 3-4 per cent crude protein. The crude protein content of the concentrate mixture should be adjusted to provide about 13-14 per cent crude protein in total ration.

- Grain portion of concentrates should be crushed else part of it may pass off undigested in the faeces. It is desirable to moisten the concentrate mixture and mix it with straws before feeding. Plentiful availability of clean drinking water must also be ensured.

- A suitable combination of berseem along with oats, maize, wheat/paddy straw and concentrates (based on the level of production) is most practical strategy of feeding dairy buffaloes during winters. The respective dry matter and crude protein contents of the above feeds are Berseem (12 and 14%), Oats (15 and 10 %), Maize (16 and 10%), Straws (90 and 4%) and Concentrates (90 and 20%) respectively. Cultivation of improved varieties of fodder crops have potential not only to improve the yield of the fodder but prolong the availability period also. The important varieties in the category are fodder Maize (African Tall, Vijay Composite), Berseem (BL 10, BL 22) and Oats (OS 6, OL 9).

⊙ If green fodder is not enough and you have to depend on straws, you can improve the quality of these straws by treating them with urea under expert guidance. By so doing you can improve the milk production of their animals by about 15 per cent and ensure a saving of about 1 kg concentrate per animal per day.

Feeding Based on Body Condition

Body condition gives as indication of how the animal has been fed over the preceding weeks/ months. The level of milk production of the dairy animal also affects it. The body condition scoring is almost a routine practice in western countries and there are some good scales available to do this in those countries. A practical scale for scoring the body condition of dairy animals suitable both for cows was developed by Prasad (1994). Later this method was tried in buffaloes as well and was found quite satisfactory (Prasad and Tomer 1996). To assess the body condition of dairy buffaloes objectively the following simplified procedure can be used with good results:

A simple body condition scoring scale suitable for buffaloes

Score	Evaluation	Description of Animal
1.	Very poor	Animal emaciated, deep cavities under tail head, no muscle cover between pelvis and skin all the bones very prominent.
2.	Poor	Animal appears weak, cavity under tail head marked, deep depression is loin area, some muscle mass evident all over the body.
3.	Moderate	Shallow cavity in tail head region seen, slight fatty tissue also evident, Pelvis can be felt easily, depression in loin region still evident. Ends of transverse

processes of lumber region can be palpated with some pressure.

4.	Good	Fatty tissue can be felt all over the animal body (chine, loin and rump region). Skin appears smooth but pelvis can be felt. Ends of transverse processes can be felt but thick layer of tissue in the region evident. Only a slight depression evident in loin area.
5.	Fat	Folds of fatty tissue present all over, pelvis felt only with firm pressure. Transverse processes difficult to palpate. No depression in the loin area visible.
6.	Very fat	Tail head buried in fat tissue, skin distended, Pelvis cannot be felt even with firm pressure, fat accumulation over transverse processes evidence. Bony structure not palpable.

Recommendation

- ◉ Monitoring condition score over the lactation cycle is very useful in focusing on management and nutritional areas for further improvement. Optimum body condition the maximum milk production may be higher than that for optimum milk production (4-4.5 v/s 3-3.5).

- ◉ Feed conversion efficiency of a fat animal is lower.

- ◉ Excessive loss of body condition (> 1.0 point) after calving is more harmful than higher condition itself. The owner must ensure that the animal does not lose body weight excessively after parturition. This has adverse effect on milk production, reproductive and health of the animal.

- ◉ The body condition of the animal should not go below 3.0

even during peak lactation. This should be ensured in high yielders by:

1. Providing adequate high quality feed.

2. Maximise DM intake.

⊙ The optimum body condition at calving would be around 4-4.5. Animals fattier than this at calving are likely to develop fatty liver syndrome during early postpartum.

References

Prasad, S.(1994). Body Condition Scoring and Feeding Management in Relation to Milk Yield, Feed Efficiency and Incidence of Diseases in Crossbred Dairy Cows. Ph.D. thesis submitted to NDRI Deemed University, Karnal.

Prasad, S. and Tomer, O.S. (1996). Effect of body condition class on important physical and production parameters of Murrah buffaloes. Indian J. Dairy Science **49:** 189-194.

CHAPTER 28

Role of Buffaloes as Draught Animal in Different Agro-Climatic Regions

S.V. Singh and R.C. Upadhyay

National Dairy Research Institute, Karnal -132 001 (Haryana)

Buffaloes (*Bubalus bubalis*) have been categorized mainly in two types viz. riverine and swamp depending upon variation in their habitat and genetic makeup. The body size of river buffaloe is generally large and they are found in India, Pakistan and some of the west Asian countries. The females produce moderate quantities of milk and males are good for draught and meat production. These buffaloes are used as a source of farm power in India, Indonesia, Philippines, Taiwan, China, Vietnam, Sri Lanka, Thailand, Burma, Bangladesh, North Eastern Africa and other countries (Cockrill, 1974). Swamp buffaloes found mainly in South East Asia and some parts of North Eastern States of India, are primarily used as draught animal for ploughing and puddling, transport, pulling logs, sledge, cart etc.

India possess the best buffalo breeds with high potential for milk production viz. Murrah, Nilli Ravi, Surti and Jaffarabadi, with their origin in Northwestern states of India. There are several other breeds in India, which have their regional importance and raise the economic value of farming community e.g. Bhadawari, and Tarai in Uttar Pradesh, Nagpuri and Pandharpuri in Maharasthra, Parlakhemundi, Manda, Jerangi, Kalahandi, Sambalpuri in Orissa and Andhra Pradesh; Toda and South Kanara in Tamil Nadu and Kerala. Mehsana breed has been developed by grading up of Surti buffaloes and Murrah in Mehsana district of Gujarat. Gogavari breed is a crossbred of local nondescript buffaloes with Murrah.

Livestock Population Dynamics

India has one of the largest livestock populations in the world, accounting for about 57% of the world buffalo population and 16% of cattle population (GOI, 2002). Between 1951 and 2002, the cattle population increased from 155.3 millions to 175.1 millions, representing growth less than 1% per year. However, the buffalo stock witnessed a significant acceleration in growth during 1977 and 1982 compared to previous years. The turning point in the composition of draught animal population was 1977 when male cattle population declined from 73.22 million to 61.14 million (12.08 millions) between 1977 and 1982, and the corresponding decline among male buffalo population was over 1.96 millions (GOI, 1999). The declining trend however is not uniform across the different states. Agriculturally advanced states such as Punjab, Haryana, Andhra Pradesh, Kerala and Tamil Nadu witnessed a decline in male draft animal population due to mechanization, while the less progressive and hilly states such as Assam, Bihar, Madhya Pradesh, Orissa and West Bengal remained dependent on work animals.

Land Holding Pattern and Energy Requirement:

Indian agriculture is characterized by small fragmented land, hill farming, shifting cultivation, Tal and Diara land (waterlogged land), suitable for cultivation by human and animal power only. There are about 106 million operational holdings possessing 165 million ha land. The average size of farm holdings in India is 1.57 ha with 79% farm holding having farm size of 1.34 ha only, which is further divided into 2-6 parcels.

Intensive cultivation as a result of introduction of high yielding varieties in mid 1960's after green revolution and increased cropping intensity (1.33%) required higher energy inputs and better management practices. In order to complete timely field operations, agro-processing and farm transport require additional farm energy. Draught animals are the major source of motive power (tractive and rotary) for majority of farmers. Tillage, irrigation and threshing operations are arduous to perform, and therefore are being replaced gradually by mechanical power (Singh *et al.*, 1977 and Singh, 1999), however, other field operations continued to be performed by draught animals. In states like Punjab and Haryana, where mechanization occurred at a faster pace, use of draught animals declined. However, the working buffaloes could not be replaced due to their significance in normal day-to-day transport system.

The buffaloes of different agro- climatic regions, vary in their body size and their draught characteristics. The different buffalo breeds, their characteristics and their habitat are presented in Table-1.

Table-1: Buffalo Distribution and Habitat

Buffalo breed	Draught characteristics	Habitat
Murrah	Medium draft, good for transport	Haryana and U.P.
Bhadawari	Can stand heat better during field operation	Agra, Etawah, and Gwalior regions
Jaffarabhadi	Good for heavy draught	Gir forest
Nili Ravi	Good for heavy draught	Valleys of Sutluj and Ravi in Punjab
Nagpuri	Slow in movement but good for heavy draft	Nagpuri Region

Spatial Distribution and Growth of Draught Buffaloes

Male buffaloes are mainly used for farm produce, transport and field operations but in eastern and southern regions where low temperature and high humidity conditions prevails, they are used for field operations. The population of working buffaloes increased in most of the states of India, which accounted for about 84.74% of the total draught buffaloes, except in Punjab, Tamil Nadu, Andhra Pradesh, Rajasthan and Maharashtra. Male buffalo population and distribution is being presented for different regions (Table-2). Over all, draught buffaloes registered a negative growth of 0.16% per annum.

Table-2: Distribution and growth of male buffaloes in different regions.

(In millions)

States	1972	1977	1982	1987	1992	Growth %
Northern region	1.95	1.17	2.05	2.19	2.31	0.85
Himachal Pradesh	0.01	0.01	0.01	0.03	0.00	-
Uttar Pradesh	1.58	1.80	1.83	2.03	1.90	0.93
Jammu & Kashmir	0.05	0.04	0.04	0.04	0.06	0.91
Punjab	0.26	0.23	0.10	0.07	0.13	-3.40
Haryana	0.06	0.08	0.11	0.12	0.22	6.70
Eastern region	1.98	2.10	2.00	1.85	2.34	0.84
Manipur	0.01	0.01	0.04	0.03	0.03	5.65
Bihar	0.70	0.77	0.75	0.70	0.87	1.09
Orissa	0.58	0.59	0.53	0.32	0.61	0.25
West Bengal	0.50	0.50	0.53	0.63	0.60	0.92
Assam	0.19	0.23	0.16	0.17	0.23	0.96
Southern region	2.08	1.93	1.63	1.15	0.98	-3.69
Tamil Nadu	0.33	0.27	0.26	0.25	0.02	-13.08
Andhra Pradesh	1.26	1.18	1.02	0.66	0.61	-3.56
Karnataka	0.27	0.27	0.25	0.20	0.27	-
Kerala	0.22	0.22	0.09	0.04	0.08	-4.93
Western region	1.57	1.67	1.53	1.49	1.71	0.43
Madhya Pradesh	1.12	1.12	1.11	1.06	1.31	0.79
Gujarat	0.03	0.02	0.03	0.02	0.04	1.45
Rajasthan	0.13	0.17	0.10	0.10	0.08	-2.40
Maharashtra	0.30	0.30	0.30	0.30	0.28	-0.34
All India	7.58	7.93	7.32	6.78	7.34	-0.16

Buffalo Development in Different Agro-climatic Regions

The climatic conditions, soil, cropping pattern and other topographical conditions in different parts of India are different, and these conditions affect the type of animals. As per the Planning Commission, Government of India, there are 15 agro climatic regions, however some of these have been clubbed together for the purpose presenting the distribution of working buffaloes.

Temperate Himalayan Region

This region experiences heavy rainfall in most parts, snowfall during winter and flood during summer. The temperate Himalayan regions are distributed from mountainous area and terai of Himalaya from Kashmir to Arunachal Pradesh. In this region, lightweight buffaloes are found which are used for ploughing and carting. Area has a shortage of feeds and fodders

Dry Northern Region

Hot tropical region including arid zones consisting Punjab, Delhi, Haryana, Eastern Rajasthan, Western U.P; M.P; and Gujarat is the home tract of well defined breed of buffaloes i.e. Murrah, Nili Ravi, Mehsana, Zaffarabadi, Surti and Nagpuri. Males are mainly used for different types of farm operations i.e. carting, ploughing, inter-cultivation and other farm and farm allied operations.

Wet Western Region

Part of eastern U.P; Bihar, W.B; lower Assam, Part of Meghalaya, Tripura, Mizoram, Manipur, Nagaland, Orissa and part of A.P. comes under this particular region. Paddy is the main cereal crop. Mostly entire male buffaloes and in some part of the region infertile / low producing female buffaloes are also used for agricultural operations. The buffaloes are mostly nondescript type and are low producing.

Southern Region

Lower part of M.P; A.P; Maharashtra, T.N; and Karnataka comes under this region. The climate is generally dry and rainfall is low. Mainly nondescript buffaloes are found and are the main draught animals for field operations and other adjunct work.

Coastal Region

The east and west border of Southern India are known as coastal region. Rainfall is about 250-500 cms/year. Local animals are of non-descript type and majority of them are unproductive and poor in health. However, in coastal Andhra Pradesh large number of good Murrah buffaloes are found.

Draught Buffalo: Utilization

Farmers use draught buffaloes for field operations such as tillage, weeding, interculture, digging and threshing besides transport in different agro-climatic regions. These animals are also used for secondary operations such as water lifting, sugarcane crushing and oil extraction. Generally males are used for draught operations; however, females are also used for draught work due to economic considerations. Females contribute only about 6.25% to the total DAP (Sharma, 1982). The decline in number of draught animals some times force use of females for timely completion of farm activities, Therefore, there is a need to review the objectives and priorities of research and exploring the possibilities of uses of alternative animals like; females (low producing and unfertile) for work as they contribute in terms of work at farm, reproduce, give milk, calf, dung and draught power(Dolbert, 1981; Mathewman *et al.*, 1988)

Why Female Buffaloes for Work?

In traditional agricultural system a major constraint to production and draught animal use is now the small size of land holding. The man: land ratio of the Asia and Pacific region is lowest and decreasing further. Under conditions of small farm size, unproductive female buffaloes have an important role to play as a means of increasing the efficiency of the use of resources (Singh and Upadhyay, 1997). The opinion for the use of female for agriculture work also came from Mahatma Gandhi who emphasized that religious scruples needed not prevent the use of females for work. Data indicates a constant increase in the number (0.37 to 1.03 millions) of working buffaloes from 1971 to 1997 in different agro climatic regions of India. In this direction several studies have been conducted at NDRI, Karnal to evaluate and increase the working efficiency of she buffaloes. Working buffaloes travel at an average speed of about 3.0km/hr and develop 0.49 hp during winter and 0.45 hp during summer while ploughing; the respective area ploughed was 0.224 ha/6hr during summer and 0.245 ha/6 hr during winter season (Singh and Upadhyay, 1997).

To improve the working efficiency of female buffaloes work rest cycle was developed. The work: rest: work schedule during a day could be 2:1:2 or 2:2:2 hours during summer and hot humid season (Singh and Upadhyay, 1997). In comfortable environments, animals may be able to work for longer durations. If animals are distressed or environmental conditions are impinging, animals will have to be given frequent rest pauses.

Draught Power and Speed of Draught Animals

The work output of working animals depends on the breed, physical condition, harnessing device, loading characteristics, environmental conditions, feeds and feeding methods. This can be assessed by measuring the draught, speed and physiological responses of animals. Studies have been conducted on draughtability of animals, speed and power developed are presented in Table-3. It was observed that speed and draft power both increased as the body weight increased.

Table-3: Draught and speed in relation to body weight of animals.

Body weight of animals (Kg)	Average speed (Km/hr)	Average draft (Kg)	Power (kW)
200-300	1.54+0.07	26.41+4.54	0.15
300-400	2.09+0.49	34.56+10.15	0.27
400-500	2.46+0.69	42.57+9.40	0.39
>500	2.37+0.40	40.07+11.55	0.35

Draughtability Characteristics and Fatigue

The fatigue levels of animals have been defined by quantitative parameters (body temperature, pulse rate and speed) and qualitative systems such as frothing, uncoordination of legs, excitement, inhibition of progressive movements and tongue protrusion. Based on these parameters Upadhyay and Madan (1985) developed a score card to assess the draughtability of animals. For oxen and buffaloes the maximum points were assigned 40 and for camel and donkey 30 and 32 respectively. The animal is considered fatigued at 50% of maximum score points (Table-4).

Table-4: Fatigue scorecard for working buffaloes.

Parameters	Score scale					Total
	1	2	3	4	5	
Respiration rate/min	$*R_0+15$	R_0+30	R_0+45	R_0+60	R_0+75	5
Heart rate / min	$*H_0+10$	H_0+20	H_0+30	H_0+40	H_0+50	5
Rectal Temp.(°C)	$*T_0+0.5$	$T_0+1.0$	$T_0+1.5$	$T_0+2.0$	$T_0+2.5$	5
Frothing	First appearance	Dribbling of saliva start	Continuous dribbling	Froth on upper lips	Full mouth frothing	5
Leg Uncoordination	Strides uneven	Occasional dragging of feet	Movement of legs uncoordinated and frequent dragging of feet	No coordination in fore and hind legs	Unable to move because of uncoordination	5
Excitement	Composed	Disturbed	Nostrils dilated and bad temperament	Movement of eye ball prominent with excitement	Furious and trying to stop	5
Inhibition of progressive movement	Brisk	Free movement	Slow walking	Very slow	Stop walking	5
Tongue protrusion	Mouth closed	Occasional opening of mouth	Frequent appearance of tongue	Continuous protrusion of tongue	Tongue fully out	5

$*R_0$, $*H_0$ and $*T_0$ represent respiration rate, heart rate and rectal temperature, respectively (Upadhyay and Madan, 1985).

The safe response values without getting the animal fatigue as reported by Upadhyay and Madan (1985) were R_0+30 H_0 +20 and T_0+1.0 in buffaloes. These measurable parameters contribute 6 points from the total 16 safe score points and subjective parameters contribute 10 points, more than quantifiable parameters. The tolerance limits for different animals have been investigated and reported by Srivastava and Ojha (1987), Srivastava (1987 and 1993) and Upadhyay (1993). The limiting physiological response values due to working load (Respiration rate, body temperature, heart rate) are given in Table-5.

Table-5: The range of fatigue parameters of draught animals due to load.

Animal	Respiration rate/min	Heart rate / min	Rectal Temp. (°C)
Buffaloes	R_0+(40-50)	H_0+(10-33)	T_0+(1.8-3.2)
Bull	R_0+(15-75)	H_0+(10-50)	T_0+(0.5-2.5)
Camel	R_0+(4-8)	H_0+(12-18)	T_0+(0.7-1.7)
Donkey	R_0+(15-50)	H_0+(15-45)	T_0+(1.0-3.0)

Source: Annual report of AICRPs on utilization of Animal energy. CIAE, Bhopal.

The buffaloes are stressed to variable levels due to adverse environmental conditions. The responses of draught buffaloes have been presented in table without load during the daytime in different seasons.

Table-6: Variations of physiological responses of working buffaloes without load during different seasons.

Animal	Respiration rate/min	Heart rate / min	Rectal Temp.(°C)
Summer	22-38	40-60	37.5-39.2
Winter	22-33	40-50	35.5-37.8
Hot & hot humid	22-35	40-60	35.5-37.7

Source: Annual report of AICRPs on utilization of Animal energy. CIAE, Bhopal.

Performance Limiting Factors

The animal, equipment, soil and environmental conditions limit the work output and improvement either by selection and modifications in soil texture and by modifying the working schedule of animal, working capacity could be increased (Fig.1).

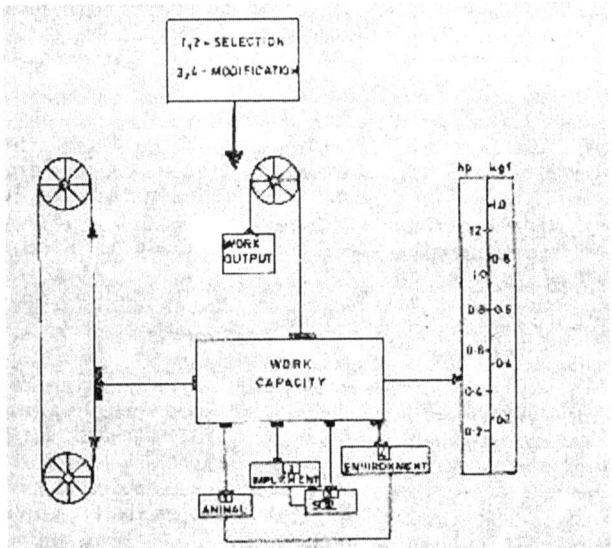

Fig. 1. Factor'saffecting work capacity

Structure and dimension of animal

Heavy animals have more pulling capacity than light weight animals (Goe, 1983, Upadhyay, 1987). Therefore large body size of buffaloes are best suited for heavy work like load pulling. The environment and genetic interaction become limiting for performance. Performance of the working buffaloes may be achieved by proper training, feeding and management (Goe, 1983, Upadhyay, 1984 and Pearson and Lawrence, 1987). Animals have to adjust to particular climatic condition and to work during summer as the adaptation of animals is two fold i.e. for work and climate.

Physiological Factors

Extreme cold and heat utilize available body energy for heat production and loosing respectively. During winter, metabolism does not increase to the same level as during summer conditions. The heat of work is used to maintain the body temperature instead of causing of stress. Depending upon the level of work, muscle metabolism increase several folds, therefore adjustments are made in the uptake and transport of oxygen by the blood and its utilization in muscle tissues. Muscular activity for longer duration depends on the oxygen supply and may be a limiting factor for performance. Increased muscular activity produce greater heat in muscles, which needs to be dissipated by different thermoregulatory processes to sustain work by maintaining body temperature (Nangia *et al.*, 1988, Upadhyay, 1987, Singh and Upadhyay, 1997).

Cardiovascular Function

During work, the cardiovascular activity increases in mammals (Nodel, 1980, Bell *et al.*, 1983, Upadhyay and Madan, 1986, Upadhyay, 1988 and Singh and Upadhyay, 1999). Cardiac activity increases due to both intensity and speed (Singh and Upadhyay, 1997 and Parks and Manohar, 1983). The pulmonary system supplies oxygen to the working muscle. Frequency and depth of respiration change to meet oxygen demands of work. Pulmonary ventilation increases to only a limited extent during work and which limit work over prolonged hour in buffaloes. The reasons relate to inefficiency of gas exchange and respiratory muscle fatigue as a result of inadequate perfusion when respiratory muscle compete for blood flow with working muscle and thermoregulatory (Rowell, 1974).

Animal Fatigue Under Environmental Stress

The working buffaloes get fatigued due to work loads and adverse environmental conditions. The draught capacity of buffaloes varies depending upon the body size and weight from 8-15%. Higher speed of work reduces the tractive efforts, decrease work output and induces fatigue early (Fig.2). The sustained capacity of buffaloes depends on Animal sex, age, body conformation, size, dimension, work output, environmental condition, physiological factors and types of implements. Buffaloes in general are able to work for 6-8 hours continuously at light work, 4-6 hours at sub-maximal work and 3-4 hours at heavy work. During summer direct exposure to solar radiation

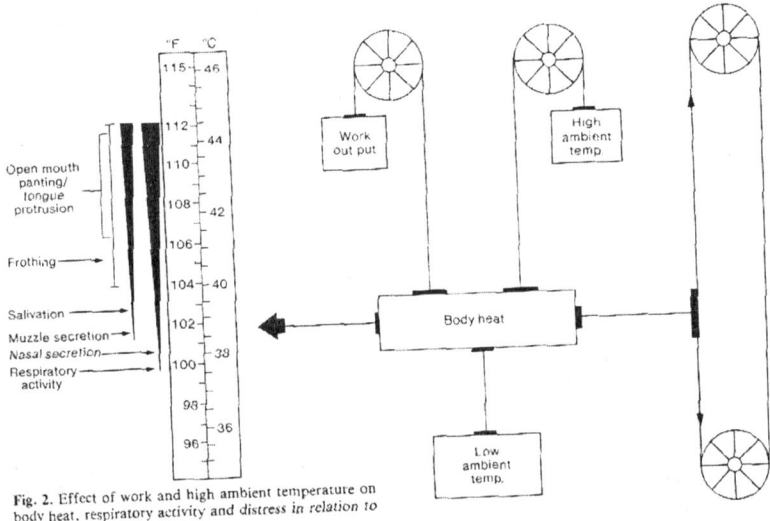

Fig. 2. Effect of work and high ambient temperature on body heat, respiratory activity and distress in relation to

Fig. 2. Effect of work and high ambinet temperature on body heat, respiratory activity and idstress in relation to temperature.

reduces work output and onset of fatigue is in about 3 hours at sub-maximal work.

The load carrying capacity of a buffalo is 1.5-2.0 ton in a pneumatic tyred bullock cart i.e. about 3-4 times of their body weight. These loads can be pulled for 2-3 hours continuously and 6-8 hours in a day during winter (Kapoor, 1985) and 5-6 hours in summer with rest pauses in between, however heavy loads are pulled by buffaloes to an extent of 4-5 ton at a slow speed of travel.

The heat load during work in summer increases (Nangia *et al.*, 1980; and Upadhyay, 1987) body temperature which rises to 43°C. This level of body temperature indicates that buffaloes store heat of work more than other working species. This higher body temperature induces certain behavioural symptoms before complete exhaustion or heat stroke develops in buffaloes. Behavioural manifestations like excessive frothing, dribbling of saliva, tongue protrusion take places in order to prevent any damage to brain and body. These symptoms must be observed as indicative of necessity of rest (Upadhyay and Madan, 1985, Upadhyay, 1987).

In some of the buffaloes body temperature may rise without

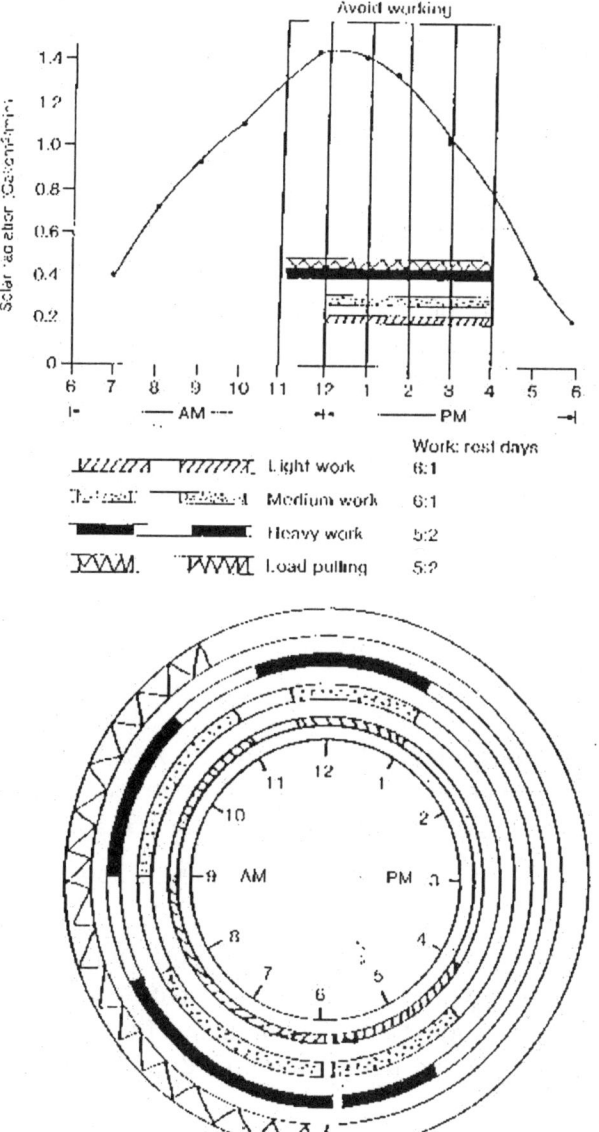

Fig. 3. Work-rest cycle for working animals in tropical climate.

adequate development of associated thermoregulatory processes and may be lethal, particularly during summer conditions. To obtain higher output from buffaloes under stressful climatic conditions, adequate ameliorative measures should be taken. Animals should be protected from direct solar radiation and must be provided frequent rest pause(s). Further to reduce heat load, the water may be sprinkled over the body or allow wallowing to reduce heat load. These processes help in quick recovery of animals to normal conditions.

Conclusion

Buffaloes are used in different agricultural farm operations in India. Since they have higher pulling capacity, buffaloes are preferred for heavy work. In spite of mechanization of agricultural operations, replacement of buffalo power is negligible vis a vis bullock power. Buffaloes exhibit distress during summer and are unable to cope thermal heat loads, therefore, require protection and alleviation of heat by artificial means or changing the working schedule for getting maximum out put in terms of working efficiency. Buffaloes are efficient draught animals as they may pull 5 to 7 times of their body weight and sustain for 6-8 hours at a moderate speed of 3-4 km/ hr. Their importance as draught animal in different agro climatic regions of India remains in many more years to come.

References

Bell, A.W; Hales, J.R.S; King, R.B. and Fawcett, A.A. 1983. Influence of heat stress on exercise induced changes in regional blood flow in sheep. Journal Applied Physiology **55**: 1916-1923.

Cockrill, W.R. 1974. The Husbandry and Health of Domestic Buffalo. FAO Rome.

Dolbert, F. 1981. Rural versus urban benefits from cattle production, feed resources utilization, the role of cattle in their farming system. *In:* Maximum Livestock Production from Minimum Land (Eds. Jackson *et al.*) Bangladesh Agricultural University, Mymensingh,

Goe, M.R. 1983. Current Status of Research on Animal Traction. World Animal Review **45**: 2-17.

Government of India 1999. Basic Animal Husbandry Statistic, 1999, Department of Animal Husbandry and Dairying, Ministry of Agriculture.

Government of India 2002. Basic Animal Husbandry Statistic, 2002, Department of Animal Husbandry and Dairying, Ministry of Agriculture.

Mathewman, R.W; Merrit, J; Oldham, J.D; Philips, P. and Smith, A.J.1988. The effect of work on milk yield and composition in draught cows. Proceedings of VIth World Conference on Animal Production, Helsinki, p-457.

Nangia, O.P; Singh, N. and Sukhija, S.S. 1980. Effect of exercise on thermal and acid base balance in buffaloes. Tropical Animal Health and Production **1**: 185-188.

Nodel, E.R. 1980. Circulatory and thermal regulations during exercise. Federation Proceeding **39**: 1491-1497.

Parks, C.M. and Manohar, M. 1983. Distribution of blood flow during moderate and strenuous exercise in ponies. American Journal Veterinary Research **44**: 1861-1866.

Pearson, R.A. and Lawrence, P.R. 1989. Continuous monitoring of draught animal performance during work. Draught Animal News **10**: 1-4

Rowell, L.B.1974. Cardiovascular adjustments to exercise and thermal stress. Physiological Review **54**: 75-159.

Shalija and Singh, S.V.2002. Role and future potential of draft animals. Indian Journal of Dairy and Bioscience **13**: 105-109.

Sharma, K.C. 1982. Some aspects of draught cows for agriculture. In: National Seminar on Draught Animal Power System in India. Indian Institute of Management, Bangalore.

Singh, G. 1999. Characteristics and use of draught animal power in India. Indian Journal Animal Sciences **69**: 621-627.

Singh, G. 2002. Spatial distribution and use of draught animals. Indian Journal Animal Sciences **72**: 689-694.

Singh, G; De, D. and Pathak, B.S.1977. Use of Energies in Agriculture and Future needs. Proceedings Third Agriculture Science

Congress, PAU, Ludhiana, March, 12-15.

Singh, S.V. and Upadhyay, R.C. 1997. Comparative performance and draught ability of buffaloes and cows during summer and winter. Indian Journal Animal Science **67**: 430-432.

Singh, S.V. and Upadhyay, R.C. 1997. Role of female buffaloes and cows for draught purposes- An appropriate technology: A review. Indian Journal Animal production and Management **13**: 1-7.

Srivastava, S.N.L. and Ojha, T.P.1987. Utilization of Draught Animal Power. Technical Bulletin, Central Institute of Agricultural Engineering, Bhopal (M.P.).

Srivastava, S.N.L. 1993. Status of research on engineering aspects of draft animal power. Journal of Rural Energy **2**: 15-32.

Upadhyay, R.C. 1993. Work rest cycle for draught animals. Journal of Rural Energy **2**: 58-67.

Upadhyay, R.C.(1984). Selection and Training of bullocks for work. Livestock Advisor **9**: 39-44.

Upadhyay, R.C. 1987. Factors limiting work capacity and fatigue assessment in draught animals. *In:* Utilization and Economics of Draught Animal Power. N.S.L. Srivastava and T.F. Ojha (Eds) CIAE, Bhopal, India.

Upadhyay, R.C. 1988. Draught Buffalo Research and Development. *In :* Buffalo Production and Health. R. Nagarcenkar (ed) II World Buffalo Congress, ICAR, New Delhi, India.

Upadhyay, R.C. and Madan, M.L.1985. Responses of buffaloes to heavy working load under tropical conditions. Livestock Production Science **13**: 199-203.

Upadhyay, R.C. and Madan, M.L. 1986. Cardiac responses to work production under tropical conditions in bovines. Indian Journal Dairy Science **39**: 388-393.

CHAPTER 29

Buffalo Productivity and Rearing Practices Prevalent in Himachal Pradesh

V.K.Sharma

College of Vet. & Animal Science,
HPKV, Palampur-176 062 (H.P)

Himachal Pradesh (HP) state came into existence on 15th April, 1948 with the inclusion of 30 small hill kingdoms. On 1st Nov., 1966 certain areas as a centrally administered territory of Punjab state included into 6 old districts of HP forming total 12 districts of new HP. HP state is situated in between 30° 22' 40" N lat. to 30° 12' 40" N lat. and 75° 47' 55" E long. to 79° 04' 20" E long. It has an area of 55673 km² and varying MSL of 460-6600 meters. Livelihood of 90 per cent of the population is agriculture, horticulture and dairy farming oriented. The state is a predominantly agro-pastral economy ha over threeforth of its working population directly engaged in these sectors. Holdings are marginal with over 83.7 per cent farmars falling in small and marginal category the net irrigated area account for onefifth of the net sown area (India, 2002). The district of Shimla and Sirmaur have alluvial soil, while the remaining ten district have forest and hill soils. Normal rainfall in the state is 18160 mm and maximum rainfall is noticed at Dharmsala in Kangra district. Himachal Pradesh is drained by a number of rivers, the most important of which are Chenab, Ravi, Beas, Sutlej and Yamuna (Anonymous, 1999). As per the census of 1992 the total animal population is 60.77 lakhs with density of 92 per km², minimum 5 animals per km² in Lauhal and Spiti districts and maximum 232 animals per Km² in Hamirpur district. The details of the increase in population trend from year 1951-1992 of different categories of the animals has been given in Table-1. District-wise distribution has been given in Table-2. The buffaloes constitute about 13.8 per cent of the total livestock population of the HP state. The entire state of HP has been divided into four agro-climatic zones as:

Zone I   Sub mountain, low –hills sub-tropical (upto 650 masl)

Zone II   Mid hills sub humid (651-1800 masl)

Zone III   High hills temperate wet (1800- 2000 masl)

Zone IV   High hills, temperate dry (above 2000 masl)

Table-1: Livestock population increase trend in Himachal Pradesh.

Sr. No.	Particulars	1951	1961	1972	1982	1992
1.	Cows	1115441	1212530	2175690	2173663	2165034
2.	Buffaloes	175136	208442	543007	616415	703542
3.	Sheep	626925	661731	1039946	1090322	1078940
4.	Goat	571697	594770	906415	1059862	118094
5.	Others	15634	14787	36517	48278	51323
6.	Total	2504833	2692271	4702455	4988540	5116933
7.	% Buffaloes	6.99	7.74	11.55	12.35	13.75

Table-2: District-wise distribution of Livestock in HP.

Sl. No.	District	Cows	Buffaloes	Others	Total	% Buffaloes
1.	Bilaspur	60959	8346	168033	237338	3.52
2.	Chamba	243673	35765	443586	723024	4.95
3.	Hamirpur	60809	94352	82676	237837	39.67
4.	Kangra	399398	147858	317212	918468	16.10
5.	Kinnaur	20935	3	91970	112908	0.002
6.	Kulu	159993	772	168637	328302	0.24
7.	Lauhal & Spiti	8117	-	55323	63440	-
8.	Mandi	438050	108416	406024	952490	11.38
9.	Shimla	327690	22450	232228	582368	3.85

Sl. No.	District	Cows	Buffaloes	Others	Total	% Buffaloes
10.	Sirmaur	234405	40059	148349	422813	9.47
11.	Solan	144613	74785	105528	324926	23.02
12.	Una	67492	91736	53791	213019	43.06
	Total	2165034	703542	2248357	5116933	13.75

Buffalo Rearing Practices in Himachal Pradesh

There has been a lot of improvement in the buffalo production systems of the rural area of HP during this decade. The farmers of the state are following the following feeding practices:

a) **Stall Feeding:** Small as well as large farmers producing enough dry and green roughage for feeding their buffaloes. The main sources of dry roughages are wheat and paddy straws. Under this system of feeding buffaloes are normally fed mixture of chaffed straws and stovers with the addition of concentrate with the crushed cotton, lathyrus and moth seeds etc. and variable quantities of chaffed green fodders. Common green leguminous and non-leguminous forages available for feeding buffaloes in larger parts of HP are maize, oats, berseem, lucerne, local grasses and barley. Flour, brans and cakes are frequently mixed in their feed for increasing the feed intake. Some of the farmers also use compound feeds, UMB and mineral mixture for feeding their animals.

b) **Migratory Grazing:** This system is prevalent among economic weaker section of landless rural families and migratory Gujjar families. They migrate their buffaloes from upper hill areas to lower areas during the months of winter, which again back to their destinations to hills during the summer. In this system, animals are unable to get sufficient amount of nutrients in most of the cases and they are liable to suffer from nutritional deficiencies.

c) **Grazing As Well Stall Feeding:** In this system of feeding buffaloes are allowed for 5-7 hours of grazing daily on the spent and depleted grasslands besides available fallow land in the nearby areas either in private or Government sector. In this system, especially during the rainy and early winter season, animals are able to get reasonably sufficient amount of green biomass available in the grazing areas but in the lean periods animals waste their energy in search of feed. During this period lactating buffaloes are usually provided with concentrate mixture including mineral mixture, UMB and green fodder, if available. Feeding of concentrate depend upon their availability. Farmers are also feeding soaked or boiled/ cooked cotton seed and or cotton seed cake, other oil cakes, brans or pulses byproducts or grains including weed seeds, flour of cereal grains, broken rice, ground mixture of minor millets and pulse milling by-products etc. After morning milking buffaloes are let loose and carried to fallow land available in nearby areas or nearby rivers. This provides light exercise and wallowing in water.

Table-3: **Milk production in HP (1998).**

Sr.No.	Catagory	Thousands tones Lit.	% of total
1.	Cows	310.57	43.50
2.	Buffaloes	359.79	50.39
3.	Goats	43.60	6.11
	Total	713.96	

Feeding Practices to Different Categories

a) **Feeding Of Dry Animals:** For non-pregnant dry animals, mostly poor quality straws alone with very small quantity of concentrate is offered.

b) **Feeding Of Pregnant Buffaloes:** Under very good management conditions, a buffalo calved at the interval of 14-18 months interval and has mostly lactation length of about 400-470 days. The dry period vary from 90-300 days mostly. During this period they are given the concentrate at the rate of 1.0-1.5 kg and increased to 2.5 to 3.0 kg in the last quarter of the pregnancy. They are also provided with the 30-40 gm of the mineral mixture daily.

c) **Feeding For Milk Production:** Normally, the milk yield ranges between 4 - 8 l/d. They are fed wheat straw, green fodder and 2kg concentrate daily. Wheat straw is mixed with green leguminous or non-leguminous fodder and fed *ad.lib.* For high milk yielder the amount of concentrate mixture is increased up to 4 kg daily depending upon the milk yield.

d)   **Calf Feeding:** Immediately after parturition the calf is cleaned and colostrums is fed to the calf. Generally calves are allowed to suckle directly from the mothers. Calves are removed immediately from the mother and help in standing. On an average 2 kg colostrums/milk is fed to the calf daily for about 8-10 days. Alone with milk they are given some amount of calf starter in some of the cases for sparing the milk from the feeding of calves. Some of farmers are also using the milk replacers at early stage. They are given a dry mixture of cereal flour, oil cakes, vegetable oil, mineral mixture and sometimes vitamins are added only in case of female calves. As yet no systematic calf rearing practices are being followed by majority of the farmers. After a period of 3 months of age, calves are fed concentrate and roughage. After 3-6 months of age female calf is given 1.5 kg concentrate mixture alone with good quality of green fodder or hay.

Table-4: Forage, grass and seed production, distribution by the H.P. Department of Animal Husbandry.

Sr. No.	Particular	1980-81	1990-91	1997-98	% increase
1.	Forage seed production (Q)	113	96	141	19.85
2.	Grass seed production (Kg)	-	145	233	37.8
3.	Forage seed distribution (Q)	38.5	180	408	90.6
4.	Grass root slips distribution (Lakh No.)	3.1	4.3	8.4	63.0
5.	Forage demonstration units	132	2800	4100	300.6
6.	Area under grass Nursery (Hectare)	0.5	1.0	4.0	87.5
7.	Silage Production & distribution(Q)	4800	4500	5500	12.7
8.	Forage production centres	4	4	5	20.0

Efforts Being Carried Out By State Animal Husbandry Department And Scientific Research Carried Out By Agriculture University

1)   Efforts are being done by the Department of Animal Husbandry as well as by the Veterinary & Animal Sciences College, Palampur, Agriculture University of the state to boost the milk production from presently 359.79 thousand tonnes to 713.98 thousand tonnes. The share of buffalo milk is 50.4 per cent (Table-3). There are 3 frozen and liquid semen laboratory has been established by the A.H. deptt. of the state. From these semen labs. the semen is transported to 955 different artificial insemination centres and in 729 centres there is facility to inseminate buffaloes also. To improve the fodder production and quality of the fodder in the state a continuous programme has been devised. Tremendous improvements in the production of the fodder for enhancing the milk production of the buffaloes in the state are being done. The increase in respect to fodder/forage seed production, fodder seed distribution, improved grass root slips distribution, number of fodder production centre, grass nursery production area, silage production and fodder seed production institutes has gone up to 69.8, 90.6, 63.1, 300.6, 87.5, 12.7 and 20.0 per cent respectively in the year 1997-98 to that of the year 1980-81 (Table-4). The Animal husbandry sector is contributing about 11-12 per cent in the total economy of the state, which was only 1-2 per cent in the year 1950.

2)   Scientific works have been carried out in the University and the following are the Conclusions of the various studies carried out:

a)   Forty five per cent CP of the total ration can be replaced with *Leucaena leucocephala* (Su-babool) leaf meal and by doing so fat yield and FCM of the buffalo milk can be increased. This also reduced the cost of the ration up to 34 per cent. With the feeding of *Panicum maximum* (Guinea grass), milk production is lowered and this lowered milk production can be covered up with the addition of oil cakes such groundnut cake, mustard cake or sunflower cake to the tune of 200, 250 or 300 g/d, respectively. Soybean straw can be fed to the yearling to meet out the maintenance requirements (Garg and Kumar, 1994).

b) With the supplementation of micronutrients copper and cobalt to the anaestrous buffaloes the reproduction efficiency in the Paonta Valley (Sirmaur Distt.) can be increased significantly, (Chadda *et. al.*, 1994).

c) *Sorghum bicolor* (M.P. Chari) can meet out the maintenance requirements where as Teosinte (Mak chari) can meet out upto 40 per cent DM requirement along with addition of concentrate and grains to supply the protein and energy requirements of the Murrah heifers. *Sorghum helepence* (Baru grass) should not fed more than 30 per cent of DM requirement to heifers. Urd straw can be fed to a limited quantity to heifers due to its poor palatability and high lignin contents, (Garg and Kumar, 1995).

d) *Bracharia decumbence* (Signal grass) at the pre-flowering stage can be fed to Murrah heifers as a sole source of roughage to meet the maintenance requirements. For attaining daily weight gain supplementation of 200 g mustard cake and 30 g of mineral mixture is necessary. Similarly, *Sorghum helepence* can be fed as a sole source of forage (at 50 per cent flowering stage) along with 400 g mustard cake and 30 g of mineral mixture supplementation for good results. *Panicum maximum* (Guinea grass) can be fed as a sole source of roughage to the growing heifers alone with the supplementation of 400 g mustard cake and 30 g of mineral mixture. It was also reported that DM consumption of milch buffaloes increased to 30 per cent as compared to non-lactating buffaloes, (Garg, 1996).

e) *Shorea robusta* (Sal) tree leaves can be fed to the buffalo heifers as a sole source of roughage with the supplementation of 100 g mustard cake per day for the optimum growth. Teosinte can be fed at the rate of 29 per cent of DM requirement to maintain average body gain of 327 g/d. *Leucaena leucocephala* leaf meal can

CHAPTER 30

A Status of Buffalo Based Economy in Asia

K. Kareemulla and B.S. Meena*

National Research Centre for Agro-forestry, Jhansi-284 003 (UP)
*Indian Grassland and Fodder Research Institute,
Jhansi-284 003 (UP)

The global human population is over six billion of which around 40 per cent is dependent on agriculture. Almost two-third of the population involved in agriculture are dependent on livestock in some way or the other. In the Asian continent due to greater reliance on agriculture and higher human population density, the per capita income is much below the global average USD 5140 (table-1). The agricultural productivity measured as value added per worker in Asian countries ranged between USD 296 to 3756 in the year 2000. The proportion of people living below the national poverty lines was 25 per cent (Sri Lanka) to 36 per cent (Bangladesh). All this dismal progress persists despite rich natural resources base including diverse livestock like buffaloes and others.

Table-1: Key Indicators of Development of some Asian Countries

Country	Population (in millions)	Population density (per km²)	Per capita income (USD)	% population below poverty line	Value added in agrl. Productivity (per worker at 1995 prices)
Bangladesh	133.4	1025	370	35.6	296
China	1271.9	136	890	4.6	321

Country	Population (in millions)	Population density (per km²)	Per capita income (USD)	% population below poverty line	Value added in agrl. Productivity (per worker at 1995 prices)
India	1033.4	348	460	35.0	397
Iran	64.7	40	1750	-	3756
Pakistan	141.5	183	420	34.0	630
Sri Lanka	19.6	304	830	25.0	753
World	6132.8	47	5140	-	-

According to statistics provided by the International Dairy Union, the world output of milk products totalled 564 million tons in 1999, with an average annual consumption of 94 kilograms per capita. With China excluded, Asia's average consumption per capita has gone up to 40 kilograms, and the figure for China is below 7 kilograms. As people lead a better life, there will be a dramatic rise in the dairy consumption, which translates into a huge space for market expansion (AIR, 2001). The per capita consumption of dairy product in some of the higher income Asian countries like Taiwan, Singapore and Hong Kong increased at 10 to 14 kg per annum. While in a lower income country like India, it was 66 kg in the year 1990. This indicates the immense potential for growth of the dairy industry (Gujaral, 2004). In order to exploit this potential the native livestock of the Asian continent, especially the buffalo can offer a wide array of prospects. A comparative analysis of buffalo contribution in terms of milk and meat by major Asian countries and their competitiveness with special reference to the status of their feeding is presented below.

Buffalo Stock in Asia

Asia harbours an estimated buffalo population of 165.44 million and almost 59 per cent of this is held in India (table-2). Pakistan has a buffalo stock of almost 15 per cent of Asia's total buffalo stock followed China (13.76%). Thus the three most populous countries of the Asian continent own almost 87 per cent of the Asian buffalo wealth. An indicator of the concentration of livestock as related to human inhabitation is the density of buffaloes per hundred human population. Based on this indicator it was found that Pakistan had highest buffalo density (17.5) followed by India (9.3) and the least density was observed in China with1.7 (figure-1). This indicates the degree of importance of buffalo as a livelihood resource in these nations. Higher buffalo density meant greater reliance and vice versa. While rural subsistence on livestock production system is the predominant form in this part of the globe, the urbanization and continued preference for fresh raw milk has paved way to commercial peri-urban and urban dairies (Younas and Yaqoob, 2004). The herd sizes of such commercial dairies especially in India and Pakistan ranged between 100 and 500 preferably, buffaloes.

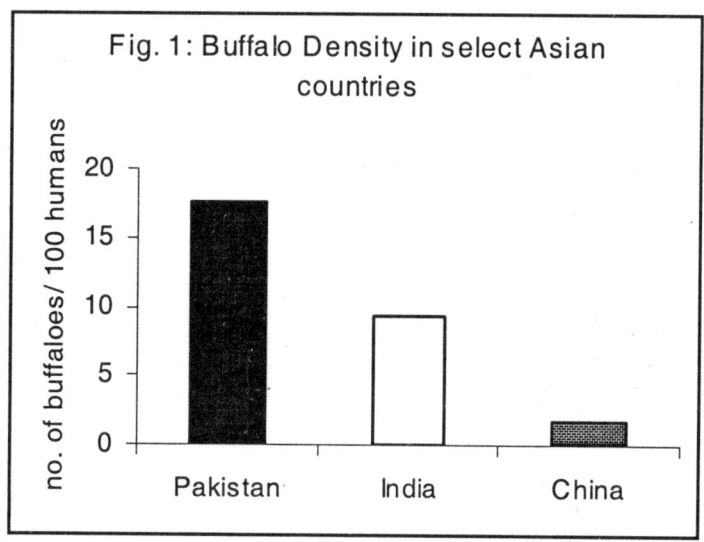

While the total animal stock indicates the strength and potential, the share of milch animals pronounces the production capability. The proportion of milch buffaloes to the total stock of buffaloes was the highest at 39.11 per cent in Pakistan followed by 35.09 per cent in India. Similar proportion in case of China was relatively lower at 23.07 per cent (table-2). Higher share of milch buffaloes indicates two things namely a healthy sex ratio and better management for economic gain. In order to understand this phenomenon, an illustration of the changes in sex ratios of buffaloes in the post green revolution period in India is discussed. While the population of male and female cattle increased by 3 and 22 per cent, respectively during 1967 to 1987, the buffaloes stocks increased by 47 and 57 per cent, respectively. This indicates the balanced preference for buffalo by the farmers and it also signifies the role of the latter in Indian agriculture in general and dairy in particular. Relatively higher share of milch animals in Pakistan could be due to better breeds like Nili-Ravi and Kundi besides the dominance of irrigated agriculture.

Table-2: **Buffalo Stock and share of Milch animals in Asia (2003)**

Country	Total stock (,000 heads)	Share (%)	Milch animals (,000 heads)	Proportion of milch animals (%)
India	96,900	58.57	34000	35.09
Pakistan	24,800	14.99	9700	39.11
China	22,759	13.76	5250	23.07
Nepal	3,750	2.27	na	
Others	17,237	10.42	na	
Asia	165,447	100.00	na	

Feeding Status in Asia

In South Asia, limited grazing, tethering and cut and carry feeding are more common. The principle aim of this process is to make available those nutrients, which buffalo lack due to limited access to grazing. The reverse is true throughout most parts of South East Asia where buffaloes derive most of their nutrient needs for maintenance and production from grazing approximately six to eight hours per day. They are also fed with limited supplements in the evening such as cereal straws and cakes (coconut cake or groundnut cake), brans (rice and wheat) or leguminous forages (*Leucaena leucocephala*,

Gliricidia maculata and *Manihot esculenta* Crantz) and salt (Mudgal, 1999) .

In South Asia chronic annual feed shortages for animals and under-nutrition are common. However, there has been a significant trend towards reduced feed deficits, which is probably reflective of improved feeding systems, more efficient use of all available feeds and increasingly intensive systems of production. Further opportunities exist for reducing this feed deficit through more intensive use of non-conventional feed resources.

By comparison, the annual availability of feed for ruminants is generally inadequate in most countries in humid South East Asia, such as Sri Lanka, Malaysia and the Philippines. In these situations, feeding strategies could be more selective, and use a variety of traditional feeds such as Guinea (*Panicum maximum*) or Napier (*Pennisetum purpureum*) grasses, agro-industrial by-products or a variety of shrubs and tree legumes.

In countries where animal feed resources are scarce, International Fund for Agriculture Development (IFAD) policy has been to provide loans/foreign exchange for the purchase of feed from external sources. In the Crop and Livestock Rehabilitation project in Korea, funds were provided for the procurement of raw feed material from China, for on-lending to feed mills. Despite efforts to ensure feed supply, shortages of high-quality feed and feed additives were still a constraint to feed production. IFAD's experience has shown that reliance on imported is not the best solution to problems of feed scarcity in developing countries for a number of reasons like access to markets for livestock products such as milk or access to livestock production areas where fodder can be marketed. Poor farmers will produce fodder if there are other benefits deriving from the activity.

Buffalo Milk Production

The Asian buffalo milk production was estimated at 70.39 million MT in the year 2003 (figure-2). India's share in this production was around 68 per cent. The contribution of Pakistan to Asia's buffalo milk production was 26.31 per cent. Despite lower contribution in terms of milk productivity of buffaloes Pakistan occupies the first rank with 1909 kg per buffalo in 2003 (Table-3 and figure-3). This could be due to the higher potential of the breeds supplemented by abundant green forage. India was way behind with 1407kg which might be due to twin factors namely preponderant rainfed farming coupled with disintegrating non-descript buffalo breeds. The productivity was considerably lower at 505 kg in China. The lower productivity in China is obviously due to the meagre potential of the Native swamp buffaloes.

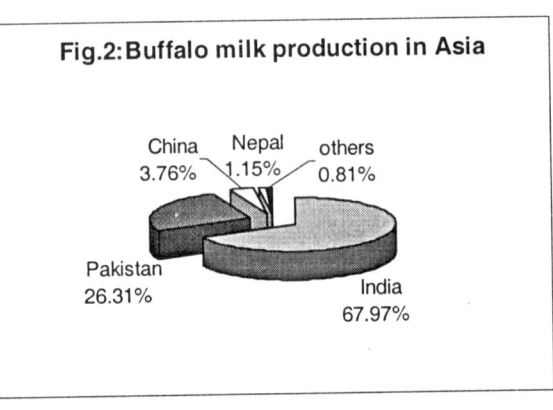

Fig.2:Buffalo milk production in Asia

China 3.76% | Nepal 1.15% | others 0.81% | Pakistan 26.31% | India 67.97%

Table-3: **Average Milk Yield of Buffaloes in Asia (2003)**

Country	Average milk yield (Kg/animal)
Pakistan	1,909
India	1,407
Vietnam	1,000
Turkey	969
Nepal	842
Sri Lanka	648
China	505
Bangladesh	407
Asia	1,389

Buffalo Meat Production

FAO estimates of buffaloes slaughtered in Asian countries during

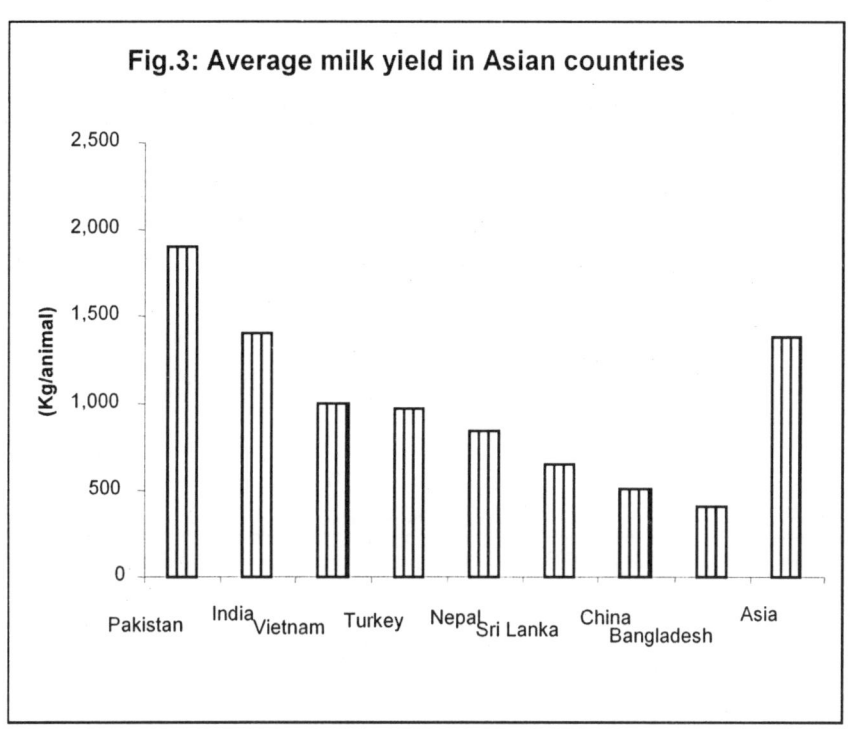

Fig.3: Average milk yield in Asian countries

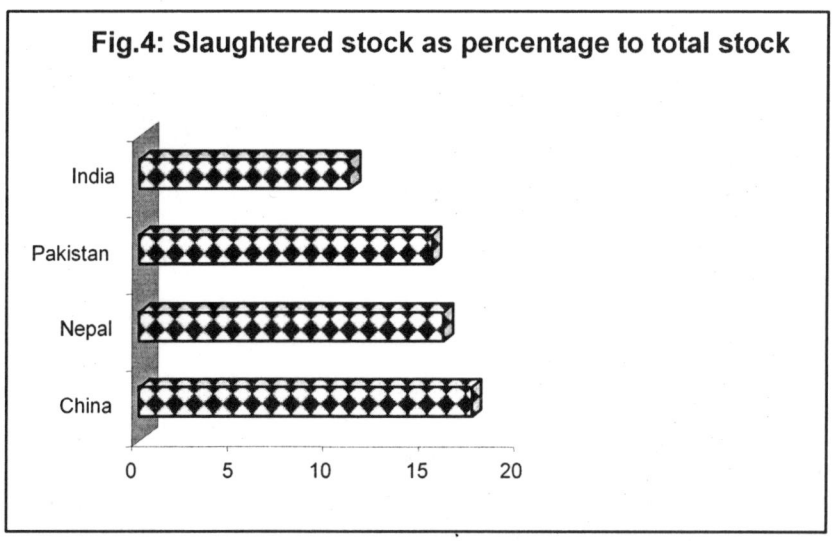

Fig.4: Slaughtered stock as percentage to total stock

2001 are given Table-4. The total buffaloes slaughtered in Asia during 2001 were 20.94 million and over half of them were accounted in India. The shares of China and Pakistan were 19 and 18 per cent, respectively in the total slaughtered buffaloes of the continent. Buffaloes are slaughtered for their meat, which is consumed both domestically and partly exported. The proportion of slaughtered buffaloes to the total stock in Asia was 12.66 per cent. Among the nations, such proportion was the lowest at 11.00 per cent in India indicating a stabilized culling rate that will not adversely affect the net population (figure-4). Table- 4: Buffaloes slaughtered in Asia for meat (2001)

Country	Buffaloes Slaughtered (Heads)	Percentage
India	10,660,000	50.90
China	3,955,700	18.89
Pakistan	3,800,000	18.15
Nepal	595,000	2.84
Vietnam	460,000	2.20
Philippines	425,000	2.03
Thailand	210,000	1.00
Indonesia	209,000	1.00
Laos	166,360	0.79
Myanmar	132,000	0.63
Other Asian countries	328,880	1.57
Asia	20,941,940	100.00

The carcass weights that influence the meat production across the countries is given in Table-5. The average buffalo carcass weight was the highest in Thailand at 253.0 kg per animal followed by 215 kg in Vietnam. This could be due to the hefty nature of the swamp buffaloes prevalent in these countries. The average buffalo carcass weight in India was only 138 kg.

Table- 5: Average carcass weight of buffaloes in Asia (2003)

Country	Carcass Weight (kg/animal)
Thailand	253.0
Vietnam	215
Philippines	190.6
Turkey	182.1
Cambodia	160
Iran, Islamic Rep of	150
India	138
Pakistan	133.9
China	100.2
Bangladesh	76.1
Asia	137.1

Given the statistics on the number of buffaloes slaughtered and the average carcass weight, the estimates of buffalo meat production follow. The total buffalo meat production in Asia during 2003 was pegged at 2.87 million MT (Table-6). Of this about 51.23 per cent is produced in India followed by 17.73 per cent in Pakistan and 13.80 per cent in China.

Table-6: Buffalo meat production in Asia (2003)

Country	Production (mt)	Share (%)
India	1,471,080	51.23
Pakistan	509,000	17.73
China	396,250	13.80
Nepal	130,000	4.53
Vietnam	98,900	3.44
Others	365197	12.72
Asia	2,871,527	100.00

Buffalo Meat Trade

The Asian countries trade bovine meat including buffalo meat and beef. The Asian countries put together were net importers as the imports (1.79 million MT) far exceed the exports (0.37 million MT) in the year 2001 (table-7 & 8). Among the exporting countries, India tops with 0.24 million MT of bovine meat whereas, in the importing countries Japan (0.88 million MT) and Republic of Korea (0.23 million MT) are prominent. The only country that figures among both exports and imports of bovine meat is China, which exports almost 50 per cent of its import. The growth rate of buffalo meat export in India was highest at 25.85 per cent per annum during the period 1990 to 2001. Among the importing countries Philippines and Israel had higher growth rates at 58.30 and 16.57, respectively. These potentials can be tapped by exporting countries like India.

Table- 7: Exports of Bovine meat from Asian countries (mt)

Country	1990	2001	Share % in 2001	Annual growth rate	Percentage of exports to imports
India	63,385	243,611	65.38	25.85	n.a.
China	226,531	82,723	22.20	-5.77	50.33
Mongolia	12,000	12,381	3.32	0.29	n.a.
Others	25,672	33,917	9.10	2.92	n.a.
Asia	327,588	372,630	100.00	1.25	n.a.

Table-8: Imports of Bovine meat in Asian countries (mt)

Country	1990	2001	Share in 2001(%)	Annual growth rate
Japan	506,383	884,937	49.41	6.80
Korea, Republic of	120,678	238,782	13.33	8.90
China	106,702	164,359	9.18	4.91

Country	1990	2001	Share in 2001(%)	Annual growth rate
Malaysia	56,194	114,581	6.40	9.45
Philippines	13,598	100,808	5.63	58.30
Israel	28,236	79,712	4.45	16.57
Asia	1,138,779	1,791,024	100.00	5.21

Competitiveness of Industry

Table- 9: Buffalo milk Producer Prices in Asia (2001)

Country	Price (USD/t)
Sri Lanka	290
India	284
Iran	145
Nepal	323
Pakistan	280
Philippines	255
Turkey	235
China	99

One of the major indicators of any industry is the unit production cost (price). The producer's price of buffalo milk in the leading Asian countries is given in table -9. The highest price was in case of Nepal at USD 323 per tonne. In India it was 284. Among the Asian countries China had the lowest producers price for buffalo milk at USD 99 in 2001. The producer prices of buffalo meat during 2001, across the major Asian countries are indicated in Table-10. A comparison of such prices indicates that the buffalo meat production was costlier in Turkey (USD 1512.75) and was cheaper in India (USD 147.09). Thus it makes sense for Turkey to rely more on imports and save the resources spent on buffalo meat production for more remunerative avenues.

Table-10: Producer Price of Buffalo meat in select Asian countries (2001)

Country	Unit Producer Price (USD/t)
Sri Lanka	258.37
India	147.09
Indonesia	538.79
Iran	937.42
Cambodia	1099.85
Nepal	515.99
Pakistan	209.00
Philippines	739.22
Thailand	353.59
Turkey	1512.75
China	260.99

Conclusion

The role of buffalo in the Asian economy has been emphasized by their concentration in countries like India and Pakistan. Higher milk productivity in countries like Pakistan is due to feed security attributed to conducive irrigated agriculture. Lower producer prices for buffalo meat in countries like India indicate the resilience and the much needed competitive strength in the wake of globalization. In view of the fact that the livelihoods of millions are intertwined with livestock like buffaloes, the same need to be conserved and nurtured in the right path.

References

AIR. 2001. China Dairy Products Market Pub: Asian Information Resources: www.chinareference.com

FAO. 2004. Agricultural Database. Food and Agriculture Organization. www.fao.org

Muhammed Younas and Muhammed Yaqoob. 2004. Rural Livestock Production in Pakistan. www.pakissan.com

Raman Gujral.2004. Profitability of Dairy farming unit.www.techno-preneur.net

V.D. Mudgal. 1999. Milking buffalo in *Smallholder Dairying in the Tropics* (eds.) Falvey L. and Chantalakhana C. International Livestock Research Institute, Nairobi, Kenya. 462 pp.

Index